T0315240

PLANT GENOME: BIODIVERSITY AND EVOLUTION

Volume 1, Part B
Phanerogams (Higher Groups)

Plant Genome
Biodiversity and Evolution

Series Editors
A.K. Sharma and Archana Sharma
Department of Botany
University of Calcutta, Kolkata
India

Advisory Board

J. Dolezel, Laboratory of Molecular Cytogenetics and Cytometry, Institute of Experimental Botany, Sokolovska, Czech Republic

K. Fukui, Department of Biotechnology, Graduate School of Engineering, Osaka University, Osaka, Japan

R.N. Jones, Institute of Biological Sciences, Aberystwyth, Ceredigion, Scotland, UK

G.S. Khush, 416 Cabrillo Avenue, Davis, California 95616, USA

Ingo Schubert, Institüt für Pflanzengenetik and Kulturpflanzenforschung, Gatersleben, Germany

Canio G. Vosa, Scienze Botanische, Pisa, Italy

Plant Genome
Biodiversity and Evolution

Volume 1, Part B
Phanerogams (Higher Groups)

Editors

A.K. SHARMA and A. SHARMA

CRC Press
Taylor & Francis Group
Boca Raton London New York

CRC Press is an imprint of the
Taylor & Francis Group, an **informa** business

A SCIENCE PUBLISHERS BOOK

First published 2005 by Science Publishers, Inc.

Published 2021 by CRC Press
Taylor & Francis Group
6000 Broken Sound Parkway NW, Suite 300
Boca Raton, FL 33487-2742

ISBN 13: 978-1-57808-353-4 (hbk)

Visit the Taylor & Francis Web site at
http://www.taylorandfrancis.com

and the CRC Press Web site at
http//www.crcpress.com

Cover illustration reproduced by kind premission of Dr Katsuhiko Kondo, et. al.

Library of Congress Cataloging-in-Publication Data

Plant genome : biodiversity and evolution / editors, A.K. Sharma and A. Sharma.
    p. cm.
  Includes bibliographical references.
  Contents: v. 1, pt. B. Phanerogams
  ISBN 1-57808-353-2
    1. Plant genomes. 2. Plant diversity. 3. Plants--Evolution. I. Sharma, Arun Kumar, 1923- II. Sharma, Archana, 1932-

QK981.P53 2003

581.3'5--dc21   2003042378

Preface to the Series

The term *genome*, the basic gene complement of an individual, is almost synonymous with the chromosome complement of both nucleus and organelles. Refinements in cellular, genetic and molecular methods in recent years have opened up unexplored avenues in genome research. The modern tools of gene and genome analyses, coupled with analysis of finer segments of gene sequences in chromosomes utilizing molecular hybridization, are now applied on a wider scale in different groups of plants, ranging from algae to angiosperms. This synergistic approach has made the study of biodiversity highly fascinating, permitting a deep insight into the molecular basis of genetic diversity. Simultaneous to the enrichment of fundamentals in systematics and phylogeny, the plant system, because of its inherent flexibility, has permitted genetic engineering and horizontal transfer of genes with immense importance in agriculture, horticulture and medicine.

Despite the fact that the data on plant genomics with its impact on the assessment of biodiversity and evolution show a logarithmic increase, a comprehensive series on the aspect covering all groups of plant kingdom is sadly lacking. In view of this lacuna, the present series on Plant Genomics: Biodiversity and Evolution has been planned. It aims to cover, in successive volumes, *comprehensive reviews, concepts and discussion on the results of genome analysis and their impact on systematics, taxonomy, phylogeny and evolution of all plant groups.* We have not gone out of our way to seek original articles, but in course of reviews and discussions, research articles, if any, are welcome.

A.K. Sharma
Archana Sharma
Series Editors

Preface to this Volume

This volume is the second in the series on 'Higher Plants', otherwise termed Phanerogam or Spermatophyte. The earlier volume, 1A, covered several taxa of Phanerogams and one of Bryophyte. A concurrent series (volume 2A, 2B and so on) cover the lower groups, represented by Algae, Fungi, Bryophytes and Lichens. Bryophyte, because of its importance in evolution of seed plants *vis-a-vis* higher groups, has been included in both volumes.

The present volume includes articles dealing with certain fundamental issues of genomics as well as phylogeny and evolution of certain economic agricultural and medicinal crops. Of these, economic crops include coffee, coconut and papaya; medicinals cover *Artemisia* and *Costus*; fodder and agricultural crops include *Phleum*, *Lolium* and *Triticale*; horticultural species include *Orchis* and *Allium* and forest plant is represented by *Populus* species.

The book starts with an article on the assessment of amino acid and nucleotide parameters in the study of phylogeny of the basic angiosperms. This is followed by genetic diversity, phylogeny and affinities of Costaceae and *Orchis* and its allies. The evolution in Graminaceous genera is represented by *Phleum*, pasture grasses and wheat rye grass. The special features are the evolution of heterochromatin in *Phleum* genomics, marker technology as a tool in *Lolium* and other genera, and heterochromatin as a marker in the analysis of evolution in *Triticosecale*. The genome study and phylogeny of the medicinals - *Artemisineae* and *Costaceae*—have been presented utilizing chromosomal features, nuclear DNA and molecular pattern. Similarly, chromosomal and molecular details as aids in evolution have been elaborated in four horticultural and commercial crops, namely, coconut, coffee, papaya and potato. The genomics of forest plant is covered in species of *Populus*.

In general, this volume contains discussions on molecular patters of basic angiosperms as well as genomics and phylogeny of different families covering several commercial, medicinal and agricultural crops. Those working on molecular genetics, evolution and phylogeny of flowering plants, including species of medicinal and agricultural value will find this book interesting.

December 2004   **Arun Kumar Sharma**
                                                 Archana Sharma

Contents

List of Contributors

Francois Anthony
IRD, BP 64501, 34394 Montpellier cedex 5, France. Telephone : +33 (0) 467416289; Fax: +33 (0) 467416320; E-mail: anthony@mpl.ird.fr

Saverio D'Emerico
Dipartimento di Biologia e Patologia Vegetale, Sezione di Biologia Vegetale, Universita di Bari, Via Orabona 4, 70125 Bari, Italy. E-mail demerico@botanica.uniba.it

Mark P. Dupal
Primary Industries Research Victoria, Plant Biotechnology Centre, La Trobe University, Bundoora, Victoria 3086, Australia

John W. Forster
Primary Industries Research Victoria, Plant Biotechnology Centre, La Trobe University, Bundoora, Victoria 3086, Australia. Telephone: +61-3-9479-5645; Fax: +61-3-9479-3618; E-mail: john-forster@dpi.vic.gov.au

Teresa Garnatje
Institute Botanic de Barcelona (CSIC-Ajuntament de Barcelona), Passeig del Migdia s/n. 08038 Barcelona, Catalonia, Spain

Sally Garvie
Department of Applied and Molecular Ecology, University of Adelaide, Glen Osmond, South Australia 5064, Australia

Godelieve Gheysen
Department of Molecular Biotechnology, Faculty of Agricultural and Applied Biological Sciences, Ghent University, Coupure Links 653, 9000 Ghent, Belgium

Kathryn M. Guthridge

Primary Industries Research Victoria, Plant Biotechnology Centre, La Trobe University, Bundoora, Victoria 3086, Australia

Kazuyoshi Hosaka

Food Resources Education and Research Centre, Kobe University, 1348 Uzurano, Kasai, Hyogo 675-2103, Japan. Telephone: +81-790-49-3121; Fax: +81-790-49-0343; E-mail: hosaka@kobe-u.ac.jp

Sharon Howlett

Primary Industries Research Victoria, Plant Biotechnology Centre, La Trobe University, Bundoora, Victoria 3086, Australia

Leonie J. Hughes

Primary Industries Research Victoria, Plant Biotechnology Centre, La Trobe University, Bundoora, Victoria 3086, Australia

Andrzej Joachimiak

Department of Plant Cytology & Embryology, Institute of Botany, Jagiellonian University, Grodzka St. 52, PL 31-044 Cracow, Poland. E-mail: a.joachimiak@iphils.uj.edu.pl

Elizabeth S. Jones

Primary Industries Research Victoria, Plant Biotechnology Centre, La Trobe University, Bundoora, Victoria 3086, Australia

Tina Kyndt

Department of Molecular Biotechnology, Faculty of Agricultural and Applied Biological Sciences, Ghent University, Coupure Links 653, 9000 Ghent, Belgium

Philippe Lashermes

Institute de Recherche pour le Developpment, "Plant resistance to pests" Unit, Montpellier, France

Ray Ming

Hawaii Agriculture Research Center, 99-193 Aiea Heights Drive, Aiea, HI 96701, USA. Telephone: 808-486-5374; Fax: 808-486-5020; E-mail: rming@harc-hspa.com

Paul H. Moore

USDA-ARS, Pacific Basin Agricultural Research Center, 99-193 Aiea Heights Drive, Aiea, HI 96701, USA

Lalith Perera

Genetics and Plant Breeding Division, Coconut Research Institute, Lunuwila, Sri Lanka. Telephone: 94 31 225-5300; Fax: 94 31 225-7391; E-mail: rescri@sri.lanka.net

Christopher Preston

Department of Applied and Molecular Ecology, University of Adelaide, Glen Osmond, South Australia 5064, Australia

Stanislawa Maria Rogalska

University of Szczecin, 70-415 Szczecin, Poland. Telephone/Fax: +48(9) 4441535; E-mail: strog@univ.szczecin.pl

Xavier Scheldeman

International Plant Genetic Resources Institute (IPGRI), Office for the Americas, A.A. 6713, Cali, Colombia

Terry Sekioka

Department of Tropical Plant and Soil Sciences, University of Hawaii, Honolulu, HI 96822, USA

Mark P. Simmons

Department of Biology, Colorado State University, Fort Collins, CO 80523, USA. Telephone: 970-491-2154; Fax: 970-491-0649; E-mail: psimmons@lamar.colostate.edu

Kevin F. Smith

Primary Industries Research Victoria, Hamilton Centre, Hamilton, Victoria 3300, Australia

Chelsea Dvorak Specht

University of Vermont, Department of Botany—Pringle Herbarium, 120 Marsh Life Sciences Building, Burlington, VT 05405 USA. Telephone: (802) 656-3221; Fax: (802) 656-0440; E-mail: chelsea@nybg.org

Joan Valles

Laboratori de Botanica, Facultat de Farmacia, Universitat de Barcelona, Av. Joan XXIII s/n, 08028 Barcelona, Catalonia, Spain. Telephone: +34-934024490, Fax: +34-934035879, E-mail: joanvalles@ub.edu

Bart Van Droogenbroeck

Department of Molecular Biotechnology, Faculty of Agricultural and Applied Biological Sciences, Ghent University, Coupure Links 653, 9000 Ghent, Belgium

Zuoheng Wang

Department of Statistics, University of Florida, Gainesville, FL 32611 USA.

Jiasheng Wu

College of Life Sciences, Zhejiang Forestry University, Lin'an, Zheijiang 311300, P.R. China

Rongling Wu

Department of Statistics, University of Florida, Gainesville, FL 32611, USA. Tel: (352) 392-3806; Fax: (352) 392-8555; E-mail: rwu@stat.ufl.edu

Francis T. Zee

USDA-ARS, Pacific Basin Agricultural Research Center, Tropical Plant Genetic Resource Management Unit, Hilo, HI 96720, USA

Yanru Zeng

College of Life Sciences, Zhejiang Forestry University, Lin'an, Zhejiang 311300, P.R. China

Errata

In the Contents page the title of Chapter 1 should read as below:

Amino Acid Versus Nucleotide Characters for phylogenetic inference of the "Basal" Angiosperms

Mark P. Simmons

Amino Acid Versus Nucleotide Characters for Phylogenetic Inference of the "Basal" Angiosperms

MARK P. SIMMONS

Department of Biology, Colorado State University, Fort Collins, CO, USA

ABSTRACT

When nucleotide and amino acid characters for phylogenetic inference are compared, six factors need to be considered. In this study, an empirical example based on eight protein-coding genes from the "basal" angiosperms was used to investigate how these six factors contribute to the relative performance of nucleotide and amino acid characters for phylogenetic inference. Nucleotide characters were generally found to outperform amino acid characters with respect to the tree-based measures examined. Of the 167 clades resolved on the amino acid jackknife trees for each of the eight protein-coding genes analyzed separately, 22 conflicted with clades resolved by the nucleotide jackknife trees for the respective genes. Independent evidence supports the nucleotide resolution in 15 cases and the amino acid resolution in four. Of the 15 cases for which independent evidence supported nucleotide resolution, five could be explained by composite amino acid characters and/or convergent amino acid character states. Although the nucleotide characters generally outperformed the amino acid characters, a wide diversity in their relative performance is evident on comparing the eight genes relative to one another. Therefore, one cannot make blanket statements regarding which type of character coding is the better of the two.

Address for correspondence: Department of Biology, Colorado State University, Fort Collins, CO 80523, USA. Tel: 970-491-2154, Fax: 970-491-0649, E-mail: psimmons@lamar.colostate.edu

Key Words: amino acid characters, composite characters, convergence, nucleotide characters, phylogenetic inference

Abbreviations: APS: amount of possible synapomorphy; *atp1*: mitochondrial gene encoding the ATPase alpha subunit; *atpB*: plastid gene encoding the ATP synthase beta chain; *cox1*: mitochondrial gene encoding cytochrome C oxidase polypeptide 1; *matR*: mitochondrial gene encoding maturase; *phyA*: nuclear gene encoding phytochrome A; *phyC*: nuclear gene encoding phytochrome C; *rbcL*: plastid gene encoding the ribulose-1,5-bisphosphate carboxylase/oxygenase large subunit; *rps2*: mitochondrial gene encoding ribosomal protein S2; TBR: tree-bisection-reconnection branch swapping

INTRODUCTION

When conducting phylogenetic or gene-tree analyses using protein-coding loci, one must decide how to code the DNA sequences as characters. DNA sequences are generally coded as either nucleotide or amino acid characters, though other approaches have been explored [e.g. 2]. When phylogenetic analyses have been conducted separately on both nucleotide and amino acid sequences, the tree topologies and branch-support values have almost always been incongruent in parts of the trees [e.g. 1, 2, 84]. The choice between alternative methods of coding these sequences can dramatically affect the inferred tree topology, regardless of whether parsimony, likelihood, or distance methods are used for tree construction. Hence, the selective use of nucleotide or amino acid characters needs to be justified for each analysis based on the particular qualities of the sequences used.

Many ideas have been expressed about which type of characters should be used to infer phylogenetic relationships at "deeper" levels (i.e., ancient cladogenic events). Slowly evolving (or "conservative") characters have generally been favored over faster evolving ones [e.g. 2, 3, 6, 65, 66, 85]. This has led to the preferential use of amino acid characters instead of nucleotide characters [e.g. 10, 60, 85], replacement substitutions without silent substitutions [e.g. 32, 61], transversions without transitions [e.g. 60, 62], and first and second codon positions without third codon positions [e.g. 10, 23, 90]. Silent substitutions and third codon positions have been considered useful only for resolving phylogenetic relationships among closely related taxa [e.g. 23, 32]. These discussions have focused on the rate of change in characters to determine their relative merit. However, when nucleotide and amino acid characters are compared, at least six factors need to be considered.

First, amino acid characters, by combining three separate nucleotide characters, constitute composite characters. Composite characters link otherwise discernible states from different characters together to form new states [92]. This causes loss of hierarchic information when unordered character states are used and can create putative synapomorphies that are not present in the separate characters [80]. Each putative synapomorphy may be viewed as an outcome of a synapomorphy at one codon position, acting in concert with a symplesiomorphy at a different codon position, to form what appears to be a separate synapomorphy. However, that novel "synapomorphy" may actually delimit a paraphyletic group. For example, given the matrix of six taxa (A – F) in Fig. 1.1a and their known phylogeny in Fig. 1.1b, taxa C and D represent a paraphyletic group. The first nucleotide character has a synapomorphy for the clade (C, D, E, F), and the third nucleotide character has a synapomorphy for the clade (E, F). Among nucleotide characters 1 – 3, there is no unique character state that could be interpreted as a synapomorphy for taxa C and D. However, that codon specifies a unique amino acid character state for taxa C and D, namely isoleucine. In the context of other characters, that isoleucine may be interpreted as a synapomorphy for taxa C and D, erroneously supporting them as a monophyletic group, as in Fig. 1.1c.

Second, because the genetic code is degenerate, those amino acids specified by more than one codon can appear to be convergently derived on a tree, whereas the underlying nucleotides are not [78]. As a result, even in the absence of multiple hits, amino acid characters may support groupings not supported by nucleotide-sequence characters as well as support contradictory groupings. Except for RNA editing, the source of all changes at the amino acid level is actual change at the nucleotide level. Therefore, in the absence of RNA editing and multiple hits, any groupings apparently supported by changes in the amino acid sequence not supported by the underlying nucleotide characters are unfounded.

Third, silent substitutions, which evolve faster than replacement substitutions and therefore contain most potential phylogenetic signals (i.e., character covariation [5, 24]), are ignored by coding amino acid characters [2]. Saturation curves [12, 13] and pairwise percentage divergences have been extensively used to determine the presence or absence of phylogenetic signals at different codon positions [e.g. 10, 23, 60]. However, this use of saturation curves and pairwise percentage divergences has recently been challenged on three grounds: 1) globally

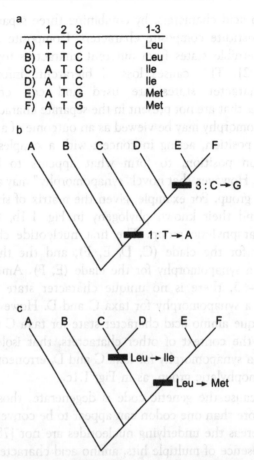

Fig. 1.1: Contrived example of how amino acid characters can create putative synapomorphies not present in their corresponding nucleotide characters. (a) Data matrix of a single codon, coded as both nucleotide and amino acid characters, for six taxa. (b) Known phylogeny for the six taxa with synapomorphies for the nucleotide characters mapped onto the tree. (c) One possible optimization of the amino acid character onto a tree with terminals C and D as sister groups.

homoplastic characters can be locally informative [47, 91]; 2) branch lengths are not considered by saturation curves and pairwise percentage divergences [50, 98]; and 3) frequency of change, without consideration of other factors such as rate heterogeneity among sites and character-state space, cannot be used to determine the phylogenetic information content of a class of characters [8]. Empirical studies have reported greater phylogenetic signals (measured by resolution, branch support, and congruence with independent evidence) at putatively saturated third codon positions than at first or second codon positions combined [e.g. 47, 50, 59, 83, 91].

The above three factors confer advantages to coding nucleotide characters rather than amino acid characters. The fourth through sixth factors are advantages of coding amino acids rather than nucleotides. The fourth factor is that amino acid characters have an increased potential character-state space relative to nucleotide characters, which makes convergence less likely to occur [2, 33]. Amino acid characters have five times the potential character-state space as nucleotide characters (i.e., 20 vs. four possible character states). However, there are functional constraints on the protein that effectively limit the character-state space for amino acid characters [18, 64, 68]. For example, in their study of 35 empirical matrices, Simmons et al. [76] found that the percentage increase in character-state space for parsimony-informative amino acid characters relative to parsimony-informative nucleotide characters ranged from –5% to 160%, with an average of only 50.4%. However, given relatively few alternative character states, as is the case with empirical nucleotide and amino acid characters, small changes in character-state space can lead to dramatic improvements in the ability to accurately reconstruct the phylogeny (Simmons et al., in litt.).

Limited character-state space has been implicated in causing increased homoplasy for nucleotide characters as a whole [36, 63], for first and/or second codon positions relative to third codon positions [44, 67], and for transitions relative to transversions [11]. As Steel and Penny [87] demonstrated, if the character-state space is large enough, parsimony is statistically consistent. Frohlich and Parker [33] also noted that increase in character-state space for amino acids reduced the likelihood of long-distance effects (i.e., optimization of a character state in one clade affects the most parsimonious optimization of the character state in another clade because of missing data in the intervening terminals; [58]). However, long-distance effects only occurred for sequence data when nucleotides or residues were misaligned.

Fifth, amino acid characters are less sensitive to changes in base composition. Convergent changes in base composition can cause the same problem as long-branch attraction [27], wherein unrelated lineages that derive similar GC contents can be resolved as sister groups [51]. Amino acids are less sensitive to this artifact because the shifts in GC content are concentrated at third codon positions [e.g. 42, 49, 69, 73, 95], at which most substitutions are silent. Convergent changes in base composition have been implicated as a source of error in phylogenetic analyses based on nucleotide characters [e.g. 51, 52, 96]. Because amino acids are less sensitive than nucleotides to this artifact, many workers have chosen to use amino acid characters instead of nucleotide characters [e.g. 22, 41, 43, 56]. However, amino acid composition is also affected by changes in base composition [e.g. 4, 30, 46, 75, 94]. As with nucleotide characters, base composition shifts have been implicated as a source of error in phylogenetic analyses based on amino acid characters [9, 39, 53].

Solutions applied to address heterogeneous nucleotide composition among terminals include the frequency-dependent significance test [88], use of LogDet transformation [40, 54 but also see 30, 76], use of models that assume neither homogeneity nor stationariness [34, 35], increased terminal sampling [14], and excluding terminals with heterogeneous nucleotide composition relative to other sampled taxa [94]. However, recent simulation studies have indicated that heterogeneity must be extreme for phylogenetic reconstruction methods to fail, suggesting that this is less of a problem than previously believed [16, 76].

Sixth, because amino acid character-state changes are only caused by replacement substitutions, amino acid characters are not as subject to saturation as the faster evolving silent substitutions [e.g. 10, 60, 85]. As cited above for the third factor, saturation curves, which are based on pairwise comparisons, do not discriminate between presence and absence of phylogenetic signal. This is because saturation curves only take into account overall distance, not taxonomic sampling. By subdividing long branches, multiple hits may be resolved on the different, shorter branches, dispersing "noise" across the tree, and thereby increasing the accuracy of the phylogenetic analyses [45]. This "noise" may be properly considered a phylogenetic signal from homoplastic characters [91]. The more terminals in an analyses for a given clade, the more phylogenetic signal homoplastic characters may contribute [72]. However, because of extinction events, it is not always possible to subdivide long branches. In such cases, one wants to use slower evolving characters to avoid long-branch attraction [28], and amino acid characters are preferable to nucleotide characters.

In this study, an empirical example was used to investigate how these six factors contribute to the relative performance of nucleotide and amino acid characters for phylogenetic inference. The Zanis et al. [100] data matrix of 104 "basal" (i.e., early derived) angiosperm taxa for 11 genes was selected as an empirical example for four reasons. First, it is taxonomically well sampled for inference of the relationships among "basal" angiosperms. Second, a broad taxonomic scale was sampled, including outgroups from all major extant lineages of seed plants and ingroups from both the monocotyledons and dicotyledons. Third, the matrix includes many (eight) protein-coding genes with which to compare nucleotide and amino acid characters. Fourth, a reference tree not based on protein-coding loci is available for the study (i.e., the three rDNA genes sampled by Zanis et al. [100]).

MATERIALS AND METHODS

The methods used in this study largely follow those of Simmons et al. [83, 84].

Matrix Conversion

The Zanis et al. [100] data matrix was obtained in NEXUS format. The 11 genes include five mitochondrial loci (*atp1*, *cox1*, *matR*, *rps2*, and mtSSU rDNA), two plastid loci (*atpB* and *rbcL*), and four nuclear loci (*phyA*, *phyC*, 18S rDNA, and 26S rDNA). The original, 104-taxon matrix was pared down to the 87 taxa sampled for 18S rDNA (the best-sampled of the three rDNA genes) to ensure that all taxa sampled in the protein-coding gene trees would also be present in the reference tree that is based on the three rDNA genes (see below).

The following modifications were performed on the 87-taxon data matrix using MacClade [57]. Nucleotide positions from the 5' and 3' ends of each protein-coding gene that represent partial codons were removed. The *matR* and *rps2* alignments were modified so as to maintain an in-frame alignment, following Zurawski and Clegg's [101] alignment criterion. The *cox1* and *rps2* RNA editing sites excluded in Zanis et al.'s [100] analysis were scored as missing data. Following Zanis et al. [100], two ambiguously aligned regions of 13 and 15 codons respectively were deleted from *rps2*. A singleton gap, inferred to represent a sequencing error, was removed from *atpB*. A long, parsimony-uninformative gap was removed from *matR*. Stop codons that were not near the 3' terminus of a gene were assumed to represent sequencing errors. As such, the expected erroneous nucleotide (based on codons from other taxa) was recoded as missing data. Stop codons that were at (or near) the very end of genes were recoded as missing data, as were any codons 3' of the stop codon so as not to include untranslated regions in this study. Stop codon positions were removed from *atpB* and *rbcL*.

Nucleotide sequences were translated into amino acid sequences based on the standard genetic code using MacClade. Although gap characters should normally be included in sequence-based phylogenetic analyses [35a, 81, 82], they were not included here because they would have been identical for both nucleotide- and amino acid-based matrices and therefore uninformative for the purpose of this study.

Matrix Statistics

To compare the potential amount of phylogenetic information contained in the nucleotide and amino acid characters, the maximum possible number of steps minus the minimum possible number of steps for each character was used as a measure of the "amount of possible synapomorphy" [25: 418]. This statistic was obtained using the "Tree Scores" option in PAUP* [89]. A matrix of nucleotide characters with roughly the same amount of possible synapomorphy as the matrix of all 961 parsimony-informative amino acid characters was created using WinClada [70] by randomly removing characters from the matrix of all parsimony-informative nucleotide characters.

To compare the character-state space of nucleotide and amino acid characters, the average number of character states per parsimony-informative character was determined, after removal of polymorphisms (which were expected to artificially increase the character-state space for amino acid characters caused by missing data in the nucleotide sequences). This was calculated for each gene by simply adding the minimum possible number of steps for parsimony-informative characters to the number of parsimony-informative characters and dividing by the number of parsimony-informative characters.

Tree Searches

All tree searches were conducted using equally weighted parsimony. Jackknife tree searches [26] were performed in PAUP* using the standard 37% deletion probability and emulating "Jac" resampling. One thousand jackknife replicates were performed, with each replicate running ten tree-bisection-reconnection (TBR) searches and up to ten trees held per search. All clades with ≥ 50% support were resolved in the jackknife trees. Searches for most parsimonious trees were performed using 1,000 TBR searches with a maximum of 100 trees held per search.

Tree-based Measures of Performance

The relative performance of the nucleotide and amino acid characters for each of the eight protein-coding genes was determined using four criteria: (1) number of clades resolved; (2) average number of terminals per clade resolved; (3) resolution of well-supported clades based on independent evidence; and (4) jackknife support for the well-supported clades based on independent evidence. The 33 clades with ≥ 95% support in the jackknife tree for the three rDNA genes were used as the independent evidence for criteria three and four. Although the rDNA trees are based on nucleotide characters, there is no reason to believe that the same selective constraints act on these rDNA genes as the protein-coding genes [86]. Therefore, there is no reason to expect the rDNA gene tree to be biased in favor of the nucleotide-based gene tree topologies over the amino acid-based gene tree topologies.

Conflicting Resolution between Nucleotide- and Amino Acid-based Topologies

For each of the eight protein-coding genes, conflicting resolution between nucleotide- and amino acid-based topologies was identified using the respective jackknife trees. Independent evidence was used to determine which, if either, of the conflicting topologies is supported. The primary source of independent evidence was the three-gene rDNA jackknife tree for the 87 taxa from Zanis et al. [100] sampled for 18S rDNA. Secondary sources were morphology-based phylogenetic analyses [19, 48, 55] and jackknife trees constructed based on all ten of the genes from Zanis et al. [100] that were not involved in the conflicting resolution. For the latter approach, both nucleotide- and amino acid-based jackknife trees were constructed independent of one another. When none of these sources yielded well-supported resolution for the taxa in question, nucleotide-based analyses for which much greater taxonomic sampling was performed for the clade in question [17, 77 (in a simultaneous analysis with morphological characters); Lars Chatrou, pers. comm. 2003], or nucleotide-based analyses for which many slowly evolving genes were sampled [37], were used. Note that by taking this approach, the assumption is made that the independent evidence represents the unknown true phylogeny and that no lineage sorting, unrecognized paralogy, horizontal transfer, or introgression events [20, 21] account for the topological discrepancies among the gene trees based on the different loci.

RESULTS

Supplementary data, including an Excel file of the results, data matrices, jackknife trees, and consensus trees are posted in zip files at: http://www.biology.colostate.edu/Research/.

Of the 11,277 nucleotide characters for the eight protein-coding genes, 31% (3,487) were parsimony informative for the 87 taxa included in the analysis. Of the 3,759 amino acid characters, 26% (961) were parsimony informative (Table 1.1). The nucleotide matrices for individual genes were found to have from 2.1 (*matR*) to 5.7 (*rbcL*) times the number of parsimony-informative characters as their corresponding amino acid matrices. The nucleotide matrix for all eight genes together was found to have 4.6 times the amount of possible synapomorphy as the amino acid matrix (23,348 vs. 5,070; Table 1.1). The nucleotide matrices for

Table 1.1: Number of parsimony-informative characters, amount of possible synapomorphy, and average character-state space for each matrix

Gene	Character coding	Number of parsimony-informative chars.	Amount of possible synapomorphy	Average character-state space[b]
all eight	DNA	3,487	23,348	2.71
analyzed	AA	961	5,070	3.39
together	same APS[a]	741	5,068	2.72
atpl	DNA	361	2,323	2.63
	AA	84	613	2.95
atpB	DNA	486	3,494	2.63
	AA	88	426	3.36
cox1	DNA	103	220	2.29
	AA	23	56	2.26
matR	DNA	682	2,696	2.55
	AA	322	1,573	3.31
phyA	DNA	639	4,738	2.93
	AA	163	770	3.74
phyC	DNA	625	4,976	2.94
	AA	153	635	3.87
rbcL	DNA	505	4,699	2.61
	AA	89	892	3.12
rps2	DNA	86	202	2.36
	AA	39	105	2.95

[a]Nucleotide characters with nearly the same amount of possible synapomorphy as all amino acid characters.
[b]For parsimony-informative characters only.

individual genes were found to have from 1.7 (*matR*) to 8.2 (*atpB*) times the amount of possible synapomorphy as their corresponding amino acid matrices. Twenty-three percent fewer (741) randomly selected nucleotide characters have roughly the same amount of possible synapomorphy (5,068 vs. 5,070) as all 961 parsimony-informative amino acid characters.

For parsimony-informative characters from all eight protein-coding genes taken together (excluding polymorphisms), the average number of character states per nucleotide character was 2.71, and the average number for amino acid characters was 3.39 (Table 1.1). This represents a 25% increase in character-state space for amino acid characters relative to their corresponding nucleotide characters. Individual genes were found to have anywhere from a 1% decrease (*cox1*) to a 32% increase (*phyC*) in character-state space for amino acid matrices relative to their corresponding nucleotide matrices.

When all eight protein-coding genes were analyzed together, the nucleotide characters resolved 39% more clades than the amino acid characters (82 versus 59; Table 1.2). However, when reduced to the same amount of possible synapomorphy as the amino acid characters, the nucleotides resolved 3% fewer clades than the amino acids. The nucleotide matrices for individual genes resolved from 10% (*phyA*) to 307% (*atpB*) more clades than their corresponding amino acid matrices.

On average, nucleotide characters from all eight protein-coding genes analyzed together resolved 32% larger clades than the amino acid characters. On average, within the ingroup (i.e., the angiosperms), nucleotide characters from all eight protein-coding genes analyzed together resolved 62% larger clades than the amino acid characters. However, when reduced to the same amount of possible synapomorphy as the amino acid matrix, the nucleotides resolved an average of 18% smaller clades than the amino acid characters within the ingroup. The nucleotide matrices for individual genes resolved from 3% smaller (*matR*) to 209% larger (*atpB*) clades than the amino acid characters, on average, within the ingroup.

The total number, percentage, and average jackknife support of the clades resolved by each of the eight genes for the 33 clades with ≥ 95% jackknife support on the rDNA jackknife tree is presented in Table 1.3. When all eight protein-coding genes were analyzed together, the nucleotide characters resolved all 33 of the clades whereas the amino acid characters resolved 81.8% of the clades. Likewise, when reduced to the same amount of possible synapomorphy as the amino acids, the

Table 1.2: Total number of clades resolved and the average number of terminals per clade in the jackknife tree for each matrix

Gene	Character coding	Number of clades resolved	Avg. number of terminals per clade	Avg. number of terminals per clade within the ingroup
all eight	DNA	82	11.1	9.7
analyzed	AA	59	8.4	6.0
together	same APS	57	7.7	5.1
atp1	DNA	48	8.5	5.7
	AA	26	11.1	2.3
atpB	DNA	61	8.0	7.1
	AA	15	7.2	2.3
cox1	DNA	12	4.4	2.5
	AA	7	5.6	2.2
matR	DNA	57	8.3	6.0
	AA	50	8.8	6.2
phyA	DNA	22	4.3	4.3
	AA	20	4.2	4.2
phyC	DNA	26	4.1	4.1
	AA	18	2.6	2.6
rbcL	DNA	67	5.2	4.1
	AA	25	5.9	2.8
rps2	DNA	11	8.2	4.8
	AA	6	9.3	2.3

nucleotides resolved 4% more of the 33 clades than did the amino acids. The nucleotide matrices for individual genes resolved from the same number (cox1) to 271% more (atpB) clades as did the corresponding amino acid matrices.

For the clades resolved by both nucleotide and amino acid characters when all eight protein-coding genes were analyzed together, the nucleotide characters provided an average of 99.1% jackknife support whereas the amino acid characters provided an average of 94.9%. However, when reduced to the same amount of possible synapomorphy, the nucleotides provided an average of 1% lower jackknife support than did the amino acids for the clades resolved by both matrices. The nucleotide characters for individual genes provided from 3% (matR) to 36% (atpB) higher average jackknife support relative to the corresponding amino acid characters for the respective clades resolved by both character codings.

Table 1.3: Total number, percentage, and average jackknife support of the clades resolved by each of the eight genes for the 33 clades with ≥ 95% jackknife support on the rDNA jackknife tree

Gene	Character coding	Number of clades resolved	Percentage of clades resolved[a]	Avg. jackknife support for clades resolved[b]	Avg. jackknife support for all clades
all eight	DNA	33	100	99.1	96.0
analyzed	AA	27	81.8	94.9/95.8[c]	77.6
together	same APS	28	84.8	94.8	78.1
atp1	DNA	26	83.9	95.6	76.2
	AA	11	35.5	78.3	27.8
atpB	DNA	26	92.9	100	83.5
	AA	7	25.0	73.4	18.4
cox1	DNA	3	30.0	100	30.0
	AA	3	30.0	83.3	25.0
matR	DNA	26	86.7	96.2	81.9
	AA	25	83.3	93.0	77.5
phyA	DNA	9	75.0	98.3	72.5
	AA	8	66.7	92.7	58.7
phyC	DNA	10	83.3	99.1	80.2
	AA	9	75.0	91.2	68.4
rbcL	DNA	25	75.8	99.9	73.8
	AA	14	42.4	82.9	35.2
rps2	DNA	6	60.0	99.0	54.7
	AA	4	40.0	87.5	35.0

[a] Relative to the number of clades that could be resolved given the taxonomic sampling for that gene.
[b] For clades resolved by both nucleotide and amino acid characters for each gene.
[c] Relative to all nucleotide characters and to nucleotide characters with the same amount of possible synapomorphy respectively.

When all eight protein-coding genes were analyzed together, the nucleotide characters provided an average of 96% jackknife support for all 33 of the well-supported rDNA clades, whereas the amino acid characters provided an average of only 77.6% jackknife support. When reduced to the same amount of possible synapomorphy, the nucleotides provided slightly higher (0.6%) support relative to the amino acids. The nucleotide characters for individual genes provided from 6% (matR) to 354% (atpB) higher average jackknife support relative to the corresponding amino acid characters for all 33 of the well-supported rDNA clades.

Of the 167 clades resolved on the amino acid jackknife trees for each of the eight protein-coding genes analyzed separately, 13% (22) conflicted with clades resolved by the nucleotide jackknife trees for the respective genes (Table 1.4). Independent evidence (i.e., phylogenetic analyses based on other loci and/or morphological evidence) supports the nucleotide resolution in 68% (15) of the cases, supports the amino acid resolution in 18% (4) of the cases, and is equivocal or supports neither the nucleotide or amino acid resolution in 14% (3) of the 22 cases. There were no cases of conflicting resolution between the nucleotide and amino acid jackknife trees for *cox1* or *rps2*. For *atp1*, *atpB*, and *rbcL*, the nucleotide resolution was supported by independent evidence for all cases of conflicting resolution. In contrast, for both *matR* and *phyA*, independent evidence supported both the nucleotide and amino acid resolutions in one or more cases each.

For the 15 cases in which the nucleotide resolution was supported over the amino acid resolution by independent evidence, only five cases could be attributed to composite characters and/or convergent character states (Table 1.4). These five cases could be explained by eight instances of convergent states and one instance of a composite character. When the composite and convergent character states were recoded as missing data for the taxa in question and the amino acid jackknife trees were recalculated, none of the five conflicting clades was resolved. For the four cases in which the amino acid resolution was supported over the nucleotide resolution by independent evidence, there was no apparent convergence in nucleotide composition at parsimony-informative characters for the taxa in question.

DISCUSSION

Consistent with Simmons et al. [83, 84], nucleotide characters were generally found to outperform amino acid characters with respect to the tree-based measures of performance examined (Tables 1.1 to 1.4). Of the 15 cases of conflicting resolution for which independent evidence supported the nucleotide-based resolution, five could be explained by composite characters and/or convergent character states. The remaining ten cases, for which no composite characters or convergent character states were identified as "synapomorphies" for the conflicting resolution, may be explained by silent substitutions being more informative than replacement substitutions. This is because the only thing distinguishing the nucleotide matrix from the amino acid matrix, other than composite

Table 1.4: Clades in conflict between the amino acid and the nucleotide jackknife trees

Amino acid topology	Nucleotide topology	Independent evidence	Composite characters or convergent character states	Resolution after removal of problematic amino acids
1 atp1 54 [Xanthorhiza, Tetracentron]	53 [Tetracentron, Trochodendron]	support DNA topology rDNA jackknife tree (100%)	none	N/A
2 atp1 100 [Aristolochia, 62 [Piper, Peperomia]]	100 [Peperomia, Piper, 97 [Anemopsis, Saururaceae]]	support DNA topology rDNA jackknife tree (100%)	none	N/A
3 atp1 77 [Pinus, Gnetales], 61 [Angiosperms, Podocarpus]	[Podocarpus, Pinus, Gnetales]	support DNA topology rDNA jackknife tree (97%)	none	N/A
4 atpB 61 [Welwitschia, Spathiphyllum]	100 [Welwitschia, Gnetum]	support DNA topology rDNA jackknife tree (100%)	2 convergent states 143: Glu (GAG, GAA) 320: Lys (AAG, AAA)	loss of resolution
5 atpB 64 [Lardizabala, Akebia, Sargentodoxa]	100 [Sargentodoxa, Akebia, Lardizabala]	support DNA topology rDNA jackknife tree (100%)	none	N/A

Contd.

Table 1.4 (Contd.)

Amino acid topology	Nucleotide topology	Independent evidence	Composite characters or convergent character states	Resolution after removal of problematic amino acids
6 atpB				
77, 54: Brasenia, Nymphaea, Nuphar	96: Nymphaea, Nuphar, Brasenia, Cabomba	support DNA topology rDNA jackknife tree (99%)	none	N/A
7 atpB				
56, 65: Pinus, Podocarpus, Gnetales, Angiosperms	56, 77: Gnetales, Pinus, Podocarpus	support DNA topology rDNA jackknife tree (97%)	none	N/A
8 matR				
50: Asimina, Annona, Cananga	70: Cananga, Annona, Asimina	support DNA topology matK + rbcL + trnL-F (Lars Chatrou pers. comm. 2003)	none	N/A
9 matR				
54, 59: Nelumbo, Platanus, Grevillea	56: Grevillea, Ceratophyllum	support AA topology all 11 genes except mat R coded as DNA jackknife tree (97%)	N/A	N/A
10 matR				
71: Croomia, Dioscorea, Asparagus	52: Asparagus, Dioscorea, Croomia	no independent, well -supported evidence		

Contd.

Table 1.4 (Contd.)

Amino acid topology	Nucleotide topology	Independent evidence	Composite characters character states	Resolution after amino acids
11 *matR* 100 ⌐ Triglochin └ Alisma 59 ⌐ Pleea └ Tofieldia	100 ⌐ Alisma └ Triglochin 78 ⌐ Tofieldia └ Pleea	support AA topology rDNA jackknife tree (89%)	N/A	N/A
12 *matR* 56 ⌐ Aristolochia └ Lactoris 65 ⌐ Asarum └ Saruma	62 ▼ Aristolochia 54 ⌐ Lactoris 100 ⌐ Asarum └ Asarum	support AA topology morphology (85% bootstrap) [48]	N/A	N/A
13 *phyA* 56 ⌐ Annona 82 ⌐ Eupomatia ⌐ Magnolia └ Degeneria	87 ⌐ Annona ⌐ Eupomatia 84 ⌐ Magnolia └ Degeneria	neither supported morphology & 12 loci (90% bootstrap) [77]		
14 *phyA* ⌐ Lactoris ⌐ Saruma └ Aristolochia	⌐ Lactoris ⌐ Saruma 68 ⌐ Aristolochia ▼	support AA topology morphology (85% bootstrap) [48]	N/A	N/A

Contd.

Table 1.4 (Contd.)

Amino acid topology	Nucleotide topology	Independent evidence	Composite characters or convergent character states	Resolution after removal of problematic amino acids
15 phyC				
65 [Lardizabalaceae / Nelumbo	54 [Xanthorhiza / Trochodendral. — 59 [Lardizabalaceae / Nelumbo	support DNA topology morphology (2 branches) [19]	2 convergent states, 1 composite character 105: Leu (TTG, CTG) 260: Lys (AAA, AAG) 262: Lys (composite)	loss of resolution
16 phyC				
58 [Xanthorhiza / Pleea	54 [Xanthorhiza / Trochodendral. — 59 [Lardizabalaceae / Nelumbo	support DNA topology morphology (2 branches) [19]	2 convergent states 118: Ala (GCG, GCA) 156: Val (GTA, GTT)	loss of resolution
17 phyC				
55 [Acorus — 56 [90 [Nymphaea / Cabomba	71 [Acorus / Spathiphyllum — 56 [Dioscorea / Pleea	support DNA topology rDNA jackknife tree (91%) 17 plastid loci (5 branches) [37]	2 convergent states 198: Ser (AGC, TCT) 285: Cys (TGC, TGT)	loss of resolution
18 phyC				
59 [Lactoris — 100 [Piper / Saururaceae	89 [Saruma / Aristolochia — 52 [Lactoris	support DNA topology all 11 genes except phyC coded as AA jackknife tree (72%) atpB + rbcL (70% jackknife) [17]	none	N/A

Contd.

Table 1.4 (Contd.)

Amino acid topology	Nucleotide topology	Independent evidence	Composite characters or convergent character states	Resolution after removal of problematic amino acids
19 phyC				
Aristolochia / Saruma / 59 Lactoris ▼ (59, 89)	Saruma / Aristolochia / Lactoris (52)	ambiguous support DNA: atpB + rbcL (70% jackknife): [17] AA: morphology (85% bootstrap) [48] AA: rDNA jackknife tree (66%)		
20 rbcL				
Ranunculaceae / Euptelea (62, 53)	Cocculus / Podophyllum / Ranunculaceae (91)	support DNA topology morphology (2 branches) [19] [55]	none	N/A
21 rbcL				
Nymphaea / Nuphar / Cabombaceae (64)	Cabombaceae / Nymphaea / Nuphar (73)	support DNA topology morphology (4 Bremer sppt.) [19]	1 possible convergent state 225: Val (GTC, GTA)	loss of resolution
22 rbcL				
Trimenia / Austrobaileya / Schisandra / Illicium / Amborella / Nymphheales (52, 64)	Austrobaileya / Trimenia / Schisandra / Illicium (96, 80, 73)	support DNA topology rDNA jackknife tree (99%)	none	N/A

characters and convergent character states, is the inclusion of silent substitutions in the nucleotide matrix. In contrast, for the four cases of conflicting resolution for which independent evidence supported the amino acid-based resolution, replacement substitutions are inferred to have accurately tracked the phylogenetic relationships among the taxa in question, whereas the majority of the silent substitutions failed to do so.

Although the nucleotide characters generally outperformed the amino acid characters, a wide diversity in their relative performance is evident on comparing the eight protein-coding genes relative to one another. If one accepts these eight genes, which have been sampled from all three plant genomes, as representative of all protein-coding plant genes, the earlier study [83, 84] was based on two genes (*atpB* and *rbcL*) that represent the extreme in terms of how nucleotide characters outperformed amino acid characters. In contrast, amino acid characters performed very similar to nucleotide characters from *matR* for all tree-based measures of performance examined.

One may expect amino acid characters to perform relatively better than nucleotide characters given lower taxonomic sampling (when examining the relationships among taxa in the same clade; [83]). This was not the case for *matR*, for which 80 taxa were sampled, nearly the same number of taxa as for *atpB* [79]. Nucleotide and amino acid characters from these two genes differed dramatically with respect to the tree-based measures of performance. However, genes for which relatively few (26-35) taxa were sampled do show some convergence in tree-based measures of performance for the nucleotide and amino acid characters.

Based on the differential relative performance of nucleotide and amino acid characters for the eight protein-coding genes examined here, one cannot make blanket statements regarding which of the two types of character coding is better. This is in part because most contemporary phylogenetic analyses sample a wide array of taxonomic diversity [e.g. 15, 47] as opposed to a paucity of exemplars [e.g. 38]. As a result, each taxon is very closely related to some other taxa in the analysis (e.g. the two species of *Ceratophyllum* included in this study), yet only distantly related to others. As Reed and Sperling [74: 286] noted: "A character may be both 'good' and 'bad' [within a given phylogenetic analysis,] depending on what level of divergence it is being used to resolve."

One approach to incorporating signal from both nucleotide and amino acid characters into a simultaneous analysis [71] is to apply the nonredundant coding of dependent characters method [31]. This method represents a modification and extension of that proposed by Agosti et al. [1], who introduced the simultaneous use of nucleotide and amino acid characters derived from the same sequence for phylogenetic analyses based on protein-coding genes. Agosti et al. [1] considered their method an objective way for up-weighting congruent, redundant characters (i.e., nonsynonymous substitutions may be given twice the weight of synonymous substitutions) with the presumed advantage that slowly evolving replacement substitutions are more reliable than faster evolving silent substitutions. However, as demonstrated by both Simmons et al. [84] and this study, silent substitutions may actually be better than replacement substitutions for phylogenetic reconstruction of some clades, and the assumption that slowly evolving nucleotide characters are necessarily more reliable than faster evolving nucleotide characters has been thoroughly refuted [7, 8, 47, 50, 91, 97, 99].

Briefly, the nonredundant coding of dependent characters method for gene-tree inference is implemented by coding nucleotide, base, amino acid, and/or class of amino acid characters from each gene and then searching for identical character-state distributions among dependent characters (e.g. the three nucleotide characters for a codon and the amino acid for which they code). The higher level characters that have identical character-state distributions with any of the lower level character(s) on which they are demonstrably dependent are then deactivated. All other characters are equally weighted in the gene-tree analysis [31]. For example, if an analysis that includes four taxa (1 – 4) has CTT at a given codon in taxa 1 and 2, and ATT at the given codon in taxa 3 and 4, the amino acid character for that codon has the same character-state distribution (leucine for taxa 1 and 2, isoleucine for taxa 3 and 4) as the first codon position nucleotide character (cytosine for taxa 1 and 2, adenine for taxa 3 and 4). Therefore, the amino acid character is redundant with the nucleotide character from the first codon position and is deactivated, rather than included, in the phylogenetic analysis of these four taxa.

Acknowledgements

I thank Lars Chatrou and Jerry Davis for helpful discussions on the phylogeny of the Annonaceae and monocots respectively, Doug Soltis for sending the data matrix from Zanis et al. [100], and John Freudenstein,

Melissa Islam, Chris Randle, and Libing Zhang for constructive feedback on the manuscript. Financial support was provided by a CSU Career Enhancement Grant.

REFERENCES

[1] Agosti D, Jacobs D, DeSalle R. On combining protein sequences and nucleic acid sequences in phylogenetic analysis: the homeobox protein case. Cladistics 1996; 12: 65-82.

[2] Albert VA, Backlund A, Bremer K, et al. Functional constraints and rbcL evidence for land plant phylogeny. Ann Mo Bot Gard 1994; 81: 534-567.

[3] Albert VA, Chase MW, Mishler BD. Character-state weighting for cladistic analysis of protein-coding DNA sequences. Ann Mo Bot Gard 1993; 80: 752-766.

[4] Anderson SGE, Sharp PM. Codon usage and base composition in Rickettsia prowazekii. J Mol Evol 1996; 42: 525-536.

[5] Archie JW. A randomization test for phylogenetic information in systematic data. Syst Zool 1989; 38: 219-252.

[6] Arnason U, Gullberg A. Relationship of baleen whales established by cytochrome b gene sequence comparison. Nature 1994; 367: 726-728.

[7] Baker RH, Wilkinson GS, DeSalle R. Phylogenetic utility of different types of molecular data used to infer evolutionary relationships among stalk-eyed flies (Diopsidae). Syst Biol 2001; 50: 87-105.

[8] Björklund M. Are third positions really that bad? A test using vertebrate cytochrome b. Cladistics 1999; 15: 191-197.

[9] Black WC, Roehrdanz RL. Mitochondrial gene order is not conserved in arthropods: prostriate and metastriate tick mitochondrial genomes. Mol Biol Evol 1998; 15: 1772-1785.

[10] Blouin MS, Yowell CA, Courtney CH, Dame JB. Substitution bias, rapid saturation, and the use of mtDNA for nematode systematics. Mol Biol Evol 1998; 15: 1719-1727.

[11] Broughton RE, Stanley SE, Durrett RT. Quantification of homoplasy for nucleotide transitions and transversions and a reexamination of assumptions in weighted phylogenetic analyses. Syst Biol 2000; 49: 617-627.

[12] Brown WM, George M, Wilson AC. Rapid evolution of animal mitochondrial DNA. Proc Natl Acad Sci USA 1979; 76: 1967-1971.

[13] Brown WM, Prager EM, Wang A, Wilson AC. Mitochondrial DNA sequences of primates: tempo and mode of evolution. J Mol Evol 1982; 18: 225-239.

[14] Chang BSW, Campbell DL. Bias in phylogenetic reconstruction of vertebrate rhodopsin sequences. Mol Biol Evol 2000; 17: 1220-1231.

[15] Chase MW, Soltis DE, Olmstead RG, et al. 1993. Phylogenetics of seed plants: an analyses of nucleotide sequences from the plastid gene rbcL. Ann Mo Bot Gard 1993; 80: 528-580.

[16] Conant GC, Lewis PO. Effects of nucleotide composition bias on the success of the parsimony criterion in phylogenetic inference. Mol Biol Evol 2001; 18: 1024-1033.

[17] Davis JI, Stevenson DW, Petersen G, et al. A phylogeny of the monocots, as inferred from *rbcL* and *atpA* sequence variation. Syst Bot 2004; 29: 467-510.

[18] Dayhoff MO, Eck RV, Park, CM. A model of evolutionary change in proteins. In: Dayhoff MO, ed. Atlas of protein sequence and structure. Washington, DC: National Biomedical Research Foundation, 1972: 89-99.

[19] Doyle JA, Endress PK. Morphological phylogenetic analysis of basal angiosperms: comparison and combination with molecular data. Int J Plant Sci 2000; 161: S121-S153.

[20] Doyle JJ. Gene trees and species trees: molecular systematics as one-character taxonomy. Syst Bot 1992; 17: 144-163.

[21] Doyle JJ. *Homoplasy* connections and disconnections: genes and species, molecules and morphology. In: Sanderson MJ, Hufford L, eds. *Homoplasy: the recurrence of similarity in evolution.* San Diego: Academic Press, 1996: 37-66.

[22] Durnford DG, Deane JA, Tan S, et al. A phylogenetic assessment of the eukaryotic light-harvesting antenna proteins, with implications for plastid evolution. J Mol Evol 1999; 48: 59-68.

[23] Edwards SV, Arctander P, Wilson AC. Mitochondrial resolution of a deep branch in the genealogical tree for perching birds. Proc R Soc Lond Ser B 1991; 243: 99-107.

[24] Faith D, Cranston P. Could a cladogram this short have arisen by chance alone? Cladistics 1991; 7: 1-28.

[25] Farris JS. The retention index and the rescaled consistency index. Cladistics 1989; 5: 417-419.

[26] Farris JS, Albert VA, Källersjö M, et al. Parsimony jackknifing outperforms neighbor-joining. Cladistics 1996; 12: 99-124.

[27] Felsenstein J. Cases in which parsimony or compatibility methods will be positively misleading. Syst Zool 1978; 27: 401-410.

[28] Felsenstein J. Parsimony in systematics: biological and statistical issues. Ann Rev Ecol Syst 1983; 14: 313-333.

[29] Foster PG, Hickey DA. Compositional bias may affect both DNA-based and protein based phylogenetic reconstructions. J Mol Evol 1999; 48: 284-290.

[30] Foster PG, Jermiin LS, Hickey DA. Nucleotide composition bias affects amino acid content in proteins coded by animal mitochondria. J Mol Evol 1997; 44: 282-288.

[31] Freudenstein JV, Pickett KM, Simmons MP, Wenzel JW. From basepairs to birdsongs: phylogenetic data in the age of genomics. Cladistics 2003; 19: 333-347.

[32] Friedlander TP, Regier JC, Mitter C. Phylogenetic information content of five nuclear gene sequences in animals: initial assessment of character sets from concordance and divergence studies. Syst Biol 1994; 43: 511-525.

[33] Frohlich MW, Parker DS. The mostly male theory of flower evolutionary origins: from genes to fossils. Syst Bot 2000; 25: 155-170.

[34] Galtier N, Gouy M. Inferring phylogenies from DNA sequences of unequal base compositions. Proc Natl Acad Sci USA 1995; 92: 11317-11321.

[35] Galtier N, Gouy M. Inferring pattern and process: maximum-likelihood implementation of a nonhomogeneous model of DNA sequence evolution for phylogenetic analysis. Mol Biol Evol 1998; 15: 871-879.

[35a] Giribet G, Wheeler WC. On gaps. Mol Phylogen Evol 1999; 13: 132-143.

[36] Goodman M, Czelusniak J, Moore GW, et al. Fitting the gene lineage into its species

lineage: a parsimony strategy illustrated by cladograms constructed from globin sequences. Syst Zool 1979; 28: 132-163.

[37] Graham SW, Olmstead RG. Utility of 17 chloroplast genes for inferring the phylogeny of the basal angiosperms. Amer J Bot 2000; 87: 1712-1730.

[38] Graur D, Durent L, Gouy M. Phylogenetic position of the order Lagomorpha (rabbits, hares and allies). Nature 1996; 379: 333-335.

[39] Gu X, Hewett-Emmett D, Li W-H. Directional mutational pressure affects the amino acid composition and hydrophobicity of proteins in bacteria. Genetica 1998; 102-103: 383-391.

[40] Gu X, Li W-H. Bias-corrected paralinear and LogDet distances and tests of molecular clocks and phylogenies under nonstationary nucleotide frequencies. Mol Biol Evol 1996; 13: 1375-1383.

[41] Hasebe M, Kofuji R, Ito M, et al. Phylogeny of gymnosperms inferred from *rbc*L gene sequences. Bot Mag Tokyo 1992; 105: 673-679.

[42] Hasegawa M, Hashimoto T. Ribosomal RNA trees misleading. Nature 1993; 361: 23.

[43] Hasegawa M, Hashimoto T, Adachi J, et al. Early branchings in the evolution of eukaryotes: ancient divergence of Entamoeba that lacks mitochondria revealed by protein sequence data. J Mol Evol 1993; 36: 380-388.

[44] Hassanin A, Lecointre G, Tillier S. The 'evolutionary signal' of *homoplasy* in protein-coding gene sequences and its consequences for a priori weighting in phylogeny. CR Acad Sci III-Vie 1998; 321: 611-620.

[45] Hillis DM. Inferring complex phylogenies. Nature 1996; 383: 130-131.

[46] Jukes TH, Bhushan V. Silent nucleotide substitutions and G + C content of some mitochondrial and bacterial genes. J Mol Evol 1986; 24: 39-44.

[47] Källersjö M, Albert VA, Farris JS. *Homoplasy increases* phylogenetic structure. Cladistics 1999; 15: 91-93.

[48] Kelly LM, González F. Phylogenetic relationships in Aristolochiaceae. Syst Bot 2003; 28: 236-249.

[49] Klenk H-P, Zillig W. DNA-dependent RNA polymerase subunit B as a tool for phylogenetic reconstructions: branching topology of the archaeal domain. J Mol Evol 1994; 38: 420-432.

[50] Lewis LA, Mishler BD, Vilgalys R. Phylogenetic relationships of the liverworts (Hepaticeae), a basal embryophyte lineage, inferred from nucleotide sequence data of the chloroplast gene *rbc*L. Mol Phylogen Evol 1997; 7: 377-393.

[51] Lockhart PJ, Howe CJ, Bryant DA, et al. Substitutional bias confounds inference of cyanelle origins from sequence data. J Mol Evol 1992; 34: 153-162.

[52] Lockhart PJ, Penny D. The problem of GC content, evolutionary trees and the origins of CHL-a/b photosynthetic organelles: are the procholorophytes a eubacterial model for higher plant photosynthesis? In: Murata N, ed. Research in photosynthesis. Dordrecht: Kluwer Academic Publishers, 1992; 3: 499-505.

[53] Lockhart PJ, Penny D, Hendy MD, et al. Controversy on chloroplast origins. FEBS Lett 1992; 301: 127-131.

[54] Lockhart PJ, Steel MA, Hendy MD, Penny D. 1994. Recovering evolutionary trees under a more realistic model of sequence evolution. Mol Biol Evol 1994; 11: 605-612.

[55] Loconte H, Stevenson DW. Cladistics of the Magnoliidae. Cladistics 1991; 7: 267-296.

[56] Loomis WF, Smith DW. Molecular phylogeny of *Dictyostelium discoideum* by protein sequence comparison. Proc Natl Acad Sci USA 1990; 87: 9093-9097.

[57] Maddison DR, Maddison WP. MacClade: Analysis of phylogeny and character evolution version 4.03. Sunderland, MA: Sinauer Associates, 2001

[58] Maddison WP. Missing data versus missing characters in phylogenetic analyses. Syst Biol 1993; 42: 576-581.

[59] Manhart JR. Phylogenetic analysis of green plant rbcL sequences. Mol Phylogen Evol 1994; 3: 114-127.

[60] Meyer A. Shortcomings of the cytochrome b gene as a molecular marker. Trends Ecol Evol 1994; 9: 278-280.

[61] Meyer A, Wilson AC. Origin of tetrapods inferred from their mitochondrial DNA affiliation to lungfish. J Mol Evol 1990; 31: 359-364.

[62] Mindell DP, Shultz JW, Ewald PW. The AIDS pandemic is new, but is HIV new? Syst Biol 1995; 44: 77-92.

[63] Mishler BD, Bremer K, Humphries CJ, Churchill SP. The use of nucleic acid sequence data in phylogenetic reconstruction. Taxon 1988; 37: 391-395.

[64] Miyamoto MM, Fitch WM. Testing the covarion hypothesis of molecular evolution. Mol Biol Evol 1995; 12: 503-513.

[65] Montgelard C, Catzeflis FM, Douzery E. Phylogenetic relationships of artiodactyls and cetaceans as deduced from the comparison of cytochrome b and 12S rRNA mitochondrial sequences. Mol Biol Evol 1997; 14: 550-559.

[66] Moritz C, Dowling TE, Brown WM. Evolution of animal mitochondrial DNA: relevance for population biology and systematics. Ann Rev Ecol Syst 1987; 18: 269-292.

[67] Naylor GJP, Collins TM, Brown WM. Hydrophobicity and phylogeny. Nature 1995; 373: 565-566.

[68] Naylor GJP, Gerstein M. Measuring shifts in function and evolutionary opportunity using variability profiles: a case study of the globins. J Mol Evol 2000; 51: 223-233.

[69] Nishiyama T, Kato M. Molecular phylogenetic analysis among bryophytes and tracheophytes based on combined data of plastid coded genes and the 18S rRNA gene. Mol Biol Evol 1999; 16: 1027-1036.

[70] Nixon KC. Winclada. Ithaca, NY: Published by the author, 2002.

[71] Nixon KC, Carpenter JM. On simultaneous analysis. Cladistics 1996; 12: 221-242.

[72] Philippe H, Lecointre G, Lê HLV, Le Guyader H. A critical study of homoplasy in molecular data with the use of a morphologically based cladogram, and its consequences for character weighting. Mol Biol Evol 1996; 13: 1174-1186.

[73] Prager EM, Wilson AC. Ancient origin of lactalbumin from lysozyme: analysis of DNA and amino acid sequences. J Mol Evol 1988; 27: 326-335.

[74] Reed RD, Sperling AH. Interaction of process partitions in phylogenetic analysis: an example from the swallowtail butterfly genus Papilio. Mol Biol Evol 1999; 16: 286-297.

[75] Rodríguez-Trelles F, Tarrío R, Ayala FJ. Switch in codon bias and increased rates of amino acid substitution in the Drosophila saltans species group. Genetics 1999; 153: 339-350.

[76] Rosenberg MS, Kumar S. Heterogeneity of nucleotide frequencies among evolutionary lineages and phylogenetic inference. Mol Biol Evol 2003; 20: 610-621.

[77] Sauquet H, Doyle JA, Scharaschkin T, et al. Phylogenetic analysis of Magnoliales and Myristicaceae based on multiple data sets: implications for character evolution. Bot J Linn Soc 2003; 142: 125-186.

[78] Simmons MP. A fundamental problem with amino-acid-sequence characters for phylogenetic analyses. Cladistics 2000; 16: 274-282.

[79] Simmons MP, Carr TG, O'Neill K. Relative character-state space, amount of potential phylogenetic information, and heterogeneity of nucleotide and amino acid characters. Mol Phylogen Evol 2004; 32: 913-926.

[80] Simmons MP, Freudenstein JV. Artifacts of coding amino acids and other composite characters for phylogenetic analysis. Cladistics 2002; 18: 354-365.

[81] Simmons MP, Ochoterena H. Gaps as characters in sequence-based phylogenetic analyses. Syst Biol 2000; 49: 369-381.

[82] Simmons MP, Ochoterena H, Carr TG. Incorporation, relative *homoplasy*, and effect of gap characters in sequence-based phylogenetic analyses. Syst Biol 2001; 50: 454-462.

[83] Simmons MP, Ochoterena H, Freudenstein JV. Amino acid vs. nucleotide characters: challenging preconceived notions. Mol Phylogen Evol 2002; 24: 78-90.

[84] Simmons MP, Ochoterena H, Freudenstein JV. Conflict between amino acid and nucleotide characters. Cladistics 2002; 18: 200-206.

[85] Simon C, Frati F, Beckenbach A, et al. Evolution, weighting, and phylogenetic utility of mitochondrial gene sequences and a compilation of conserved polymerase chain reaction primers. Ann Entom Soc Amer 1994; 87: 651-701.

[86] Stanger-Hall K, Cunningham CW. Support for a monophyletic Lemuriformes: overcoming incongruence between data partitions. Mol Biol Evol 1998; 15: 1572-1577.

[87] Steel M, Penny D. Parsimony, likelihood, and the role of models in molecular phylogenetics. Mol Biol Evol 2000; 17: 839-850.

[88] Steel MA, Lockhart, PJ, Penny D. Confidence in evolutionary trees from biological sequence data. Nature 1993; 364: 440-442.

[89] Swofford DL. PAUP*: Phylogenetic analyses using parsimony (*and other methods). Sunderland, MA: Sinauer Associates, Inc., 1998.

[90] Swofford DL, Olsen GJ. 1990. Phylogeny reconstruction. In: Hillis DM, Moritz C, eds. Molecular systematics. Sunderland, MA: Sinauer Associates, Inc., 1990: 411-501.

[91] Wenzel JW, Siddall ME. Noise. Cladistics 1999; 15: 51-64.

[92] Wilkinson M. A comparison of two methods of character construction. Cladistics 1995; 11: 297-308.

[93] Wilson K, Cahill V, Ballment E, Benzie J. The complete sequence of the mitochondrial genome of the crustacean *Penaeus monodon*: are malacostracan crustaceans more closely related to insects than branchiopods? Mol Biol Evol 2000; 17: 863-874.

[94] Wilson WC, Ma H-C, Venter EH, et al. Phylogenetic relationships of bluetongue viruses based on gene S7. Virus Res 2000; 67: 141-151.

[95] Wirth T, Le Guellec R, Veuille M. Directional substitution and evolution of nucleotide content in the *cytochrome oxidase* II gene in earwigs (dermapteran insects). Mol Biol Evol 1999; 16: 1645-1653.

[96] Woese CR, Achenbach L, Rouviere P, Mandelco L. Archael phylogeny: reexamination of the phylogenetic position of *Archaeoglobus fulgidus* in light of certain composition-induced artifacts. Syst Appl Microbiol 1991; 14: 364-371.

[97] Yang Z. Maximum-likelihood models for combined analyses of multiple sequence data. J Mol Evol 1996; 42: 587-596.

[98] Yang Z. On the best evolutionary rate for phylogenetic analysis. Syst Biol 1998; 47: 125-133.

[99] Yoder AD, Vilgalys R, Ruvolo M. Molecular evolutionary dynamics of cytochrome *b* in strepsirrhine primates: the phylogenetic significance of third-position transversions. Mol Biol Evol 1996; 13: 1339-1350.

[100] Zanis MJ, Soltis DE, Soltis PS, et al. The root of the angiosperms revisited. Proc Natl Acad Sci USA, 2002; 99: 6848-6853.

[101] Zurawski G, Clegg MT. Evolution of higher-plant chloroplast DNA-encoded genes: implications for structure-function and phylogenetic studies. Annu Rev Pl Physiol 1987; 38: 391-418.

Phylogenetics, Floral Evolution, and Rapid Radiation in the Tropical Monocotyledon Family Costaceae (Zingiberales)

CHELSEA DVORAK SPECHT
University of Vermont, Department of Botany–Pringle Herbarium, Burlington, VT 05405 USA

ABSTRACT

A phylogenetic analysis of the pantropical monocotyledon family Costaceae (Zingiberales) was developed using both molecular (ITS, *trn*L-F, *trn*K, *mat*K) and morphological data. Character sampling was designed to provide phylogenetic signal at the species level. Taxon sampling within the Costaceae was selected to encompass geographical and morphological diversity of the family in order to test the monophyly of taxonomic groups proposed by Schumann [39] and Maas [22, 23] and to define trends in character evolution. The results indicate that the generic taxonomy of Costaceae need to be revised to reflect the phylogenetic hypothesis. Evolutionary trends in floral morphology show that close associations with pollinators have evolved several times from generalist ancestors with an open floral form and that these shifts in floral type to reflect pollination syndromes are associated with species radiations. Bee pollination has evolved once in the family, arising in Africa from an open-flowered ancestor. Bird pollination has evolved multiple times: once from an open-flowered ancestor in Southeast Asia and several times from a bee-pollinated ancestor in the Neotropics. Based on dating exercises using a local

Address for correspondence: University of Vermont, Department of Botany—Pringle Herbarium, 120 Marsh Life Sciences Building, Burlington, VT 05405 USA. Tel: (802)656-3221, Fax: (802)656-0440, E-mail: cspecht@urm.edu

molecular clock technique, these shifts to pollination specific floral forms are found to be associated with rapid species radiations, creating clades with significantly more extant species than sister clades. Potential mechanisms for rapid evolution of these clades are discussed in light of the molecular evidence.

Key Words: Costaceae, rapid radiation, key innovation, local molecular clock, pollination syndrome

INTRODUCTION

The order Zingiberales has long been regarded in the taxonomic literature as a natural lineage within the monocotyledons [5, 7, 50]. The monophyly of the order has been supported by recent cladistic analyses including both morphological and molecular data [4, 16, 17, 19, 32, 45]. This order contains mostly tropical plants and includes the economically important banana and ginger as well as a variety of prominent tropical ornamentals such as *Heliconia* and *Strelitzia* (bird of paradise plant). Two informal groups within the Zingiberales are often recognized based on shared morphological and anatomical characters: the "banana families" (Musaceae, Heliconiaceae, Strelitziaceae, Lowiaceae), which include bananas and the bird of paradise, and the "ginger families" (Zingiberaceae, Costaceae, Cannaceae, Marantaceae), which contain ginger and the prayer-plant family, among others [6]. While the "banana families" form a paraphyletic grade at the base of the Zingiberales, the "ginger families" together form a derived monophyletic lineage (Fig. 2.1).

The flowers of the ginger group are highly modified compared to those of the banana group, with fertile stamen number reduced to one (Zingiberaceae and Costaceae) or one-half (Marantaceae and Cannaceae) (see Fig. 2.2). In Costaceae and basal Zingiberaceae, the petals are reduced in size while the five remaining infertile stamens (androecium) develop as petaloid structures, fusing in various ways to form a prominent "labellum" that dominates the floral display [13, 14]. The form of the labellum has been used extensively for taxonomic purposes within the families, with sections and subsections defined based upon labellum shape, size, and color.

As the main morphological character involved with overall shape and appearance of the floral display, the labellum is often modified to advertise nectar rewards for the attraction of pollinators. The shape and form of the labellum are linked with a preferred pollinator and combined with a suite of characters, including labellum color and bract color, that

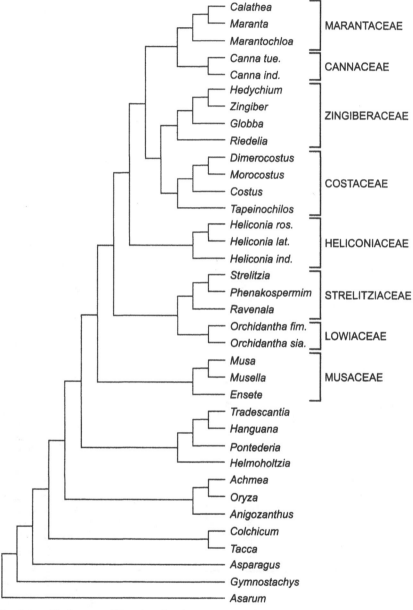

Fig. 2.1: Phylogeny of Zingiberales from Kress et al. [19]. Single most parsimonious tree from total combined analysis of 4 data sets (*rbc*L, *atp*B, 18S rDNA, and morphology).

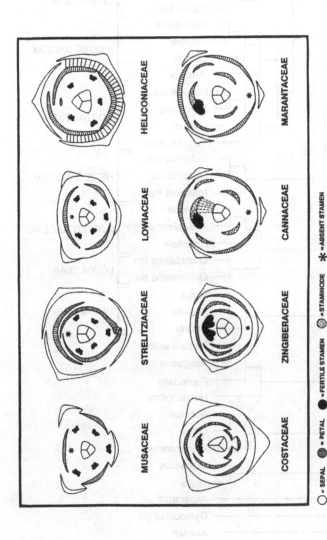

Fig. 2.2: Floral diagrams of the eight families of Zingiberales. Adopted from Kress [16] and Kirchoff [14a, 13b]. In the banana familes (top row), the five separate stamen are filamentous in structure. The sixth stamen is either absent (Musaceae, Strelitziaceae, Lowiaceae) or petaloid (Heliconiaceae). In the ginger families (bottom row), the stamen are petaloid and represented as staminodes. Cannaceae and Marantaceae have 1/2 fertile stamen, 1 absent stamen, and four remaining staminodes that do not fuse. In Costaceae and Zingiberaceae, fusion of staminodes results in formation of the labellum. In Zingiberaceae, two stamen fuse, one is lost, one is fertile, and two are infertile (staminodes). In Costaceae, 5 staminodes fuse to form the labellum and the remaining fertile stamen is petaloid in structure.

create a multicharacter "pollination syndrome" [12, 46] or specific morphology designed to attract a specific pollinator. In Costaceae, the labellum has been modified to attract bird or bee pollinators, with pollinator specificity contributing to reproductive isolation of species differing in pollination syndrome [12]. The diversity of floral morphology and pollination syndromes in the family Costaceae and the increased diversity in clades associated with pollinators indicate that floral evolution and pollinator attraction have been important factors in the diversification of Costaceae [22, 36, 48].

Transitions between floral types may play an active role in evolution, prompting speciation under certain environmental or ecological conditions. Such questions require an evolutionary framework in which to analyze developmental and distributional data. A phylogenetic hypothesis for the family Costaceae is necessary in order to explore potential roles of pollination attraction in species-level diversification.

Molecular sequence data provide phylogenetically informative characters that are independent of floral morphology and may corroborate or refute suggested taxonomic or evolutionary relationships based on composite floral characteristics. In addition, a molecular phylogeny by nature has a temporal component, enabling ages to be estimated for each of the nodes using an appropriate molecular clocklike algorithm. Studies concerning absolute rates of diversification can then be done using the knowledge of phylogenetic relationships combined with clade age and current clade diversity. Comparisons among clades with and without pollination associations can then be undertaken to assess the role of this key innovation in speciation.

A molecular-based phylogeny of Costaceae has been developed and a new taxonomy for the family established, based on the resultant phylogenetic hypotheses. Further investigations of species-level evolution are presented here, investigating the association of speciation rates with shifts in pollination syndromes.

COSTACEAE (MEISN.) NAKAI

Costaceae, a member of the ginger families clade, is one of the most easily recognizable groups within the Zingiberales, distinguished from other families within the order by well-developed and sometimes branched aerial shoots that have a characteristic monostichous (one-sided) spiral phyllotaxy [15]. Its close relationship with Zingiberaceae is evidenced by its former placement as a subfamily within the Zingiberaceae family, largely based on similarities of inflorescence and floral characters. Tomlinson [50] suggested that while these types of characters may indicate common ancestry, they are not sufficient to overcome the mounting morphological and anatomical differences that warrant independent familial rank of the two lineages. Molecular evidence has likewise shown that the two lineages are distinct [19, 44].

Anatomically, the monophyly of Costaceae and its separation from Zingiberaceae are supported by multicellular, uniseriate, unbranched hairs with the base never sunken as found in the Zingiberaceae [49, 50]. In addition, the hypodermis is always well developed with one or more layers below each surface in contrast to the Zingiberaceae, which either lacks a hypodermis or has only a single hypodermal layer below each surface. Finally, the Costaceae completely lack oil cells, which are abundant in all parts of the Zingiberaceae. Nakai [27a] cited the nonaromatic vegetative body, spirally arranged leaves, and absence of anther appendages to separate the Costaceae from the Zingiberaceae. Taken together, these characters indicate the uniqueness of the Costaceae lineage and provide anatomical and morphological synapomorphies for the family.

The well-developed, sometimes branched aerial stem, the distinctive monostichous spiral phyllotaxy, and the fusion of five staminodes into a labellum [14] versus three staminodes in Zingiberaceae, form the suite of characters most commonly cited as unique to and defining of Costaceae.

Floral Structure and Pollination Syndromes

In flowering plants and more specifically in Costaceae, combinations of floral color and structure are of major importance for the attraction of pollinators [8, 12, 36]. Referred to as "pollination syndromes," these are suites of floral and inflorescence characters that function to attract specific pollinators and indicate dependence of the plant on pollinator visitation for reproductive success and potentially providing isolation mechanisms to promote speciation and prevent hybridization. Transitions in these suites of characters can reflect correlated evolution among various floral and vegetative traits. Flowers of Costaceae are generally large and showy with a large delicate labellum, a structure formed from the fusion of five sterile staminodes, that dominates the floral ensemble [14]. This labellum and the overall floral structure that it creates serve to attract insects and birds that are subsequently rewarded with nectar produced by septal glands at the top of the gynoecium [36]. Reported pollinators include hummingbirds as well as bees of genera *Euglossa, Exaerete, Eulaena, Euplusia, Trigona,* and *Chrysantheda* in the Neotropics [12] and *Xylocopa* (Indonesia), *Lithurgus* (Indonesia), and *Anthophora* (India) in the Paleotropics [22].

In Costaceae there are four basic labellum types (Fig. 2.3): (a) ornithophilous, with a small tubular labellum only slightly protruding

Fig. 2.3: Representative taxa of the four floral types found in the family Costaceae: (a) *Chamaecostus subsessilis* (Nees et Mart.) Specht in ed. and (b) *Dimerocostus strobilaceus* O. Kuntz var. *strobilaceus* represent the open floral form, (c) *Costus guanaiensis* Rusby var. *guanaiensis* the melittophilous (bee-pollinated) type, (d) *Costus woodsonii* Maas represents the ornithophilous (bird-pollinated) type, and (e) *Tapeinochilos palustrus* Gideon the *Tapeinochilos* type. Vegetative branching characteristic of *Tapeinochilos* and *Cheilocostus* in ed. is demonstrated by *Tapeinochilos palustrus* (e). All photographs taken by C.D.Specht from collections made in Bolivia (a, c), Costa Rica (b, d) and Australia (e, courtesy of Alan Carle's Botanical Ark living collection).

beyond the corolla, yellow, orange or red with bracts of the same color forming a conical inflorescence; (b) melittophilous, with a short and broad labellum forming a distinct limb that is white or light yellow, often "edged" (lateral lobes striped) in red or purple, bracts mainly green; (c) the open form, with a large labellum forming a narrow tube with a distinct open limb, white, red, yellow or purple, bracts green to yellow; and (d) *Tapeinochilos* type, with a small and inconspicuous labellum, included within the subtending bracts or slightly exserted, calyx color usually red, reddish-brown or dark grayish-green, bracts mostly bright red or yellow. Of these four, the first two types have been directly associated with pollination syndromes (ornithophilous = bird pollinated; melittophilous = bee pollinated). The latter two types have not been directly associated with attraction of specific pollinators, although the tubular flower and red bracts of the *Tapeinochilos*-type are suggestive of a bird-pollinated pollination syndrome. Flowers of *Tapeinochilos* have reportedly been visited by sunbirds, a diverse group of nectar-feeding birds.

Taxonomic History

Four genera placed in Zingiberaceae subfamily Costoideae [38, 39] were upheld when the subfamily was elevated to the status of family [27a]. These are *Costus* L. (ca. 90 spp.), *Tapeinochilos* Miq. (ca. 18 spp.), *Dimerocostus* O. Kuntze (2 - 4 spp.), and the monotypic *Monocostus* K. Schum. The pantropical genus *Costus* was the most diverse with respect to species numbers and floral form. In order to accommodate this diversity, Schumann (1) described five subgenera for *Costus*: *Eucostus* (=*Costus* according to the International Code of Botanical Nomenclature), *Cadalvena*, *Metacostus*, *Epicostus* and *Paracostus*. The same subgenera were maintained by Loesener [20].

Costus and Pollination Syndromes

In Maas' treatments of the Neotropical Costaceae [22, 23], a formal division between the two subgenera found in South America (*Costus* and *Cadalvena*) was maintained. In addition, subgenus *Costus* was divided into two separate sections: *Costus* subgen. *Costus* sect. *Costus* and *Costus* subgen. *Costus* sect. *Ornithophilus*. These sections were based on labellum characters [23] or, more succinctly, upon the pollination syndrome, with section *Costus* as bee pollinated and section *Ornithophilus* as hummingbird pollinated. The first section was characterized as "having a labellum with a short, rather broad tube, and a distinct, exposed limb; its color varies from white to yellow, but the lateral lobes are often striped with red to purple." The bracts of this group are typically green and all characteristics classically linked with bee pollination (Fig. 2.3). The second section is comprised of species with "a small, tubular labellum of yellow, orange, or reddish colour: the bracts are of the same colour, or rarely green." These phenotypic characteristics indicate adaptation to hummingbird pollination, thus the section was designated *Ornithophilus* (bird-loving). Maas considered the two sections to be natural groups within subgenus *Costus*, but relationships of these sections to old world species were not discussed except to infer a potential relationship between section *Ornithophilus* and genus *Tapeinochilos*.

While all Neotropical *Costus* subgen. *Costus* could be placed rather easily in one of the two sections based on floral and inflorescence characteristics, the Neotropical subgen. *Cadalvena* as well as subgenera *Metacostus*, *Epicostus*, and *Paracotus* do not share this distinction of floral forms associated with pollination syndromes. All these subgenera have an open floral form. The open floral form is not associated with any specific pollination syndrome and flowers with this form are is most often reported to be a pollination generalists.

Dimerocostus, *Monocostus*, and *Tapeinochilos*

Unlike the pantropical *Costus*, *Dimerocostus* and *Monocostus* are both restricted in distribution to the Neotropics, the former extending from Honduras in the North to central Bolivia in the South, and the latter known only from the Rio Huallaga region of central Peru. *Monocostus* is the only taxon to have a solitary flower in the axils of the leaves rather than a highly structured inflorescence of spirally arranged bracts subtending single or paired flowers. Both *Monocostus* and *Dimerocostus* share the open floral form (Fig. 2.3).

Tapeinochilos is restricted to the Paleotropics where it is found primarily in New Guinea as well as Indonesia and Queensland, Australia. While most closely resembling the bird-pollinated floral form found in *Costus* section *Ornithophilus*, the floral and inflorescence morphology of *Tapeinochilos* is distinct from any other morphology found in Costaceae (Fig. 2.3).

Current Taxonomy and Indications of Floral Evolution

Recent studies using cladistic methods to investigate phylogenetic relationships within the family show that the large genus *Costus* is not monophyletic with respect to the other genera (Fig. 2.4; [14]). Further more, the bird-pollinated and bee-pollinated sections suggested by Maas are likewise not monophyletic. This led to a revision of the taxonomy of the family and the description of three new genera in order to reflect phylogenetic relationships ([42, 43; see Fig. 2.4):

(1) *Chamaecostus* in ed. forms a Neotropical clade with *Monocostus* and *Dimerocostus* (Fig. 2.4). These plants have the open floral form.

(2) *Cheilocostus* in ed. forms a monophyletic group with *Tapeinochilos*, forming an Asian Costaceae clade. Unlike *Tapeinochilos*, however, *Cheilocostus* also has the open floral form found in *Chamaecostus*.

(3) *Paracostus* is comprised of two taxa that form a sister clade to *Cheilocostus* + *Tapeinochilos*. While one taxon is from Africa and the other from Southeast Asia, both share with *Cheilocostus* the open floral form.

Biogeographically, the genus *Costus* as revised includes taxa from African and New World tropics, with all former Asian *Costus* removed to *Cheilocostus*. With respect to floral forms, *Costus* maintains African taxa with both *Cadalvena* and melittophilous floral forms, as well as New World taxa with exclusively ornithophilous or melittophilous floral forms. Thus, all Neotropical *Costus* are associated with a particular pollination syndrome, as are several African species: A monophyletic group of African melittophilous taxa is sister to a New World *Costus* clade containing both ornithophilous and melittophilous taxa (Fig. 2.4). At the base of the *Costus* clade is a grade of African taxa with the open floral form. This includes a clade of epiphytic taxa ("epiphytic clade", Fig. 2.4). Thus, while there are approximately eleven African *Costus* with the open floral form, there are no Neotropical *Costus* with the open floral form. All Neotropical taxa with the open floral form are found in the clade comprised of *Chamaecostus*, *Monocostus*, and *Dimerocostus*.

Evolution of Floral Forms

The results of recent cladistic analyses demonstrate that certain floral morphologies are evolutionarily conserved while others exhibit

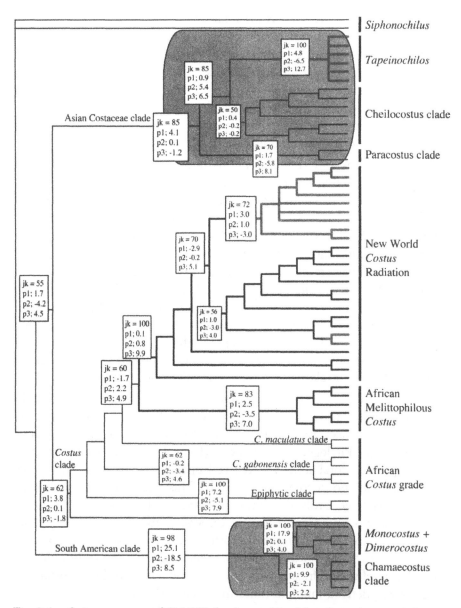

Fig. 2.4: Strict consensus of 56 MPTs for the combined four gene plus morphology data set. *jk* = jackknife, *p* = Partitioned Branch Support values (*p1* = chloroplast gene regions; *p2* = nuclear gene regions; *p3* = morphology). Thick black branches indicate melittophilous floral form, gray branches indicate ornithophilous floral form in NW *Costus* (dark gray) and *Tapeinochilos* (light gray). All other branches indicate terminals with the open floral form.

homoplasy. The open-type is plesiomorphic in the family (see Fig. 2.4; thin lines = open form), occurring as the basal floral type in all three major clades and as the only type of floral form present in the basal *Chamaecostus* clade (*Chamaecostus*, *Monocostus*, *Dimerocostus*). *Tapeinochilos*-type morphology was derived once and is unreversed in the genus *Tapeinochilos*. The two morphology types related to pollination syndrome, ornithophily, and melittophily, were independently derived within the newly revised monophyletic *Costus* lineage.

Thus, the overall trend in floral form with regard to pollination syndrome is one that moves from the generalist open floral form to more specialized forms attracting specific bird or bee pollinators: The open form is basal within Costaceae while the melittophilous and ornithophilous forms were derived several times independently from the plesiomorphic open floral form (Fig. 2.4) in *Costus* and *Tapeinochilos*. This same pattern is found in Zingiberaceae, in which the basal *Siphonochilos* and *Tamijia* both have a floral form similar to that found in basal Costaceae. This indicates that the common ancestor of the Costaceae and Zingiberaceae lineage had the open floral form and was probably also a generalist for pollinators. In both families, it appears that a close association with pollinators has driven species-level diversification based solely on relative number of taxa in clades that have specialized floral forms.

Recent studies on the evolutionary ecology of Neotropical *Costus* [12] clearly demonstrate that the New World ornithophilous and melittophilous taxa are indeed pollinated exclusively by birds and bees respectively. Phylogenetic studies (Fig. 2.4) indicate that the melittophilous floral form was derived once from an open floral form ancestor, with no subsequent reversals. This bee-affiliated floral form appears to have evolved within an African clade, with an African melittophilous taxon subsequently dispersing to the New World and giving rise to the Neotropical *Costus* radiation clade. The earliest members of the New World radiation clade maintain the melittophilous floral form of their African ancestors. Bird pollination within Neotropical *Costus* was derived multiple times from the bee-pollinated ancestor.

Based on the Costaceae phylogeny, bird pollination appears in two separate clades, the Asian Costaceae clade and the NW *Costus* radiation clade. In the first case, the ornithophilous *Tapeinochilos* is derived from an open-flowered ancestor as evidenced by its sister relationship with the open-flowered *Cheilocostus* clade and *Paracostus* as sister to *Cheilocostus* +

Tapeinochilos. Ornithophily in this clade evolved only once, and there are no reversals back to an open-flowered form within the *Tapeinochilos* clade. In the latter case, the New World *Costus* ornithophilous taxa are derived from New World melittophilous taxa that dispersed to the Neotropics from Africa. The ornithophilous *Costus* of the New World, however, do not form a monophyletic group, indicating that ornithophily evolved more than once in the genus *Costus*. As with *Tapeinochilos*, there have been no reversals either back to melittophily or to the plesiomorphic open floral form once the ornithophilous form was obtained. From the total evidence analysis presented, it appears as though ornithophily evolved three times in the *Costus* lineage. This is not the full sampling of ornithophilous taxa (12 of approximately 30) or of New World melittophilous taxa from which they are apparently derived (12 of approximately 23). In reality, there may be more independent derivations of the ornithophilous form.

The most speciose clades in the family (*Tapeinochilos* clade and New World *Costus* radiation clade) include taxa adapted to pollination by birds, a "key innovation" that may influence diversification rates. Of these two, the New World *Costus* radiation clade, which has evolved the bird pollination syndrome multiple times, has 33% more taxa than *Tapeinochilos*, indicating a potential benefit to the alternation of pollination syndromes for increased speciation rates. It is possible that an evolutionary toggle between hummingbird and bee pollination helps to drive speciation within this New World clade.

The potential association between bird pollination and increased speciation rates is also suggested in, the Neotropical *Heliconia* (Heliconiaceae), a group of approximately [75] bird-pollinated taxa. Speciation within Neotropical *Heliconia* has been extremely rapid, occurring exclusively within the past 10 million years [18], while other Zingiberalean families that do not have an exclusive bird-pollination syndrome have speciated at a much lower rate. With *Heliconia*, however, there are no bee-pollinated taxa for interfamilial comparison of speciation within clades to investigate the potential benefits of alternating between two specialized pollination syndromes.

A more detailed analysis of rates of evolution within the clades of Costaceae reveals the biogeographic and temporal processes involved in speciation within the family. It involves developing a molecular clock model for the Costaceae and looking at species diversification over time.

The first step is use of a molecular data matrix to develop a temporal context for the evolution of Costaceae. The second step compares absolute rates of diversification in various clades of Costaceae, including those with and without pollination syndromes. This enables an investigation of the role of a particular key innovation, pollination syndrome, on rates of diversification in the tropics.

COSTACEAE DIVERSIFICATION IN SPACE AND TIME: A LOCAL MOLECULAR CLOCK ANALYSIS

A matrix consisting of intron DNA sequence data (*trn*L-F and *trn*K, combined) for 36 ingroup taxa and 1 outgroup taxon was used for dating the origin of internal nodes for Costaceae. A phylogenetic analysis of this data set using the parsimony optimality criterion implemented in PAUP* [47] reflects the phylogeny presented in Figure 2.4: thus all major clades are represented and their relationships to one another reflect those found in the total evidence phylogeny.

The topology obtained from the chloroplast data set was used in a Dispersal-Vicariance Analysis (DIVA, [31]), with geographic areas scored as present or absent characters for each of the 37 taxa. The areas of distribution used for the DIVA analysis are: Central America, South America, West Africa, East Africa, Southeast Asia and Melanesia (New Guinea and Northern Australia). These areas were chosen as they represent major habitat areas for Costaceae and because defined taxonomic divisions (species and clades) do not typically span more than one of these areas except for the few widely distributed taxa. The results shown are those obtained when the maximum number of area optimizations for each node was constrained to three (maxareas = 3), the conditions which gave the most parsimonious solution (15 dispersal events) with the least number of optimal reconstructions for the basalmost node. The assigned distributions are compared with age estimates to hypothesize correlation to geological history.

Divergence Time Estimation

The local molecular clock method [53, 54] was used to analyze the molecular data and estimate node ages for the chloroplast intron data set of Costaceae (*trn*L-F and *trn*K sequence data for 37 ingroup taxa plus one outgroup taxon = *Siphonochilus decorus*). The local clock option of PAML

[51, 52] was executed with four rate classes (0-3) designated based on the results of a pairwise ratio test and branch lengths (Fig. 2.5) obtained with PAUP* v.4.10beta (ML analysis; model = F81uf, invariant sites and

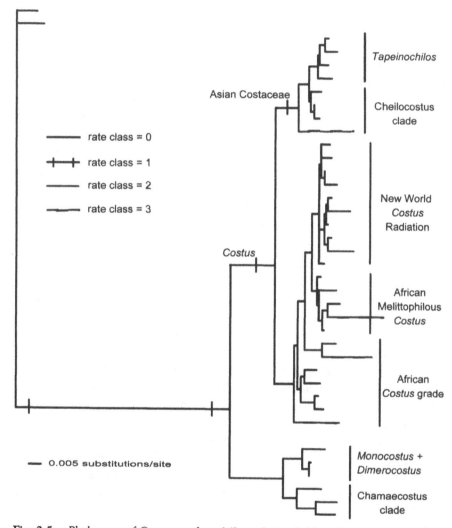

Fig. 2.5: Phylogram of Costaceae from ML analysis of chloroplast intron data for 36 taxa. Shaded branches represent branch length classes grouped together based on statistic similarity in a HYPHY analysis of pairwise relative rates. Each branch length ("rate") class was analyzed as a separate local molecular clock using the local clocks option in PAML.

gamma parameters estimated from the data). For additional details of the analysis, consult Specht 2004 [41].

Phylogenetic dating using molecular data requires at least one reference node to be fixed. A fixed calibration point is required to convert the relative rates obtained from relative comparisons of molecular sequence data to absolute dates. The ages of the remaining unfixed nodes of the phylogenetic hypothesis are estimated in this manner. Typically fossil data is used to calibrate a molecular clock, acting as a point of conversion from relative rates (branch lengths obtained from a maximum likelihood or parsimony search) to absolute rates that allow for the dating of discrete nodes. No fossils have been described for the Costaceae; hence alternative techniques must be developed to date the internal nodes of Costaceae phylogeny. One technique is to use the date of origin of the group of interest as determined in an independent molecular clock analysis. For Costaceae, the calibration point used is the split between the Costaceae and Zingiberaceae lineages as determined by a previous analysis of the order Zingiberales, for which multiple fossil calibration points are available [18].

Based on analysis of Zingiberales [18], there are two potential dates to use as the calibration point for Costaceae age estimations—that of the node connecting Costaceae and Zingiberaceae (i.e., the date of divergence or stem age) and that of initiation of speciation within Costaceae (i.e., date of diversification or crown age). These two dates are 105 million years ago (mya) and 55 mya respectively. Both dates were used as calibration points in separate analyses to test which better explains the current diversity and distribution of Costaceae. The results presented here reflect the use of the 105 mya date, the point at which Costaceae first became a separate evolutionary lineage. This allows the age of diversification of the crown group to be subsequently determined based upon molecular rates of a more thorough sampling of Costaceae taxa and a molecular data set with greater internal character variation that provides more informative characters for the model used to estimate node ages.

The chronogram reflecting the result of the PAML local clock analysis with 105 mya set as the calibration date indicates the ages of nodal diversification for Costaceae (Fig. 2.6). The ancestral distribution of the Costaceae-Zingiberaceae common ancestor is optimized by DIVA as Africa, Southeast Asia (SE) and South America (SA). The first

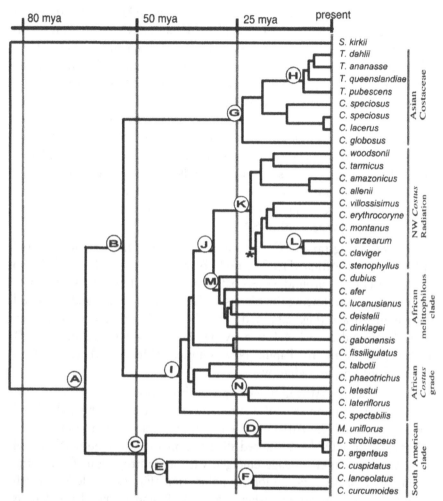

Fig. 2.6: Chronogram of Costaceae. Phylogeny of Costaceae based on trnL-F and trnK data matrix with nodes arranged according to the age estimates of the local clock method (31). Mya = million years ago. Lettered nodes are referred to in the text. * = first occurrence of hummingbird pollination.

divergence within Costaceae (Fig. 2.6; node A) dates back to the Cretaceous (~65 mya) and leads to a division between a South American and an African + SE Asian lineage. DIVA places the ancestral distribution in South America (SA) and Africa. The two clades resulting from the initial lineage split are the South American clade (including *Chamaecostus* and *Dimerocostus* + *Monocostus*) and a large clade containing the remaining species of Costaceae. The South American clade (Fig. 2.6; node C) begins to diversify around 49 +/- 6.9 mya with the separation of the *Monocostus + Dimerocostus* clade from *Chamaecostus*. The ancestral distribution was determined by DIVA to be Neotropical (SA). The *Chamaecostus* lineage (Fig. 2.6; node E) began to diversify shortly thereafter (44 +/- 6.7) while *Dimerocostus* did not diverge from *Monocostus* until much later (Fig. 2.6; node D. 20.88 +/- 4.99). Speciation within the *Chamaecostus* lineage (Fig. 2.6; node F) is estimated as having occurred as early as 21.9 +/- 4.6 mya. This is based on results of larger analyses showing that *Chamaecostus lanceolatus* and *C. curcumoides* are indeed sister taxa. The ancestral distributions for all lineages within the *Chamaecostus* clade are SA, with all extant taxa having current ranges within SA.

The remaining taxa of Costaceae, according to DIVA results, existed in SE Asia and Africa (Fig. 2.6; node B) and subsequently divided into two major lineages around 57 +/- 6.3 mya. Of these lineages, a vicariance event led to one lineage in SE Asia (Fig. 2.6; node G) and a second lineage in Africa (Fig. 2.6; node I). The SE Asian lineage began to diversify around 24 +/- 4.4 mya. At 11 +/- 2.9 mya (Fig. 2.6; node H) a dispersal event to Papua New Guinea (Melanesia) resulted in *Tapeinochilos* lineage radiation.

Meanwhile, the African lineage spread throughout tropical Africa (Fig. 2.6; node I), forming several distinct clades each maintaining the basal open floral form. The African clade began diversification around 40.7 +/- 4 mya with a DIVA-optimized distribution in Africa only. About 34 mya, however, a new lineage emerged with a melittophilous floral form, potentially as an adaptation for bee pollination (Fig. 2.6; node M), which dispersed to the Neotropics (Fig. 2.6; node J) and initiated the *Costus* radiation in South and Central America. The Neotropical lineage itself began diversifying approximately 22 mya (Fig. 2.6; node K), initially maintaining its ancestral bee-pollinated floral form but eventually acquiring a more tubular form better adapted to hummingbird pollination. The rapid diversification of this group may have resulted in part from close association with pollinators and the relative ecological success of this strategy for survival, dispersal, and diversification. In addition, the Neotropics had very little Zingiberalean flora at this time, so competition was negligible factor in diversification. Based on these data, it is not possible to date the transition from bee- to bird-pollinated forms. It can be said, however, that radiation of the New World group began around 22 mya (21.71 +/- 2 mya) and that speciation within this group was occurring rapidly at that time, with both floral forms represented ca. 20 mya.

TESTING KEY INNOVATIONS AND RAPID RADIATIONS

The study of key innovations and their role in evolution is fundamental to the study of evolutionary mechanisms. In the past, scenarios involving the role of a key adaptation on increased rates of speciation based on reported increase in number of species with the key innovation were untestable, and therefore unscientific [40], in explaining the diversity of individual groups or clades. Variability in species number per clade is readily consistent with simple stochastic models of phylogenetics since, under a null model of random speciation, all degrees of species diversity are equally likely [10, 40]. A relationship between a trait and increased diversity can only be tested if several groups possessing the same trait are considered in a comparative context [27, 55], but even multiple testing has flaws as it selects traits known to be associated with species-rich groups, thus committing a Type I statistical error [40]. However, the effect of a key innovation on species diversification can be tested if several groups are considered, and if the groups considered are sister-groups in order to remove any ambiguity about relative versus absolute size [40]. Because sister groups arose from the same ancestor and thus are of the same age, evolutive success of the group can be measured in terms of relative rather than absolute size.

Sister-group comparisons are useful for the identification of diversity correlates; however, they can only provide a relative assessment of rate variation between sister clades [33, 34]. Restriction of analytical tools to relative comparisons and sister lineages limits the extension of such studies to changes in diversification rates across a large group of organisms and to situations in which the "sister group" to a clade in question is actually a basal grade. In Costaceae, it is important not only to compare sister-group diversification rates to look for the effects of key innovations (i.e., pollinator association) on diversification, but also to compare across the entire family to determine whether those clades with increased rates of diversification are higher relative to nonsister clades and basal grades. In this way, paraphyletic groups are able to contribute to the rate of diversification analysis within the context of a larger inclusive clade. If a temporal component is available, then even more distantly related groups can be compared in a relative context. Costaceae is amenable to this study because there are several clades that possess the key innovation (i.e., pollination syndrome associations) and several grades as well as clades that maintain a basic generalist floral form. Thus, a positive association can be made between possession of the key innovation trait (pollination syndrome) and species diversity.

Fossil data has been employed in the recent past to determine absolute ages of various clades and thus enable examination of comparative diversification rate changes over time [9, 26, 30]. Given the absence of a fossil record for Costaceae, absolute rates of diversification will be obtained using the most recent phylogenetic hypothesis for Costaceae and local molecular clock analysis of clade ages. A method-of-moments estimator of the diversification process, will then enable comparative statements to be made concerning species diversification over time.

MOLECULAR CLOCKS AND DIVERSIFICATION RATES

Molecular systematics produces phylogenetic information with a temporal dimension that can be used to determine not only the relationships of taxa, but also the rate at which clades diversify. Since molecular data are available for all lineages of Costaceae, absolute diversification rates for each clade can be estimated. The following analysis uses information on the current diversity of groups in combination with statistical methods to investigate rates of diversification within Costaceae. Rates of diversification can then be compared for clades both with and without specific pollination syndromes to determine the role of pollinating specificity in diversification.

Recently, several papers have used molecular phylogenetics and molecular dating to determine both relative and absolute diversification rates [2, 26, 28, 29, 35]. Using molecular phylogenetic techniques and fossils as calibration points to date phylogenetic nodes, Magallón and Sanderson estimated the absolute diversification rates for most major angiosperm lineages [26]. This angiosperm-wide study provided a background rate of diversification for all angiosperms as well as major groups therein, including "higher monocots" and the Zingiberales. For each clade, a range of diversification rates was established using calculations of rates of diversification under two basic assumptions concerning extinction: its absence ($\varepsilon = 0.0$) or the presence of a high relative extinction rate ($\varepsilon = 0.9$) (see [42] for definition of extinction rates).

Obtaining the rate of diversification of a clade requires information about the number of lineages that originate and become extinct during an interval of time [26]. Since there is no such record for Costaceae, the rate of diversification is estimated considering only its present-day species diversity. A range is used to accommodate the potential relative extinction rate (ε). The lower bound of the extinction rate is set at zero ($\varepsilon = 0$) while the upper bound is chosen arbitrarily at $\varepsilon = 0.9$, which is the relative extinction rate bound used in previous studies [26]. This upper bound is used here to provide direct comparisons with rates determined in the angiosperm study, as well as for the mathematical, statistical, and biological reasons outlined by the authors of that study [26: 1765].

For the Costaceae analysis, lineage speciation and extinction is modeled as a stochastic birth-and-death process that is dependent on the diversification rate and the relative extinction rate [1]. In the birth-and-death model, speciation and extinction are assumed to occur at a constant rate, such that the process is homogeneous through time and leads to either increasing or decreasing diversity in an exponential manner. If extinction is negligible, a simple maximum likelihood estimate of diversification rate can be used [9]. However, in the presence of extinction, birth-and-death models using a maximum likelihood estimator tend to be biased by the failure to sample extinct clades. As an alternative, a method-of-moments estimator can take into account a variable extinction rate. Following Magallón and Sanderson [26], diversification is modeled as a stochastic, time-homogeneous, birth-and-death process that depends on diversification rate and relative extinction rate. The following formula, applied successfully to all angiosperms, is used, where r is rate of diversification given a determined value of rate of extinction (ε), t the time of diversification (age of node), and n the number of species in a given clade.

Stem group: $r = 1/t * \log[n(1 - \varepsilon) + \varepsilon]$

Crown group: $r = 1/t\{\log[1/2n(1-\varepsilon) + 2\varepsilon + 1/2(1-\varepsilon)\sqrt{n}(n\varepsilon^2 - 8\varepsilon + 2n\varepsilon + n)] - \log 2\}$

Differentiation between a stem and crown group age is less critical in Costaceae in which ages are based on molecular clock estimates rather than on placed fossil taxa. In the present analysis, a stem group age indicates the date of divergence between two lineages, while a crown group age represents the date at which diversification begins within a clade. Both dates are available for all clades of Costaceae. When all taxa are sampled for a clade, the diversification date (crown group age) is an accurate representation of complete speciation rates. However, the diversification age can underestimate the real crown group age if a basal member of the clade is not sampled for the molecular clock analysis. This is similar to the situation with fossil taxa wherein fossil age can represent either the absolute oldest date, or the age of a taxon nested within the clade of interest. To compensate for the potential of missing taxa to bias the ages estimated for dates of diversification (crown age), diversification rates are estimated using both crown ages and stem ages for clades in which there is an incomplete sampling of taxa (Table 2.1).

Using the formulas given above, Magallón and Sanderson found that the rate of diversification for all flowering plants ranges from 0.077 to 0.089 net speciation events per million years given the bounds of a high extinction rate and no extinction respectively. The higher monocots were found to have a rate higher than that of all angiosperms combined, falling between 0.0945 and 0.1144 speciation events per million years. Lastly, the Zingiberales were found to have a diversification rate of 0.0599 to 0.0798, a rate unexpectedly low compared to other speciose monocot lineages [26].

DIVERSIFICATION AND KEY INNOVATION: EFFECT OF POLLINATION SYNDROME ON RATES OF DIVERSIFICATION IN COSTACEAE

A fundamental question in the study of adaptive radiation is whether or not putative key innovations evolve at the same time as shifts occur in rate of speciation. In order to assess whether increases in diversification rate coincide with shifts in a key innovation, in this case development of a close association with specific pollinators and development of a floral and inflorescence morphology to reflect this association, a phylogeny is used along with ages of nodes to determine absolute rates of diversification for Costaceae (Table 2.1). Using the statistical models of the diversification process detailed above, the absolute rates of diversification for each major clade of Costaceae were calculated, enabling discussion of hypotheses about evolutive innovations that might be responsible for shifts in rates of diversification [3, 21, 35, 37].

Table 2.1: Absolute rate of diversification for Costaceae and clades. Rates were estimated in absence of extinction ($\varepsilon = 0.0$) and under a high relative extinction rate ($\varepsilon = 0.9$). Use of stem group or crown group age for clades with both dates is indicated. Groups in bold are from Magallón and Sanderson's published rate of diversification estimates [26]

Clade	Age in my (t)	Number of species (n)	Reference	$\varepsilon = 0.0$	$\varepsilon = 0.9$
Costaceae	65 (crown)	85		0.0577	0.0337
S.Am. clade	49 (crown)	12		0.0366	0.014
Chamaecostus	44 (crown)	8	[22, 23]	0.0315	0.0108
Monocostus + *Dimerocostus*	20 (stem)	4	[22, 23]	0.0693	0.0131
Asian clade	19 (crown)	25		0.1329	0.0617
Paracostus	56 (stem)	2	[41, 42]	0.0124	0.0017
Cheilocostus	19 (stem)	5	[24, 25]	0.0847	0.0177
	14 (crown)		A.Poulsen & J.Mood, pers.comm.	0.0654	0.0199
Tapeinochilos	19 (stem)	18	[11]	0.1521	0.0523
	11 (crown)			0.1997	0.0855
Costus	56 (stem)	53	[41, 42]	0.0709	0.0326
	40 (crown)			0.0819	0.0443
African	34 (stem)	6		0.0527	0.0119
Melittophilous	33 (crown)			0.0333	0.0106
NW *Costus*	34 (stem)	36	[22, 23]	0.1054	0.0442
Radiation	22 (crown)			0.1314	0.0660
Zingiberales	**83.5 (crown)**	**1560**	**[26]**	**0.0798**	**0.0599**
"higher"	**(crown)**		**[26]**	**0.1144**	**0.0945**
monocots incl.					
Zingiberales					

The rate of diversification for the entire Costaceae lineage based on the crown group (diversification) age of 65 mya is 0.034 – 0.058 net speciation events per million years, slightly less than the rate estimated by Magallón and Sanderson [26] for the diversification of the Zingiberales as a whole (0.059 – 0.079). In all cases where calculations were made for both stem age and crown age, differences in the resultant ranges were not significant. While both are presented in Table 2.1, only the crown ages are discussed as the basal sampling of each clade was sufficient to justify the estimated crown (diversification) date as representative of the earliest speciation within a lineage.

The South American clade, comprised of *Monocostus*, *Dimerocostus*, and *Chamaecostus* and with a crown group age of 49 mya, has a rate lower than the net Costaceae diversification rate (0.014 – 0.037). Based on the crown group age of 19 mya, the Asian Costaceae clade, comprised of *Tapeinochilos*, *Cheilocostus*, and *Paracostus*, has a diversification rate of 0.0617 – 0.1329. Even with this large range, the diversification rate is significantly higher than that found for Costaceae as a whole.

For *Costus* the rate of diversification was found to be 0.044 – 0.0819. This range overlaps that of all Costaceae, but demonstrates a more rapid diversification rate in the absence of extinction. In order to understand the rate variation for these composite groups, individual clades were further examined to determine which lineages contribute to the overall diversification rates for these clades.

South American Clade

The South American clade is comprised of genera *Chamaecostus* in ed., *Monocostus*, and *Dimerocostus*. *Chamaecostus* is a small clade comprising eight species located in the Guyana and Brazilian shields from French Guiana to northern Bolivia. The plants are all small in stature, which initially prompted their separation from the remaining Neotropical members of genus *Costus* into subgenus *Cadalvena*. In addition, they have the plesiomorphic generalist open floral form. *Chamaecostus* is sister to the *Monocostus* + *Dimerocostus* clade.

Monocostus is monotypic, with a single species located in Peru. *Dimerocostus*, whose range stretches from Honduras to Bolivia along the low foothills of the Andes, has approximately three species with some morphological variation accounted for as subspecies. The genetic diversity of *Dimerocostus* is low, indicating that use of subspecies is appropriate rather than considering the morphological variation to indicate a greater number of total species in the genus (Specht, unpubl. data). All the taxa in this clade have the plesiomorphic open floral form and are not associated with specific pollinators.

The overall rate of diversification for this clade, as mentioned earlier, is lower than that for Costaceae as a whole and, except for *Paracostus*, is the lowest found for all of Costaceae. The rate of diversification within *Chamaecostus* alone, using the 44 mya diversification (crown) age, proved indistinguishable from the entire clade (0.011 – 0.0315). The rate of the *Dimerocostus* + *Monocostus* lineage, using the stem age of 20 mya, was slightly faster in the absence of extinction (0.0693) but similar to the remainder of the clade when a high extinction rate was assumed (0.013). In general, then, speciation rates are lower than average in this generalist-pollinated clade of Costaceae.

Asian Costaceae (*Cheilocostus* and *Tapeinochilos*)

Based primarily on the superficial similarity of flower shape (tubular labellum) and bract color giving both groups a bird-pollinated aspect, *Tapeinochilos* was thought to be related to an ornithophilous group of New World *Costus* [23]. Molecular phylogenetic studies, however, identified the *Cheilocostus* clade of Asian taxa as the sister group to the monophyletic *Tapeinochilos*. Following the separation of these two lineages, genus *Tapeinochilos* acquired a pollination-specific floral morphology and underwent a localized radiation on the island of New Guinea. Diversification rates indicate the rapidity of this radiation.

The entire Asian Costaceae lineage, including *Paracostus* (sister to *Cheilocostus* + *Tapeinochilos*), had a diversification rate of 0.0617 – 0.1329, which is significantly faster than the rate for Costaceae as a whole. The *Paracostus* lineage, with just two species, separated from the remainder of Costaceae 56 mya (stem age), showed a very low rate of 0.0017 – 0.0124. Both species of *Paracostus* maintain the open, generalist floral form. While significantly faster than *Paracostus*, the *Cheilocostus* lineage also seems to have undergone diversification at a slower rate (0.019 – 0.065) than the average Costaceae lineage. Interestingly, *Cheilocostus* maintained a low diversification rate while holding (or possibly obtaining) a relatively large geographic distribution that overlaps with the easternmost distribution of *Tapeinochilos* on New Guinea. *Cheilocostus*, like Paracostus, maintains an open, generalist floral form. However, some species of *Cheilocostus globosus* can develop a more closed floral form approximating that found in bird-pollinated species. The *C. globosus* species complex has high morphological diversity while maintaining relatively low molecular diversity, although the interspecific molecular diversity is significantly higher than that within its sister species, *C. speciosus*. The ability of species in this complex to develop more pollination-specific floral morphologies and the high morphological diversity combined with increased molecular diversity of this complex as compared to its sister taxon, may attest to the ability of this clade to eventually undergo a rapid radiation event. It would be particularly interesting were a shift to rapid diversification rates accompanied by a fixed transition in floral form to an ornithophilous-type pollination syndrome.

The comparatively rapid radiation within *Tapeinochilos* (0.085 – 0.199 using the crown age of 11 mya is particularly noteworthy. This significantly higher rate of diversification is coincident with a shift from the generalist open floral form to the *Tapeinochilos* ornithophilous-type floral form. The question remains whether or not this shift in floral form is responsible for the increased rate of diversification. Reasons for a correlation could involve the use of a novel resource for pollination (in this case, sunbirds) enabling increased rates of diversification, or the fact that pollination-specific taxa are more likely to become reproductively isolated and therefore could potentially undergo more rapid speciation events. In order to demonstrate such a correlation, however, other clades with similar pollination syndrome-associated floral form must show similar increases in diversification rates.

Genus *Costus*

As mentioned above, the overall rate of diversification for *Costus* is slightly higher than that found for Costaceae as a whole. At the base of *Costus* is a plesiomorphic grade of eleven African taxa divided into single taxon or 2-3 taxa lineages (e.g. epiphytic clade) that all maintain the generalist open floral form. While it is difficult to get the rate for these species as a whole, it can be assumed that they have a diversification rate similar to that found for all *Costus* with a slight bias toward any extremely fast or extremely low rate found in any of the crown groups. In general, however, the average rate for all *Costus* approximates the rate of diversification for taxa within this basal grade.

African melittophilous clade

The first major clade to diverge from this grade is the African melittophilous clade, a group of six taxa found throughout tropical Africa that have a bee-pollinated floral form. Actual pollination of these taxa by bees has not been confirmed but the composite floral and inflorescence morphology is that of a typical bee-pollinated flower. This clade diverged from the remaining African *Costus* around 34 mya and, according to the molecular clock analysis, began to diversify shortly thereafter at 33 mya. Using the 33 mya crown age, the diversification rate for the African melittophilous clade is estimated at 0.016 – 0.033, slower than average for Costaceae. Thus, while this clade does have a pollinator-specific floral form derived from the more generalist open floral form, it does not have an increased diversification rate compared to other Costaceae. This indicates that the acquisition of pollination specificity alone is not sufficient to promote rapid diversification. It should be noted, however, that this is the largest clade of African *Costus*, indicating that obtaining a morphology with specific pollinator association does have some effect on increased speciation at least within African *Costus*. It is important to determine whether these plants are actually bee pollinated in order to make conclusive statements about rates of diversification and the acquisition of specific pollination syndromes in this clade.

New World Costus Radiation

The final clade within the genus *Costus* is the New World radiation clade, so called because it is the most speciose clade in all of Costaceae and is found exclusively in the Neotropics. The NW *Costus* radiation clade diverged from the African melittophilous clade approximately 34 mya and comprises 36 species that are exclusively bird and bee pollinated. Unlike with African melittophilous group, pollination ecology studies of NW *Costus* clearly show that they are indeed pollinated exclusively by the organisms which their floral morphologies suggest [12], indicating that the pollination syndromes are effective means of attracting specific pollinators. Because the bee- and bird-pollinated taxa are interspersed, it is not possible at this time to establish an independent rate of diversification for each of the two pollination syndromes. The ornithophilous floral form was derived multiple times from the ancestral melittophilous form and it is quite possible that each ornithophilous clade has a distinct diversification rate.

Regardless, it is clear from the current study that the rate of diversification for the NW *Costus* radiation clade is significantly faster than the rate of diversification for any other clade of Costaceae (0.66 – 0.131) except *Tapeinochilos*, which has a rate still higher than that estimated for the NW *Costus* radiation. Thus, in New World *Costus* and *Tapeinochilos*, two clades comprised exclusively of taxa with distinct pollination syndromes, the rates of diversification are significantly higher than for the remaining Costaceae.

The argument could be made for a biogeographic cause to the increase in diversification rate found in the NW *Costus* radiation clade. The NW radiation clade is sister to an African melittophilous clade, thus it could be argued that dispersal to the Neotropics is likely to be the cause of the subsequent rapid radiation with increased habitat diversity. It should be noted, however, that the South American clade (including *Chamaecostus, Monocostus* and *Dimerocostus*) already existed in the New World at the time that the radiation clade arrived. This clade, however, did not rapidly radiate in the absence of competition or with the presence of available niche space, maintaining the ancestral open floral form. This comparison thus strengthens the indication that the NW *Costus* radiation clade's rapid diversification is likely due to the acquisition of specific associations with Pollinators rather than the result of a novel colonization event.

The finding of increased rates in both the NW *Costus* radiation clade and in *Tapeinochilos* provides a positive association between increased diversification rates and the key innovation of pollination syndrome. Such increases in rate of diversification found in non-sister clades comprised exclusively of species with a particular key innovation can be considered indicative of independent radiation events caused by the acquisition of a key innovation, in this case association with pollinators and development of a floral form associated with a particular pollination syndrome.

CONCLUSION

The role of a key innovation in speciation or rates of diversification has been the subject of much attention in the last decade. With the increase in phylogenetic hypotheses combined with genomic analyses and the construction of molecular phylogenies, increasingly robust statistical techniques have been developed for estimating absolute rates of diversification for clades within a known phylogeny. In the absence of fossil data, molecular clock algorithms can be used to determine ages of nodes.

A phylogenetic hypothesis was developed for the family Costaceae and a local molecular clock algorithm used to estimate divergence times in the phylogeny. This information was then used to test the effect of key innovations on rates of diversification in various clades possessing floral morphologies associated with pollination specificity. The results indicate that acquisition of pollination-specific floral structure correlates with an increase in species diversification rate in two major clades of Costaceae. Thus, at least in Costaceae, pollinator specificity is considered an advantage for promoting increased diversification and rapid species radiation in the tropics.

Acknowledgments

Sincere thanks to D. Stevenson, R. DeSalle, and W. J. Kress for their continuing support of my research on Costaceae, P. O'Grady for ideas and discussions, A. Poulsen and J. Mood for discussions on Asian Costaceae, Ray Baker and the Lyon Arboretum, for use of their collections, and D. Barrington and C. Paris for providing the opportunity to work at the University of Vermont. This work was partially funded by the National Science Foundation (DEB-0206501 Doctoral Dissertation Improvement Grant), a Fulbright Research Fellowship, New York University's Graduate School of Arts and Science, the New York Botanical Garden Graduate Studies Program, the Lewis B. and Dorothy Cullman Program in Molecular Systematic Studies, and through a Visiting Research Fellowship and Postdoctonal Fellowship at the National Museum of Natural History, Smithsonian Institution, Department of Botany.

REFERENCES

[1] Bailey NTJ. The elements of stochastic processes with applications to the natural sciences. New York; NY: Wiley, 1964.

[2] Baldwin BG, Sanderson MJ. Age and rate of diversification of the Hawaiian silversword alliance (Compositae). Proc Natl Acad Sci, USA 1998; 95: 9402-9406.

[3] Berenbaum MR, Favret C, Schuler MA. On defining "key innovations" in an adaptive radiation: cytochrome P450s and Papilionidae. Amer Nat 1996; 148: S139-S155.

[4] Chase MW, Soltis DE, Soltis PS, et al. Higher-level systematics of the monocotyledons: an assesment of current knowledge and a new classification. In: Wilson KL, Morrison DA, eds. Monocots: Systematics and Evolution. Melbourne: CSIRO, 2000: 3-16.

[5] Cronquist A. An integrated System of classification of flowering plants. New York; NY: Columbia University Press, 1981.

[6] Dahlgren R, Clifford HT. The monocotyledons : a comparative study. New York, NY: Academic Press, 1982.

[7] Dahlgren RMT, Clifford HT, Yeo PF. The families of monocotyledons. Berlin: Springer-Verlag, 1985.

[8] Endress PK. Diversity and evolutionary biology of tropical flowers. Cambridge: Cambridge University Press, 1994.

[9] Eriksson O, Bremer B. Pollination systems, dispersal modes, life forms, and diversification rates in angiosperm families. Evolution 1992; 46(258-266).

[10] Farris JS. Expected asymmetry of phylogenetic trees. Syst Zool 1976; 25: 196-198.

[11] Gideon OG. Systematics and Evolution of the genus *Tapeinochilos* Miq. (Costaceae-Zingiberales) [PhD]. Queensland, Australia: James Cook University, 1996.

[12] Kay KM, Schemske DW. Pollinator assemblages and visitation rates for 11 species of Neotropical *Costus* (Costaceae). Biotropica 2003; 35(2): 198-207.

[13] Kirchoff BK. Floral ontogeny and evolution in the ginger group of the Zingiberales. In: Leins P, Tucker SC, Endress PK, eds. Aspects of floral development. Berlin: J. Cramer 1988a: 45-56.

[14] Kirchoff BK. Inflorescence and flower development in *Costus scaber* (Costaceae). Can J Bot 1988b; 66: 339-345.

[14a] Kirchoff BK, Kunze VH. Inflorescence and Floral development in *Orchidantha maxillarioides* (Lowiaceae) Int.J.Pl. Sci. 1995: 156: 159-171.

[15] Kirchoff BK, Rutishauser R. The phyllotaxy of *Costus* (Costaceae). Bot Gaz. 1990; 151: 88-105.

[16] Kress WJ. The phylogeny and classification of the Zingiberales. Ann Missouri Bot Garden 1990; 77: 698-721.

[17] Kress WJ. Phylogeny of the Zingiberanae: morphology and molecules. In: Rudall P, Cribb PJ, Cutler DF, Humphries CJ, eds. Monocotyledons: systematics and evolution: Royal Botanic Gardens, Kew; 1995: 443-460.

[18] Kress WJ, Specht CD. Using local clocks and dispersal-vicariance analysis to investigate the origin and diversification of the major lineages of the tropical monocot order Zingiberales. Syst Biol 2004 (submitted).

[19] Kress WJ, Prince LM, Hahn WJ, Zimmer EA. Unraveling the evolutionary radiation of the families of the Zingiberales using morphological and molecular evidence. Syst Biol 2001; 50(6): 926-44.

[20] Loesener T. Zingiberaceae. In: Prantl EA, ed. Die Naturlichen Pflanzenfamilien, 1930: 547-640.

[21] Losos JB, Miles DB. Testing the hypothesis that a clade has adaptively radiated: Iguanid lizard clades as a case study. Amer Nat 2002;160(2):147-157.

[22] Maas PJM. Costoideae (Zingiberaceae). New York, NY: Haner Publishing Co., 1972.

[23] Maas PJM. Renealmia (Zingiberoideae) and Costoideae additions (Zingiberaceae). In: Flora Neotropica. Bronx, New York: New York Botanical Garden, 1977.

[24] Maas PJM. Notes of Asiatic and Australian Costoideae (Zingiberaceae). Blumea 1979; 25: 543-549.

[25] Maas PJM, Maas van de Kamer H. Notes on Asiatic Costoideae (Zingiberaceae) II. A new *Costus* from Celebes. Notes Roy Bot Gard Edinburgh 1983; 41(2): 325-326.

[26] Magallón S, Sanderson MJ. Absolute diversification rates in angiosperm clades. Evolution 2001; 55(9): 1762-1780.

[27] Mitter C, Farrell B, Wiegmann B. The phylogenetic study of adaptive zones: has phytophagy promoted insect diversification. Amer Nat 1988; 132: 107-128.

[27a] Nakai T. Notulae ad Plant as asiae orientalis (XVI). Journal of Japanese Botany 1941; 17: 189-203.

[28] Nee S. Inferring speciation rates from phylogenies. Evolution 2001; 55(4): 661-668.

[29] Paradis E. Detecting shifts in diversification rates without fossils. Amer Nat 1998; 152(2): 176-187.

[30] Ricklefs R, Renner SS. Species richness within families of flowering plants. Evolution 1994; 48: 1619-1636.

[31] Ronquist F. DNA version 1. 1. Computer Program and manual available by anonymous FTP from Uppsala University. ftp.uu.se. 1996.

[32] Rudall PJ, Stevenson DW, Linder HP. Structure and Systematics of Hanguana, a monocotyledon of uncertain affinity. Austr Syst Bot 1999; 12: 311-330.

[33] Sanderson MJ, Donoghue MJ. Shifts in diversification rate with the origin of angiosperms. Science 1994; 264: 1590-1593.

[34] Sanderson MJ, Donoghue MJ. Reconstructing shifts in diversification rates on phylogenetic trees. Trends Ecol Evol 1996; 11: 15-20.

[35] Sanderson MJ, Wojciechowski MF. Diversification rates in a temperate legume clade: are there "so many species" of Astragalus (Fabaceae)? Amer J Bot 1996; 83(11): 1488-1502.

[36] Schemske DW. Variation among floral visitors in pollination ability: A precondition for mutualism specialization. Science 1984; 225(4661): 519-521.

[37] Schena M, Shalon D, Davis RW, Brown PO. Quantitative monitoring of gene expression patterns with a complementary DNA microarray. Science 1995; 270(5235): 467-470.

[38] Schumann K. Monographie der Zingiberaceae von Malaisien un Papuasia. Bot J Syst, Pflanzenges Pflanzengeo 1899; 27: 259-350.

[39] Schumann K. Zingiberaceae. In: Engler A, ed. Das Pflanzenreich IV. Leipzig: Englemenn; 1904.

[40] Slowinski JB, Guyer C. Testing whether certain traits have caused amplified diversification: an improved method based on a model of random speciation and extinction. Amer Nat 1993; 142: 1019-1024.

[41] Specht CD. Gondwanan Evolution or Dispersal in the Tropics? The biogeography of Costaceae. Proceedings of Monocots III Symposium, Ontario, CA. 2004a (submitted).

[42] Specht CD. Systematics and Evolution of the tropical monocot family Costaceae (Zingiberales): a multiple data set approach. Syst Bot 2004b (submitted).

[43] Specht CD, Stevenson DW. A new generic taxonomy for the monocot family Costaceae (Zingiberales) Brittania (submitted).

[44] Specht CD, Kress WJ, Stevenson DW, DeSalle R. A molecular phylogeny of Costaceae (Zingiberales). Mol Phylogen Evol 2001; 21(3): 333-45.

[45] Stevenson DW, Davis JI, Freudenstein JV, Hardy CR, Simmons MP, Specht CD. A phylogenetic analysis of the monocotyledons based on morphological and molecular character sets with comments on the placement of Acorus and Hydatellaceae. In: Wilson KL, Morrison DA, eds. Monocots: Systematics and evolution. Melbourne: CSIRO, 2000: 17.

[46] Stiles FG. Notes on the Natural History of Heliconia (Musaceae) in Costa Rica. Brenesia 1979; 15: 151-180.

[47] Swofford DL. PAUP* (Phylogenetic Analysis Using Parsimony)* and other methods. Version 4.10 beta. Sunderland, MA. Sinauer Associates, 2001.

[48] Sytsma KJ, Pippen RW. Morphology and pollination biology of an intersectional hybrid of Costus (Costaceae). Sys Bot 1985; 10(3): 353-362.

[49] Tomlinson PB. Studies in the systematic anatomy of the Zingiberaceae. J Linn. Soc (Bot) 1956; 55: 547-592.

[50] Tomlinson PB. Phylogeny of the Scitamineae—morphological and anatomical considerations. Evolution 1962; 16: 192-213.

[51] Yang Z. PAML: a program package for phylogenetic analysis by maximum likelihood. CABIOS 1997; 13: 555-556. http://abacus.gene.ucl.ac.uk/software/paml.html.

[52] Yang Z. PAML: a program package for phylogenetic analyses by maximum likelihood. Version 3.14beta, 2003. http://abacus.gene.ucl.ac.uk/software/paml.html.

[53] Yang Z, Yoder AD. Comparison of likelihood and Bayesian methods for estimating divergence times using multiple gene loci and calibration points, with application to a radiation of Cute-Looking Mouse lemur species. Syst Biol 2003; 52(5): 705-715.

[54] Yoder AD, Yang Z. Estimation of primate speciation dates using local molecular clocks. Mol Biol Evol 2000; 17(7): 1081-1090.

[55] Zeh DW, Zeh JA, Smith RL. Ovipositors, amnions, and eggshell architecture in the diversification of terrestrial arthropods. Quart Rev Biol 1989; 64: 147-168.

Cytogenetic Diversity in *Orchis* s.l. and Allied Genera (Orchidinae, Orchidaceae)

SAVERIO D'EMERICO

Dipartimento di Biologia e Patologia Vegetale, Sezione di Biologia Vegetale, Università di Bari, Via Orabona 4, 70125 Bari, Italy.

ABSTRACT

Cytogenetic relationships among species of genus *Orchis* s.l. and allied genera are considered. Chromosome numbers and karyotype structure of numerous taxa are compared based on Feulgen, Giemsa C-banding, and DAPI (4'-6-diamidino-2-phenylindole) stained somatic metaphase chromosomes. The fundamental diploid numbers observed were 2n = 36, 40, and 42. Analyses indicated that chromosome complements differed markedly in chromosome size and karyotype structure. The amount and distribution of constitutive heterochromatin varied considerably among the taxa studied; in some cases a correlation was found between asymmetrical karyotype and heterochromatin content. In some species DAPI staining revealed that most heterochromatin regions are rich in A-T base pairs. The 18S, 5.8S-25S (pTa71) and 5S (pTa794) r DNA were used in *Anacamptis* s.l. and *Steveniella-Himantoglossum* s.l. species as probes for *in-situ* hybridization to reveal localization of ribosomal genes. Our results seem to indicate that both karyomorphology and heterochromatin distribution are a major cause for a possible evolutionary pathway. Chromosomal rearrangements and evolutionary trends in representative genera of subtribe Orchidinae are discussed.

Key Words: Orchidinae, *Anacamptis* s.l., *Dactylorhiza* s.l., *Gymnadenia* s.l., *Neotinea* s.l., *Ophrys*, *Orchis* s.s., *Serapias*, *Steveniella-Himantoglossum* s.l., *Traunsteinera-Chamorchis*. Chromosome numbers, constitutive heterochromatin, karyotype structure, phyletic relationships.

Address for correspondence: E-mail: demerico@botanica.uniba.it.

INTRODUCTION

Orchidinae is the largest subtribe of family Orchidaceae, comprising 34 genera and about 361 species, widely distributed in every continent and well represented in northern temperate and some tropical areas [29, 48, 67]. Cytological knowledge of Orchidinae consists of chromosome counts of more than 194 [14]. Data indicate five principal basic chromosome numbers in this group, taxa x = 14, 16, 18, 20, and 21, and mainly diploid species. The upper number observed is therefore a major contributor to the generation of genetic diversity within and between species of Orchidaceae. Despite the great number of chromosome counts, karyomorphological information on this subtribe was scant and only recently have more details been obtained [9,15,18,22]. Numerous species of Orchidaceae differ in karyotype structures. A comparison of karyotype characteristics was used to trace the evolutionary pathways in this group of species. For karyomorphological study some indices have been proposed for estimation of karyotype asymmetry [e.g. 35, 54, 61] (Tables 3.1, 3.2, and 3.3).

In recent years, karyological studies using differential staining techniques have revealed interesting variations in heterochromatin contents in the chromosomal complements of many groups of plants [36]. Heterochromatin domains mostly consist of repetitive DNA sequences and quantitative variations in the genome can have various effects in genetic differentiation and karyotypic evolution [51,65]. The distribution and chemical composition of heterochromatin have been compared in different taxa to support taxonomical and phylogenetic correlation [25,30,33].

In orchids, C-banding and fluorescent staining have been used to characterize species of genera *Cypripedium* [68], *Cymbidium* [57], *Cephalanthera* [56], *Pleione* [62], *Chamorchis, Dactylorhiza, Epipactis, Gymnadenia, Limodorum, Listera, Orchis, Serapias,* [18,19,20,21,23,25, 26,46], and *Phalaenopsis* [40].

This paper examines the cytological, karyomorphological characteristic and the distribution of constitutive heterochromatin regions to trace the evolutionary relationships of representative Orchidinae genera.

CYTOGENETIC RELATIONSHIPS IN REPRESENTATIVE ORCHIDINAE GENERA

Previous investigations based on karyomorphology and heterochromatin distribution in *Orchis* s.l. species indicated a possible grouping of some

Table 3.1: Number of representative wild species of Orchidinae subtribe. Taxa, codes, total chromosome length in μm, range of chromosome size in μm (long/short), chromosome number, and karyotype formulae

Taxa	Code	Haploid complement μm (mean values)	Size l/s μm (mean values)	Chromosome number 2n	Formula
Barlia robertiana	BAR	35.02	4.31 to 1.87	36	32 m + 4sm
Himantoglossum hircinum	HIM	31.72	4.05 to 1.89	36, 36 + 1B	28 m + 8 sm
Anacamptis pyramidalis	PYR	41.62	4.05 to 1.20	36, 54, 63, 72	20 m + 16 sm
A. coriophora	COR	48.33	4.25 to 2.17	36, 54	14 m + 20 sm + 2st
A. laxiflora	LAX	42.00	3.50 to 1.50	36, 54	24 m + 12 sm
A. palustris	PAL	49.08	3.83 to 1.70	36	22 m + 14 sm
A. morio	MOR	48.26	4.84 to 2.05	36	30 m + 6 sm
A. papiliomacea	PAP	38.80	3.91 to 1.45	32	18 m + 12 sm + 2st
A. morio × A. papiliomacea	GEN	—	—	34, 52	—
A. collina	COL	44.73	4.00 to 2.00	36	18 m + 18 sm
Neotinea tridentata	TRI	39.50	3.67 to 1.08	42, 42 + 1B	34 m + 8 sm
N. lactea	LAC	50.33	4.00 to 1.92	42	28 m + 14 sm
Orchis mascula	MAS	39.50	3.00 to 1.17	42, 42+1B	24 m + 14 sm + 4st
O. provincialis	PRO	—	—	42	22 m + 14 sm + 6st
Serapias vomeracea	VOM	38.70	3.62 to 1.23	36	20 m + 16 sm
S. parviflora	PAR	38.47	3.15 to 1.27	36	16 m + 20 sm
S. lingua	—	—	—	72	—
Ophrys incubacea	INC	44.02	3.83 to 1.75	36	22 m + 14 sm
O. tenthredinifera	TEN	43.09	3.70 to 2.07	36, 54	30 m + 6 sm
O. fusca	FUS	41.66	3.67 to 1.47	36, 72	32 m + 4 sm
Dactylorhiza romana	ROM	37.42	3.33 to 1.00	40, 40 + 1B, 40 + 2B	16 m + 22 sm + 2st
D. saccifera	SAC	31.25	2.17 to 1.00	40	32 m + 8 sm
Gymnadenia conopsea	CON	46.16	2.67 to 1.00	40	—
G. rhellicani	RHE	38.67	2.83 to 1.16	40	—
Chamorchis alpina	ALP	41.83	3.00 to 1.33	42	20m + 20sm + 2st

Table 3.2: Taxa, intrachromosomal asymmetry index (A_1), interchromosomal asymmetry index (A_2) [54], and Giemsa C-band. Abbreviations: tb, telomeric band; scb, subcentromeric band; cb, centromeric band; lcb, large centromeric band; tcb, thin centromeric band

Taxa	A_1 (mean values)	A_2 (mean values)	Giemsa C-band
Barlia robertiana	0.24	0.33	
Himantoglossum hircinum	0.22	0.26	
Anacamptis pyramidalis	0.30	0.37	
A. coriophora	0.42	0.17	20 tb
A. laxiflora	0.24	0.26	
A. palustris	0.33	0.26	
A. morio	0.21	0.25	
A. papilionacea	0.38	0.27	4 tb + 2 scb
A. morio × A. papilionacea	0.32	0.29	2 tb + 1 scb
A. collina	0.33	0.25	
Neotinea tridentata	0.33	0.38	cb
N. lactea	0.34	0.18	cb and tb
Orchis mascula	0.39	0.26	20 lcb and 22 tcb
O. provincialis	0.44	0.32	36 tcb and 6 tb Serapias
vomeracea serapias	0.46	0.26	lcb
S. parviflora	0.39	0.27	lcb
S. lingua	—	—	lcb
Ophrys incubacea	0.32	0.22	cb and 12 tb
O. tenthredinifera	0.23	0.17	cb and 2 tb
O. fusca	0.26	0.21	cb and 2 tb
Dactylorhiza romana	0.30	0.30	Tb and B heterochromatic
D. saccifera	0.27	0.19	tb
Gymnadenia conopsea	0.44	0.30	32 cb and 6 tb
G. rhellicani	0.39	0.23	24 cb and 16 tb
Chamorchis alpina	0.38	0.22	lcb

taxa [18]. These cytological studies revealed that in this genus there are three basic chromosome numbers, x = 16, 18, and 21. Besides the chromosome number, species of genus *Orchis* s.l. vary in chromosome size; in fact, taxa with 2n = 32, 36 possess chromosomes of larger size than those with 42 [15,16,22]. Bateman et al. [2], on the basis of nucleotide sequences of ribosomal gene spacers (ITS1 and ITS2 nuclear rDNA), divided the genus into three taxonomic groups, namely *Anacamptis* s.l., including all species with 2n = 36 chromosomes, *Neotinea* s.l., and *Orchis* s.s., including 2n = 42 chromosomes taxa.

Table 3.3: Taxa, 4'-6-diamidino-2-phenyl-indole (DAPI), chromomycin A_3 (CMA), probes pTa71 (18S-5.8S-25S) and pTa794 (5S). Abbreviations: tb, telomeric band; cb, centromeric band; stb, subtelomeric band; l/a, long arm; s/a, short arm

Taxa	DAPI (rich AT)	CMA (rich GC)	pTa 71	pTa 794
Barlia robertiana			Pair 5 l/a	Pair 5 s/a
Himantoglossum hircinum			4 sites	4 sites
Anacamptis pyramidalis				
A. coriophora	20 tb			
A. laxiflora				
A. palustris				
A. morio			4 sites	2 sites
A. papilionacea	4 tb		Pair 8 l/a	Pairs 2,3 l/a
A. morio × A. papilionaceai	2 tb		3 sites	3 sites
A. collina			2 sites	2 sites
Neotinea tridentata	cb			
N. lactea	cb and tb			
Orchis mascula	tb and stb			
O. provincialis	tb and stb			
Serapias vomeracea	cb		6 sites	6 sites
S. parviflora	cb			
S. lingua	cb			
Ophrys incubacea	—	cb and tb		
O. tenthredinifera		cb		
O. fusca	—	cb		
Dactylorhiza romana	tb			
D. saccifera	tb	.		
Gymnadenia conopsea	tb			
G. rhellicani	tb			
Chamorchis alpina				

Anacamptis s.l. (Including Former *Orchis morio* Group) [2]

Genus *Anacamptis* Rich. includes terrestrial plants distributed in northern, central, and southern Europe, western Asia and southern to northern Africa [67]. Previous records indicate mainly diploid with 2n = 36 and polyploid cytotypes with 2n = 54, 63, and 72 chromosomes. Only in *Anacamptis pyramidalis* (L.) L.C.M. Rich. has intraspecific polyploidy been seen in many specimens. Moreover, one specimen from Italy had 2n = 72 + 1B at metaphase I in EMC.

Rare cases of triploidy ($2n = 3x = 54$) have been observed in *A. laxiflora* and *A. fragrans* (Pollini) R.M. Bateman (= *Orchis coriophora* L. subsp. *fragrans* Sudre) and meiotic plates at metaphase I in EMC showed numerous trivalents confirming the autopolyploid origin [9,16] (Fig. 3.1).

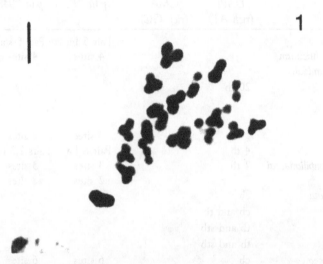

Fig. 3.1: *Anacamptis laxiflora.* EMC at MI showing 14 trivalents, 4 bivalents, 4 univalents. Scale bar = 5 μm (original by S. D'Emerico).

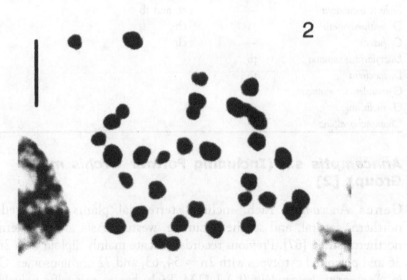

Fig. 3.2: X *Orchiserapias nelsoniana.* 22 univalents and 7 bivalents at MI in an EMC. Scale bar = 5 μm (original by S. D'Emerico).

The karyotype of the species studied showed that in spite of similarity between taxa, there are differences in the morphology of chromosomes. In addition, differences in amount and distribution of heterochromatin have been found. Species with uniformly symmetrical karyotypes, comprising mainly metacentric chromosomes, such as A. morio (Fig. 3.3), A. longicornu, A. laxiflora, and A. pyramidalis, contain little or no heterochromatin. In fact, karyomorphological similarities in these taxa are confirmed by similar reaction to differential staining technique. Whereas, species with a quite asymmetrical karyotype characterized by the predominance of submetacentric chromosomes, such as A. papilionacea (Fig. 3.4), and in particular A. coriophora and A. fragrans, exhibit a large amount of telomerically distributed heterochromatin [18].

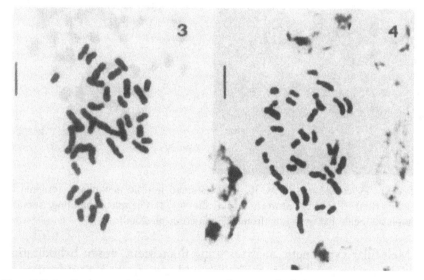

Figs. 3.3-4: Feulgen staining somatic metaphases in Orchid species. 3) *Anacamptis morio*, 2n = 36; 4) A. *papilionacea*, 2n = 32. Scale bar = 5 μm (original by S.D'Emerico)

Anacamptis collina and A. *palustris* possessed karyomorphological characters, which separate them from species of the O. *morio* cluster. In particular, A. *collina* showed a high number of chromosome pairs bearing secondary constrictions.

Species A. *papilionacea*, A. *coriophora*, and A. *fragrans* (Fig. 3.5) are characterized by the presence of telomeric blocks which react positively to both C-banding and DAPI or H33258 staining. The differential staining of heterochromatin by different staining techniques may relate to the intimate composition of chromatin to DNA organization, since different heterochromatin classes might contain different families of repeated DNA [13,39,47]. These data suggest that the DNA of heterochromatin in the latter species is rich in A-T base pairs. (Fig. 3.5)

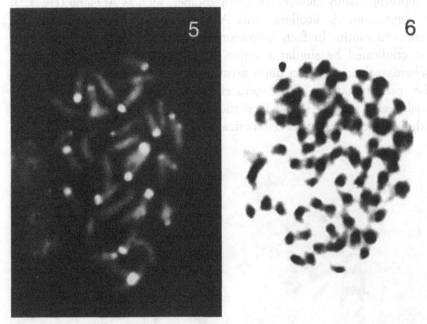

Fig. 3.5: *A. coriophora*, 2n = 36, DAPI stained mitotic metaphase. (original by S.D.'Emerico) (Fig. 3.6) *Serapias lingua*, 2n = 72, Giemsa C-banding, somatic metaphase. Scale bar = 5 μm (from D'Emerico et al. 2000).

Molecular cytogenetic analyses using fluorescent *in-situ* hybridization (FISH), evidenced differences in physical location of ribosomal genes on the chromosome of some species and indicated that repetitive DNA sequences in orchids may prove very useful for the understanding of evolutionary trends [31,42,69]. With the probes pTa71 (18S,5.8S,26S) and pTa794 (5S), in some species rDNA genes were identified on different chromosome pairs. After FISH only *A. papilionacea* showed two pairs of 5S rDNA sites, while *A. morio* and *A. collina* showed only one. Moreover, in *A. papilionacea* the signals on one pair were much more intense than those on the other pair [24] (Fig. 3.7). Repeated sequences are subject to very rapid changes in copy number. Reddy and Appels [50] and Sastri et al. [55] suggested that in some plant groups (e.g. *Secale, Hordeum*) major sites, deriving from overrepresentation of a few ancestral

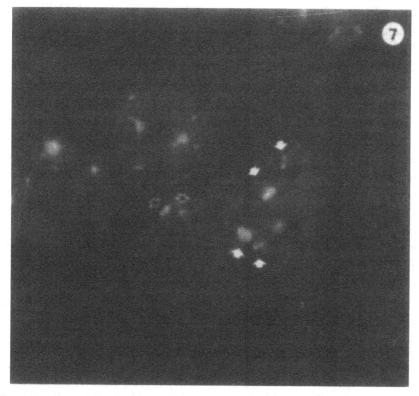

Fig. 3.7: *In-situ* hybridization to chromosomes of *Anacamptis papilionacea*. Red and green signals show sites of hybridization of 18S-25S rDNA (chromosomes are also indicated with open arrowheads) and 5S rDNA (closed arrowheads) respectively (original by S. D'Emerico).

ones, might denote a more recent origin of the major clusters in respect to the minor sites. In this connection, the presence of a major site of 5S rDNA gene clusters in *A. papilionacea* could be regarded as a further indication of a recent origin of Fig. 3.7 chromosomal rearrangements of this species.

With respect to the 18S,25S rDNA sites, pTa71 signals were present on one pair of chromosomes in *A. papilionacea* and *A. collina* and two pairs of sites were observed in *A. morio*.

Steveniella-Himantoglossum s.l. Group (Including *Comperia* and *Barlia*) [3]

This group of species is widespread from Portugal, Spain, and across the Mediterranean region, south to northern Africa, Aegean Isles, Crimea,

Syria, Turkey, Caucasus, western and northern Iran [67]. Species *Himantoglossum hircinum*, *H. adriaticum*, and *Barlia robertiana* presented a chromosome number of 2n = 2x = 36. Cases of aneuploidy with 2n = 36 + 1B are known in both *H. hircinum* and *H. adriaticum* [10,16].

A comparison of the karyotypes of *H. hircinum*, *H. adriaticum*, and *Barlia robertiana*, revealed a similar karyotype with mainly metacentric chromosomes [16,22]. This close karyological resemblance may indicate that the three species are closely related phylogenetically. The chromosomes in these species contain very little or no heterochromatin. In addition, they have low asymmetrical karyotype and properties to banding techniques similar to the *Anacamptis morio* group. However, double-target *in situ* hybridization in *B. robertiana* chromosomes revealed one chromosome pair carrying both the pTa794 and the pTa71 signals on opposite arms; the 5S rDNA sequences were located at the telomeric position on the short arm and the 18S,25S rDNA loci in the telomeric domain of the long arm [24].

In *H. hircinum*, FISH revealed localization of four 18S,25S rDNA signals and four 5S rDNA sites (data unpubl.).

Orchis s.s. (Including *Aceras* R. Br.) [2]

Orchis s.s. species are distributed in Europe, temperate Asia, northern Africa, Azores, Canary Islands, western Iran, Caucasus, and northern Scandinavia [67].

Cytological studies in *Orchis* s.s. revealed chromosome number 2n = 2x = 42, although two cases of tetraploidy (2n = 84) are recorded, in *O. patens* and *O. canariensis*. One rare case of triploidy has been observed in *O. italica* with 2n = 3x = 63; its origin seems to be attributable to the fusion of reduced and nonreduced gametes [9]. In this group, the species possess chromosomes of smaller size and consequently only in some species has it been possible to construct the karyotype. *Orchis italica*, *O. militaris*, *O. purpurea*, *O. quadripunctata*, and *O. simia* showed similar reaction to C-banding. Karyological analyses revealed that all the chromosomal chromatin had a neutral reaction to both C-banding and DAPI staining.

In *Orchis* s.s., the complex of *O. mascula* constitutes an object of divergences of taxonomic evaluation; in fact, *O. provincialis* and *O. mascula*, belonging to subgroup *O. mascula*, exhibit much morphological similarity and close phylogeny [1,48].

The karyotype of *O. provincialis* is represented by 22 metacentric, 14 submetacentric and 6 subtelocentric chromosomes. Numerous chromosomes of the complement were characterized from thin centromeric bands. Two chromosome pairs displayed a heterochromatic short arm and the chromosome pair 1 showed telomeric bands on the long arm.

The karyotype morphology of *O. mascula* consists of 24 metacentric, 14 submetacentric and four subtelocentric chromosomes. C-banded somatic metaphase chromosomes showed particular heterochromatin distribution. In 10 chromosome pairs of the complement, heterochromatin occupied the entire extent of the chromosomes, with only the euchromatic telomeric region, while 11 pairs were completely euchromatic and characterized by the presence of thin centromeric bands. C-banded metaphase I of *O. mascula* showed 10 bivalents having large heterochromatic bands and 11 bivalents with thin heterochromatic bands [25].

The chromosomes of *O. mascula* and *O. provincialis* after staining with both Hoechst 33258 and DAPI showed a bright fluorescence at the telomeric or subtelomeric regions of many chromosomes. For their reacting properties these bands should be enriched in A-T base pairs; moreover this pattern corresponded to no Giemsa C-band.

Neotinea s.l. (Including Former *O. tridentata* Group) [2]

This group is distributed from western Ireland and the Isle of Man eastward across southern Europe from Portugal and the Canary Islands, Turkey, Lebanon, northern Africa, Caucasus, Iraq, and Caspian region in western Asia [67].

Neotinea s.l. revealed the basic chromosome number x = 21 and mainly diploid species. The only polyploid species is *N. commutatum* with 2n = 4x = 84 [43]. *N. tridentata* and *N. ustulata* also showed 2n = 42 + 1B. These B-chromosomes were observed in some individuals found in stations at the boundary of their distribution area and might possibly correlate with B-chromosome presence [16].

The *Neotinea* group shows chromosomes of small size and complex karyomorphology; hence only in *N. tridentata* and *N. lactea* has it been possible to construct the karyotype. The chromosome complement of these species consists of mainly metacentric chromosomes; however, the karyotype of N. *lactea* is more asymmetrical than N. *tridentata* [15,22]. In the two analyzed species, distribution of heterochromatin differed but not its organization. *Neotinea tridentata* showed in all chromosomes small centromeric C-bands. N. *lactea* showed stronger heterochromatin bands than N. *tridentata*; in fact, Giemsa C-banding showed conspicuous bands located at telomeric positions on many chromosomes. In both species DAPI$^+$ heterochromatin always colocalized with C$^+$ bands.

Serapias

Genus *Serapias* L. comprises 20 species [49] distributed in the Mediterranean area. Cytological studies have shown that most of the species possess 2n = 36 chromosomes, with S. *lingua* having 2n = 72 (see Fig. 3.6) [6,8,12,15,27,28,44]. Using the Feulgen technique, all the species revealed a complex chromosome structure with karyotypes moderately asymmetrical and mainly submetacentric chromosomes. Giemsa C-banding showed conspicuous bands at centromeric positions in numerous chromosome pairs. Heterochromatin occupied the entire chromosome with a segment euchromatic at the telomeric position [19]. A study based on sequence divergence of the ITS of ribosomal genes confirmed little differentiation among the species [48].

Preliminary *in-situ* hybridization of *Serapias vomeracea* chromosomes allowed localization of six 5S rDNA signal, of which two were much more intense than those on the other pair. In this species 18S,25S rDNA hybridization signals were present on six chromosomes (data unpubl.).

Ophrys

Genus *Ophrys* L. includes twelve sections and approximately 140 taxa widely spread in the Euro-Mediterranean area [49]. The basic haploid chromosome number is x = 18; however, polyploidy has been recorded with 2n = 4x = 72 in O. *fusca* [34], 2n = 72, 74 in O. *dyris* and O. *vasconica* [5,41], 2n = 3x =54 in O. *tenthredinifera* [9] (Fig. 3.8), and 2n = 90 in O. Pyris [5]. Cytological studies revealed numerous cases of aneuploidy with chromosome numbers 37, 38, and 39 [9, 34]. A characteristic aneuploidy was noticed in interspecific hybrid *Ophrys bertolonii* x O. *tarentina* from Apulia, Italy, with chromosome number 2n = 38 and meiotic irregularities. Meiotic plates during metaphase I showed 17 bivalents + 1 trivalent + 1 univalent or 18 bivalents + 2 univalents. In metaphase II the count was always n = 19. However,

9

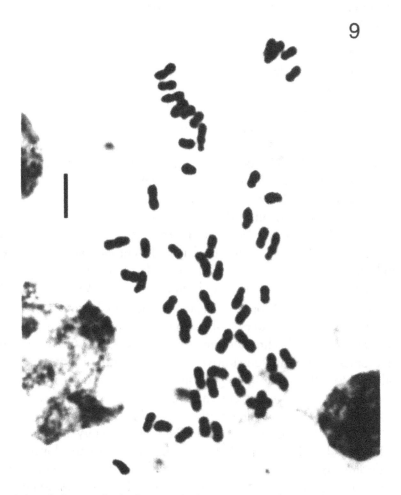

Fig. 3.8: *Ophrys tenthredinifera*. Triploid somatic metaphase with 2n = 3x = 54 chromosomes. Scale bar = 5 μm (original by S. D'Emerico).

Ophrys may be considered rather stable with a constancy of chromosome number.

Bianco et al. [7,9] provided the first karyotypes for *Ophrys* based on analyses of six species. Karyomorphology of *Ophrys* showed that its taxa have karyotypes that can be arranged in a series showing a progressive trend from symmetrical to moderately asymmetrical type. Preliminary C-banding revealed that all chromosomes possess small amounts of centromeric heterochromatin. In some chromosomes, intercalary and terminal C-bands have also been observed.

Dactylorhiza s.l. (Including *Coeloglossum* Hartm.) [2]

Genus *Dactylorhiza* Necker ex Nevski comprises approximately 50 taxa [49] distributed mostly in boreal, temperate, Mediterranean areas, northern Scandinavia, east to the Himalayas, Japan, western Siberia, and eastern North America [67].

Different ploidy levels based on x = 20, ranging from diploid, $2n = 2x = 40$ to polyploid $2n = 4x = 80$ and $2n = 6x = 120$ species have been reported by Soò [60] and Moore [45]. Chromosome number $2n = 3x = 60$ is known in *D. insularis* and *D. maculata* subsp. *meyeri* [15,32,58,66]. It is evident that polyploidy is a very common phenomenon in this group. Chromosome sizes are small in the genus and similar to those of the 42-chromosome *Orchis* s.s. and *Neotinea* s.l. [16]. Karyological analysis in *D. romana* from southern Italy showed $2n = 40$ chromosomes; moreover several specimens with $2n = 41$ and 42 have been observed. Chromosome counts of $2n = 42$ have been recorded for *D. sambucina* [11].

In a cytogenetic study of *D. romana*, characterized from individuals with yellow or purple flowers, C-banding techniques allowed the best identification of B-chromosomes [26]. These supernumerary chromosomes are rich in constitutive heterochromatin and present only in individuals with purple flowers; in fact, no yellow flower individuals ever showed supernumerary chromosomes. On the other hand, this association was present only in one station, probably as a consequence of specific ecological/adaptive conditions. The supernumerary chromosomes were large and had a morphology and size similar to the longest A-chromosomes of the complement. In addition, the heterochromatin structure in late prophase or early metaphase chromosomes comprised two main heterochromatic blocks joined by a poorly stained small gap.

In some species of genus *Dactylorhiza*, examination of the amount and distribution of heterochromatin and karyotype structure indicates an interesting variation in chromosome complement. The karyotype structure and constitutive heterochromatin of *D. sambucina*, *D. insularis*, and *D. sambucina* belonging to section *Sambucinae* and *D. saccifera* ($2n = 40$) and *D. maculata* ($2n = 80$) of the *Dactylorhiza* section [59], proved to be quite different. The latter species have smaller chromosomes and a higher average amount of heterochromatin than the former. Consequently, the data obtained for the species reported here support the basal position of the *Sambucinae* clade [1].

Gymnadenia s.l. (Including Former *Nigritella* Group) [2]

Gymnadenia R. Br. in Aiton comprises about 20 taxa distributed in Europe and central and eastern Asia [67]. The basic chromosome number is x = 20, and up to now diploids with 2n = 40, triploids with 2n = 60, tetraploids with 2n = 80, and pentaploids with 2n = 100 have been reported [64]. Only the diploid species reproduce sexually, while the polyploid are apomicts [63].

Chromosomes are small and the karyotype morphology very complex; hence only C-banding offers more information about chromosome morphologies. In this group, the chromosomes of G. *rhellicani* and G. *conopsea*, both diploids (2n = 40), revealed notable variations in the amount and distribution of constitutive heterochromatin. G. *rhellicani* showed numerous chromosomes with evident telomeric heterochromatin after Giemsa C-banding, while G. *conopsea* showed that all the chromosomes possess centromeric heterochromatin and a few telomeric C-bands. These data seem to indicate that structural rearrangements are present in the chromosome complement of G. *rhellicani* [21].

The polyploid species G. *austriaca* (2n = 80) and G. *buschmanniae* (2n = 100) show a great similarity in heterochromatin content and distribution along the chromosomes to diploid species.

Traunsteinera-Chamorchis Group [3]

This group includes three species—*Traunsteinera globosa, T. sphaerica,* and *Chamorchis alpina*—distributed in northern, western, central, and southeastern Europe south to Italy and east to Poland, north to Russia, east to the Carpathians, Turkey, and Caucasus [67]. An interesting result of previous studies was that although the three species have the same chromosome number (2n = 42), the size of their genomes and constitutive heterochromatin varies considerably (data unpubl.). In C. *alpina* the sizes of chromosomes and karyomorphology are rather different from those in other genera with 42 chromosomes, i.e., *Orchis* s.s., *Neotinea* s.l., and *Trausteinera*. In all the chromosomes of the complement of C. *alpina* it is clearly possible to observe the centromeric position and a moderately asymmetrical karyotype; chromosomes in T. *globosa* are small and similar to those of the 42-chromosome *Orchis* s.s. and *Neotinea* s.l. In addition, C-banding technique revealed in C. *alpina* conspicuous bands in numerous chromosome pairs. In many cases heterochromatin occupied the entire extent of the chromosomes, with euchromatin confined to the telomeric position [21]. On the contrary, in *T. globosa* all the chromosomes have small centromeric bands rather similar to those of *Neotinea tridentata*. Our data on *C. alpina* could be interpreted to indicate structural rearrangements in chromosome complement [4,61].

EVOLUTIONARY RELATIONSHIPS AMONG GENERA

In recent years, cytogenetic analysis has been widely applied as a phylogenetic tool in orchids and has considerably influenced the taxonomy and classification of the Orchidinae subtribe [3].

Orchis s.l. and allied genera form a complex that displays variations in chromosome number, karyotype structure and amount and distribution in heterochromatin. Some examples involving possible pathways of karyotype evolution are given based on karyomorphological data obtained from some Orchidinae genera.

Genus *Anacamptis* s.l. (2n = 36) has a diverse evolutionary pattern in species formation (Fig. 3.9); chromosome morphology and heterochromatin distribution allowed identification of different groups within the genus. *Serapias* and *Ophrys*, both with 2n = 36 chromosomes, have normally homogeneous karyomorphology. In *Anacamptis* s.l., Giemsa staining revealed variation in banding pattern between the species. In *A. papilionacea*, *A. coriophora*, and *A. fragrans*—species with asymmetrical karyotypes—have been found with a major amount of heterochromatin, mainly clusters around telomeres. The species *A. morio*, *A. longicornu*, *A. laxiflora*—with less asymmetrical karyotypes—showed Fig. 3.9 chromosomes with insignificant heterochromatin.

In *Serapias*, C-banding showed conspicuous bands around the centromeres, indicating that these species form a homogeneous cluster. In addition, the presence of a larger amount of heterochromatin and the high level of karyotype asymmetry in this genus seem to denote recent structural rearrangements. *Ophrys* species are characterized by the presence of quite homogeneous, moderately asymmetrical and scarcely heterochromatic karyotypes; heterochromatin present in thin centromeric bands possibly indicates that evolutionary forces are acting to a lesser extent in this genus.

Though an evolutionary pattern for the three genera is still rather complex, a possible scenario may be drawn. If the assumptions that chromosomal rearrangements and amount and distribution of

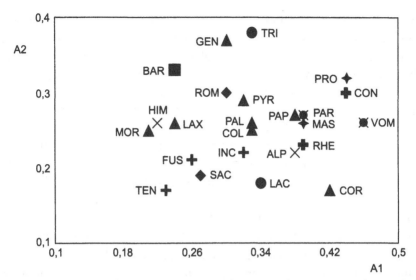

Fig. 3.9: Diagram of the values of A_1 (intrachromosomal asymmetry index) and A_2 (interchromosomal asymmetry index) of the karyotypes of the Orchidinae taxa examined [54]. Species codes given in Table 3.1. In this group as a whole, karyotypes showed a progressive trend from symmetrical (*Anacamptis morio*; $A_1 = 0.21$) to moderately asymmetrical (*Serapias vomeracea*; $A_1 = 0.46$). Interestingly, our data showed that *Anacamptis* s.l. group (triangle) is accompanied by progressive variation in the intrachromosomal asymmetry index (A_1 = range 0.21-0.42) and amount of constitutive heterochromatin, undicating that both these parameters play a possible role in karyotype evolution.

heterochromatin can play a functional role in karyotype evolution and affect the phenomenon of speciation, our data can justify the perception that *Anacamptis* s.l. is an ancestral group, which has not ended its evolution, possibly due to adaptative radiation, from which both *Ophrys* and *Serapias* evolved. The former due to adaptation to a specialized reproductive system has probably reduced its evolutionary rate, while the latter is still positively evolving. On the other hand, the karyotype structure of *Ophrys* species appears very distinct from that of *Serapias* and the hypothesis of a derivation of both genera from a common ancestor is not probable.

Based on the chromosome morphology and heterochromatic distribution in the *Dactylorhiza* (2n = 40) species studied, it can be seen that the chromosome set recalls the set in *Orchis* s.s. and *Neotinea* s.l., both with 2n = 42 chromosomes [9,15,16,22]. The hypothesis of a derivation of *Dactylorhiza* from 42-chromosome *Orchis* species indicates that the *Sambucinae* section, including *D. sambucina* and *D. romana*, is a possible primitive group from which *Dactylorhiza* genus evolved. In the latter species the presence of a variable number (from $2n = 40$ to 42) chromosomes, of an ancestral karyotype and particular heterochromatic supernumerary chromosomes, suggest that the number $2n = 40$ originated from an ancestral $2n = 42$ through chromosome number reduction and heterochromatinization of two chromosomes. Later the supernumerary chromosomes can be progressively depleted of functional genes and eventually lost. Heterochromatin elimination is reported in parasitic nematodes, crustacean copepods, and arthocladines [51]. On the other hand, the karyotypes of *D. saccifera* ($2n = 40$) and *D. maculata* ($2n = 80$) belonging to section *Dactylorhiza* represent a secondary arrangement that occurred during the rapid radiation of the genus [26].

In *Gymnadenia* s.l. karyological analyses showed a good level of similarity to the *Dactylorhiza* group in chromosome number, karyomorphology, and heterochromatin pattern.

CYTOGENETIC EVIDENCE OF INTERSPECIFIC AND INTERGENERIC HYBRIDIZATION IN SOME GENERA

Interspecific Hybrids

Cytogenetic studies on interspecific hybrids involving taxa of Orchidinae groups are very rare. In *Anacamptis* the natural interspecific hybrid *A. morio* x *A. papilionacea* (*O. x gennarii*) gave evidence of chromosome structural hybrid. The parental species possessed distinct chromosome numbers: *A. morio* had $2n = 36$ chromosomes while *A. papilionacea* exhibited $2n = 32$ chromosomes. *A. morio* x *A. papilionacea* showed 34 chromosomes. Meiotic plates in hybrid specimens at metaphase I in EMC mostly showed univalents and just a few bivalents, indicating a reduced chromosomal homology between parental genomes, while both parent species exclusively exhibited bivalents. With C-banding and fluorochrome staining, two chromosomes were characterized by the presence of large telomeric heterochromatic blocks on the short arms, which strongly resembled the ones observed in *A. papilionacea* [18]. *In-situ* hybridization of ribosomal genes allowed localization of three 18S,25S rDNA signals and three 5S rDNA sites. In *O. x gennarii* it was observed that the number of 18S,25S and 5S ribosomal sites was precisely the sum of the haploid site numbers of the two parents, and that the chromosomal domains of the hybridization sites corresponded to those of the parental species [24].

In one population two hybrid specimens were triploid. The 52-chromosomed A. *morio* × A. *papilionacea* plants showed at metaphase I in EMC 18 bivalents and 16 univalents. These triploid individuals might have originated from the fusion of an unreduced gamete with a normal haploid gamete. When the triploid hybrid karyotype was compared with the karyotypes of both putative species, it appeared that the 18 chromosomes, which can form pairs, should have originated from A. *morio* and the 16 unpaired ones from A. *papilionacea* [17]. Regarding the set corresponding to the genome of A. *morio*, there is a difference in pair n. 1; this pair is submetacentric while the equivalent pair of A. *morio* is metacentric.

Most populations of *Anacamptis morio* × A. *papilionacea* consist of diploid individuals and the origin of numerous F_1 individuals was derived by cross-fertilization between individuals of parental species. However, asexual reproduction through tuber represents a mechanism frequently observed. The chromosome number of the hybrids has showed 2n = 34, while 2n = 33 or 35 were never found; moreover, comparison of chromosome complement of numerous individuals has shown different karyomorphology. These data suggest extensive chromosomal rearrangements and not backcross with the putative species. The effects of chromosomal differences between supposed parents are largely responsible for hybrid sterility and play an important role in the absence of the introgression process.

In conclusion, the hybridization mechanism is of particular relevance in the natural hybrid *Anacamptis morio* × A. *papilionacea*, in which the formation of this interspecific hybrid involves the sympatric species A. *morio* and A. *papilionacea*. Numerous individuals showed some traits from one parent, while others were intermediate between the parental species. Moreover, a high frequency of extreme or novel characters in many hybrid specimens [52,53] was also seen.

Another case regarding an interspecific hybrid was proposed by D'Emerico et al. [16] in the formation of individual hybrids between putative parents *Anacamptis collina* and A. *morio*, both with chromosome number 2n = 36. Of four specimens that displayed characters morphologically intermediate between the parents, three showed 2n = 36 chromosomes and a chromosome complement typical of the putative species. Karyological analyses of the fourth hybrid individuals showed a chromosome number $2n = 3x = 54$. The allotriploid hybrid karyotype revealed numerous chromosome pairs typical of A. *collina*, while the haploid genome seems to belong to A. *morio*.

An interesting case was observed in the interspecific hybrid *Ophrys apulica* × *O. bombyliflora*, in which examination of ovary wall cells, normally diploid ($2n = 2x = 36$), revealed cells with polyploid nuclei and number $2n = 4x = 72$, $2n = 6x = 108$ and $2n = 8x = 144$. These tetraploid and octoploid numbers probably result from endomitosis, while the hexaploid cells could have arisen through fusion between nuclei. This process of endonuclear chromosome duplication, considered the most common mode of cell polyploidization in plants, was described in *Ophrys* for the first time by Bianco et al. [9].

In *Orchis* s.s., though *O. mascula* (L.) L. and *O. provincialis* Balbis ex Lam. et DC. display much morphological similarity and are closely related from a phylogenetic point of view [1,48], yet they represent two clearly isolated *Orchis* species that should often be found living in sympatry.

In this respect it is notable that in Sardinia, in mixed populations of *O. mascula* s.l. and *O. provincialis*, intermediate phenotypes between these species have frequently been detected and classified as *O. x penzigiana* A. Camus. The presumed hybrid specimens display intermediate morphological characters between the parental species. Like both the parental species, all examined individuals of *Orchis x penzigiana* revealed a chromosome number of $2n = 42$.

The banding patterns of *Orchis mascula* s.l. displayed the presence of particular heterochromatic distribution. C-banding showed in numerous chromosomes conspicuous bands located around centromeric positions. On the contrary, in the complement of *O. provincialis* C-banding revealed in all chromosomes a small centromeric heterochromatin (represented mainly as two dots), and three chromosome pairs with telomeric bands.

Banded metaphases were observed in *O. x penzigiana*. C-banding revealed 10 heterochromatic chromosomes and 32 euchromatic chromosomes. Some parental chromosomes could be distinguished by the difference in heterochromatin distribution. In fact, ten heterochromatic chromosomes are of *Orchis mascula* s.l. origin, whereas, 32 completely euchromatic chromosomes (21 chromosomes of *O. provincialis* origin and 11 chromosomes of *O. mascula*) cannot be distinguished [25].

Intergeneric Hybrids

Within the Orchis s.l. alliance, Anacamptis hybridizes readily with Serapias. However, specific individualization of the putative parents on a morphological base, as is normally seen in Orchiserapias, in this case gives very complex results.

X Orchiserapias nelsoniana (Anacamptis collina × Serapias parviflora) was first described by Bianco et al. [8]. Its number is 2n = 36 and the lengths of the chromosomes range from 4.38 to 1.65 μm. Meiotic plates at MI in an EMC showed predominantly univalents (Fig. 3.2), indicating very low affinity between putative parents A. collina and Serapias parviflora. However, combined cytogenetic and morphological approaches suggest as putative parents the species A. collina and S. parviflora in the formation of the intergeneric hybrid. On the other hand, the rarity of the hybrid can be partly explained, considering that S. parviflora practical of norm the autogamy and the precocity of A. collina to give the scarce superimposition of the periods of flowering of the two species.

POLYPLOIDY IN ORCHIDINAE

A high number of polyploid taxa have been described in subtribe Orchidinae for Dactylorhiza [37,38] and Gymnadenia [63,64]. Genus Dactylorhiza contains autotetraploid taxa such as D. maculata subsp. maculata (2n = 80) belonging to D. maculata group and allotetraploid taxa derived by hybridization between D. maculata s.l. and D. incarnata s.l. lineages [37]. In the Gimnadenia group on the other hand, the polyploid taxa (2n = 80 and 2n = 100) are apomictic and produce adventitious (nucellar) embryos [63]. However, some cases of triploidy (2n = 60) have been observed [21].

Contrarily, groups Anacamptis s.l., Orchis s.s., Neotinea s.l., Ophrys, Serapias, Steveniella-Himantoglossum, and Traunsteinera-Chamorchis show only a few cases of polyploidy. Within the Anacamptis group, A. pyramidalis showed three levels of ploidy, diploids (2n = 36), triploids (2n = 54), and tetraploids (2n = 72), often associated with morphological and chromatic differences. In this species, cytotypes differ in geographic distribution, with diploids occupying restricted areas and undisturbed vegetation (e.g. fringes of undergrowth), while tetraploids are the most widespread cytotypes (e.g. uncultivated land). Triploid individuals are usually found in geographically identical areas where with both cytotypes are present, and arise from a diploid × tetraploid cross as a consequence of contact between diploid and tetraploid individuals.

Autotriploid individuals (2n = 3x = 54) have been observed occasionally in *Anacamptis laxiflora* and *A. fragrans* populations, both of which are diploid with 2n = 2x = 36 chromosomes. Cytological examination suggests the fusion of reduced and nonreduced gametes as the origin of these triploid individuals.

Genus *Ophrys* is characterized by a constant diploid chromosome number, 2n = 2x = 36, whereas tetraploidy, 2n = 4x = 72, is observed only in the *O. fusca* populations. In this group also numerous cases of somatic aneuploidy with 2n = 37, 38, 39, and 40 are known. However, here the presence of aneuploidy appears to be correlated with hybridization and introgression.

In *Orchis* s.s. (2n =2x = 42) two cases of tetraploidy, 2n = 4x = 84, have been noted for *O. patens* and *O. canariensis*, while in *Neotinea* s.l. (2n = 2x = 42) a chromosome count of 2n = 84 has been recorded for *N. commutata*.

Predominantly diploid species are found in genus *Serapias* with 2n = 36 chromosomes; only in *S. lingua* does 2n = 72. Karyological analysis in *S. lingua*, based on heterochromatin content, showed some chromosomes in the karyotype with amounts of heterochromatin characteristic of the chromosome complement of *S. parviflora*. Let it be noted here that the interspecific hybrid *Serapias x todaroi* Tin. originated from hybridization between *S. lingua* (2n = 72) and *S. parviflora* (2n = 36). The chromosomal number of the individual hybrid is 2n = 54. Profit sharing of *S. lingua* in hybridization is also revealed by the fact that it is the only species of genus *Serapias* to have 2n = 72 chromosomes.

DISTRIBUTION OF HETEROCHROMATIN IN REPRESENTATIVES OF ORCHIDINAE

Orchidinae species have shown a striking range of variation in heterochromatin content (see Tables 3.2 and 3.3). The most detailed studies in this respect pertain to *Anacamptis* s.l., *Orchis* s.s., *Neotinea* s.l., *Serapias*, *Dactylorhiza* s.l., *Ophrys*, *Steveniella-Himantoglossum* s.l., and *Traunsteinera-Chamorchis* group.

The karyotypes of nine *Anacamptis* species examined in an earlier study [18], show that most (*A. morio*, *A. longicornu*, *A. laxiflora*, *A. palustris*, *A.*

collina) have negligible or no heterochromatin. On the other hand, some do (A. *papilionacea*, A. *coriophora*, and A. *fragrans*). Staining with DAPI showed bright blocks at telomeric regions corresponding to Giemsa C-bands [18].

In *Serapias*, heterochromatin distribution revealed low levels of divergence among the species and the presence of large centromeric heterochromatin might indicate recent structural rearrangements in the chromosomal complement of this genus [4]. In this connection, the reduced timescale might account for the small interspecific variation of nuclear and morphological characters observed in the genus [19].

Karyological analysis in *Orchis* s.s. demonstrated that in species *O. mascula* and *O. provincialis* most DAPI-positive sites did not colocalize with Giemsa C-bands. In addition, DAPI showed bright fluorescence at the telomeric or subtelomeric regions of many chromosomes. Consequently, the fluorescent data was able to indicate that these telomeric regions contain AT-rich satellite DNA. On the other hand, in group *Neotinea* DAPI⁺ heterochromatin always colocalized with Giemsa C-bands, indicating an organization quite different from that observed in the *Orchis mascula* subgroup.

Interestingly, in *Ophrys* species preliminary karyological analyses revealed that CMA⁺ heterochromatin colocalized with C⁺-bands. All the species studied exhibited similar banding patterns when stained with C-banding and CMA.

CONCLUSIONS

Given its wide variability, the Orchidaceae family is an interesting group in which to monitor biodiversity at the global level.

As for the cytogenetic diversity of Orchidinae species, two points seem clear. First, our investigation showed wide degree of variability in chromosome number and karyomorphology in a number of orchid species. Second, observational studies have documented that a major amount of constitutive heterochromatin in some groups, sometimes plays a role in the evolution of species differences. Finally, due to the sensitivity of most orchids to environmental changes, as a consequence of their breeding system, karyomorphological variation in orchid species could be proposed as a biodiversity indicator in a given environment [20].

REFERENCES

[1] Aceto S, Caputo P, Cozzolino S, Gaudio L, Moretti A. Phylogeny and evolution of *Orchis* and allied genera based on ITS DNA variation: morphological gaps and molecular continuity. Mol Phylogen Evol 1999; 13: 67-76.

[2] Bateman RM, Pridgeon AM, Chase MW. Phylogenetics of subtribe *Orchidinae* (Orchidoideae, Orchidaceae) based on nuclear ITS sequences. 2. Infrageneric relationships and reclassification to achieve monophyly of *Orchis* sensu stricto. Lindlejana 1997; 12: 113-141.

[3] Bateman MR, Hollingsworth MP, Preston J, et al. Molecular phylogenetics and evolution of Orchidinae and selected Habenariinae (Orchidaceae). Bot J Linn Soc 2003; 142: 1-40.

[4] Bernard J, Miklos GLG. Functional aspects of satellite DNA and heterochromatin. Int Rev Cytology 1979; 58: 1-14.

[5] Bernardos S, Amich F, Gallego F. Karyological and taxonomic notes on *Ophrys* (Orchidoideae, Orchidaceae) from the Iberian Peninsula. Bot J Linn Soc 2003; 142: 395-406.

[6] Bianco P, Medagli P, D'Emerico S, Ruggiero L. Numeri cromosomici per la flora italiana. Inform Bot Ital 1987; 19: 322-332.

[7] Bianco P, D'Emerico S, Medagli P, Ruggiero L. Karyological studies of some taxa of the genus *Ophrys* (Orchidaceae) from Apulia (Italy). Caryologia 1989; 42: 57-63.

[8] Bianco P, D'Emerico S, Medagli P, Ruggiero L. X *Orchiserapias nelsoniana* Bianco, D'Emerico, Medagli et Ruggiero, hybr. nat. nov. della Puglia. Webbia 1990; 44: 315-322.

[9] Bianco P, D'Emerico S, Medagli P, Ruggiero L. Polyploidy and aneuploidy in *Ophrys*, *Orchis* and *Anacamptis* (Orchidaceae). Pl Syst Evol 1991; 178: 235-245.

[10] Capineri R, Rossi W. Numeri cromosomici per la flora italiana. Inform Bot Ital 1987; 19: 314-318.

[11] Cauwet-Marc AM, Balayer M. "Les orchidées du bassin méditerranéen". Contribution à l'étude caryologique des espéces des pyrénées-orientales (France) et contrées limitrophes. II. Tribu des *Ophrydeae* Lindl. Bull Soc Bot France, Lett Bot 1986; 133: 265-277.

[12] Cauwet-Marc AM, Balayer M. Les genres *Orchis* l., *Dactylorhiza* Necker ex Newski, *Neotinea* Reichb. et *Traunsteinera* Richb: caryologie et proposition de phylogénie et d'évolution. Bot Helv 1984; 94: 391-406.

[13] Cuadrado A, Jouve N. Mapping and organization of highly repeated DNA sequences by means of simultaneous and sequential FISH and C-banding in 6x-*Triticale*. Chromosome Res 1994; 2: 331-338.

[14] D'Emerico S. Orchideae, Cytogenetics. In: Pridgeon AM, Cribb PJ, Chase MC, Rasmussen FN, eds. Genera Orchidacearum 2: Orchidoideae. Oxford, UK: Oxford University Press, 2001b; Part 1: 216-224.

[15] D'Emerico S, Bianco P, Medagli P. Karyological studies on Orchidaceae. Tribe *Ophrydeae*, subtribe *Serapiadinae*. Caryologia 1992; 45: 301-311.

[16] D'Emerico S, Bianco P, Medagli P. Cytological and karyological studies on Orchidaceae. Caryologia 1993; 46: 309-319.

[17] D'Emerico S, Bianco P, Pignone D. Cytomorphological characterization of diploid and triploid individuals of *Orchis x gennarii* Reichenb. Fil. (Orchidaceae). Caryologia 1996; 49: 153-161.

[18] D'Emerico S, Pignone D, Bianco P. Karyomorphological analyses and heterochromatin characteristic disclose phyletic relationships among 2n = 32 and 2n = 36 species of *Orchis* (*Orchidaceae*). Pl Syst Evol 1996; 200: 111-124.

[19] D'Emerico S, Pignone D, Scrugli A. Giemsa C-band in some species of *Serapias* L. (Orchidaceae). Bot J Linn Soc 2000a; 133: 485-492.

[20] D'Emerico S, Pignone D, Scrugli A. Karyomorphology and evolution in Italian populations of three Neottieae species (Orchidaceae). Cytologia 2000b; 65: 189-195.

[21] D'Emerico S, Grünanger P. Giemsa C-banding in some *Gymnadenia* species and *Chamorchis alpina* from the Dolomites (Italy). J. Europäischer Orchideen. 2001a; 33: 405-414.

[22] D'Emerico S, Bianco P, Medagli P, Ruggiero L. Karyological studies of some taxa of the genera *Himantoglossum*, *Orchis*, *Serapias* and *Spiranthes* (Orchidaceae) from Apulia (Italy). Caryologia 1990; 43: 267-276.

[23] D'Emerico S, Grünanger P, Scrugli A, Pignone D. Karyomorphological parameters and C-bands distribution suggest phyletic relationship within the subtribe *Limodorinae* Bentham (Orchidaceae). Pl Syst Evol 1999; 217: 147-161.

[24] D'Emerico S, Galasso I, Pignone D, Scrugli A. Localization of rDNA loci by fluorescent *in situ* hybridization in some wild orchids from Italy (Orchidaceae). Caryologia 2001c; 54: 31-36.

[25] D'Emerico S, Cozzolino S, Pellegrino G, Pignone D, Scrugli A. Heterochromatin distribution in selected taxa of the 42-chromosome *Orchis* s.l. (Orchidaceae). Caryologia 2002a; 55: 55-62.

[26] D'Emerico S, Cozzolino S, Pellegrino G, Pignone D, Scrugli A. Karyotype structure, supernumerary chromosomes and heterochromatin distribution suggest a pathway of karyotype evolution in *Dactylorhiza* (Orchidaceae). Bot J Linn Soc 2002b; 138: 85-91.

[27] Del Prete C. Numeri cromosomici per la flora italiana: Inform Bot Ital 1977; 9: 135-140.

[28] Del Prete C. Contributi alla conoscenza delle orchidaceae d'Italia. VI. Tavole cromosomiche delle orchidaceae italiane con alcune considerazioni citosistematiche sui generi *Ophrys*, *Orchis* e *Serapias*. Inform Bot Ital 1978; 10: 379-389.

[29] Dressler RL. Phylogeny and classification of the orchid Family. Portland, OR: 1993. Timber Press.

[30] Flavell RB. Repetitive DNA and chromosome evolution in plants. Phil Trans 1986; ser. B 312: 227-242.

[31] Galasso I, Schmidt T, Pignone D, Heslop-Harrison JS. The molecular cytogenetics of *Vigna unguiculata* (L.) Walp: the physical organization and characterization of 18S-5.8S-25S rRNA genes, 5S rRNA genes, telomere-like sequences, and a family of centromeric repetitive DNA sequences. Theor Appl Genet 1995; 91: 928-935.

[32] Gathoye JL, Tyteca D. Contribution à l'Etude Cytotaxonomique des *Dactylorhiza* d'Europe Occidentale. Mém Soc Roy Bot Belgique 1989; 11: 30-42.

[33] Greilhuber J. Evolutionary changes of DNA and heterochromatin amounts in the *Scilla bifolia* group (Liliaceae). Pl Sys Evol 1979; suppl. 2: 263-280.

[34] Greilhuber J, Ehrendorfer F. Chromosome numbers and evolution in *Ophrys* (*Orchidaceae*). Pl Syst Evol 1975; 124: 125-138.

[35] Greilhuber J, Speta F. C-banded karyotypes in the *Scilla hohenackeri* group, *S. persica* and *puschkinia* (*Liliaceae*). Pl Syst Evol 1976; 126: 149-188.

[36] Guerra M. Patterns of heterochromatin distribution in plant chromosomes. Genet Mol Biol 2000; 23, 4: 1029-1041.

[37] Hedrén M. Plastid DNA variation in the *Dactylorhiza incarnata/maculata* polyploid complex and the origin of allotetraploids *D. sphagnicola* (*Orchidaceae*). Mol Ecol 2003; 12: 2669-2680.

[38] Hedrén M, Fay MF, Chase MW. Amplified fragment length polymorphisms (AFLP) reveal details of polyploid evolution in *Dactylorhiza* (*Orchidaceae*). Amer J Bot 2001; 88: 1868-1880.

[39] Jamilena M, Ruiz Rejón C, Ruiz Rejón M. Repetitive DNA sequences families in *Crepis capillaris*. Chromosoma 1993; 102: 272-278.

[40] Kao YY, Chang SB, Lin TY, et al. Differential accumulation of heterochromatin as a cause for karyotype variation in *Phalaenopsis* Orchids. Ann Bot 2001; 87: 387-395.

[41] Kullemberg B. Studies in *Ophrys* pollination. Zoologiska Bidrag Uppsala 1961; 34: 1-340.

[42] Maluszynska J, Heslop-Harrison JS. Physical mapping of rDNA in *Brassica* species. Genome 1993; 36: 774-781.

[43] Mazzola P, Crisafi F, Romano S. Numeri cromosomici per la flora italiana. Inform Bot Ital 1981; 13: 182-188.

[44] Mazzola P, Miceli G, Not R. Numeri cromosomici per la flora italiana. Inform Bot Ital 1982; 14: 275-279.

[45] Moore RJ. Flora Europaea. Check-list and Chromosome Index. Cambridge, UK: Cambridge University Press, 1982.

[46] Pellegrino G, Cozzolino S, D'Emerico S., Grünanger P. The taxonomic position of the controversial taxon *Orchis clandestina* (*Orchidaceae*): Karyomorphological and molecular analyses. Bot Helv 2000; 110: 101-107.

[47] Pignone D, Galasso I, Rossino R, Mezzanotte R. Characterization of *Dasypyrum villosum* (L.) Candargy chromosomal chromatin by means of *in situ* restriction endonucleases, fluorochromes, silver staining and C-banding. Chromosome Res. 1995; 3: 109-114.

[48] Pridgeon AM, Bateman RM, Cox AV, Hapeman JR, Chase MW. Phylogenetics of subtribe *Orchidinae* (*Orchidoideae*, *Orchidaceae*) based on nuclear ITS sequences. 1. Intergeneric relationships and polyphyly of *Orchis* sensu lato. Lindleyana 1997; 12: 89-109.

[49] Quentin P. Synopsis des Orchidées Européennes. Cahiers de la S.F.O., no. 2. Paris.

[50] Reddy P, Appels R. A second locus for the 5S multigene family in *Secale* L.: sequence divergence in two lineages of the family. Genome 1989; 32: 456-467.

[51] Redi AC, Garagna S, Zacharias H, Zuccotti M, Capanna E. The other chromatin. Chromosoma. 2001; 110: 136-147.

[52] Rieseberg LH. The role of hybridization in evolution: old wine in new skins. Amer J Bot 1995; 82: 944-953.

[53] Rieseberg LH, Ellstrand NC. What can morphological and molecular markers tell us about plant hybridization? Crit Rev Pl Sci 1993; 12: 213-241.

[54] Romero Zarco C. A new method for estimating karyotype asymmetry. Taxon 1986; 35: 526-530.

[55] Sastri DC, Hilu K, Appels R, et al. An overview of evolution in plant 5S DNA. Pl Syst Evol 1992; 183: 169-181.

[56] Schwarzacher T, Schweizer D. Karyotype analysis and heterochromatin differentiation with Giemsa C-banding and fluorescent counterstaining in *Cephalanthera* (Orchidaceae). Pl Syst Evol 1982; 141: 91-113.

[57] Schweizer D, Nagl W. Heterochromatin diversity in *Cymbidium* and its relationships to differential DNA replication. Exper Cell Res 1976; 98: 411-423.

[58] Scrugli A. Numeri cromosomici per la flora italiana. Inform Bot Ital 1977; 9: 116-125.

[59] Soó Rde. Die Geschichte der Erforschung der Gattung *Orchis* (sensu lato), besonders von Dactylorhiza. J Naturwiss Vereins Wuppertal 1968; 21/22: 7-19.

[60] Soó Rde. Genus *Dactylorhiza*. In: Tutin TG, Heywood VH, Burges NA, et al. Flora Europaea Cambridge, UK: Cambridge University Press, 1980: (5)

[61] Stebbins GL. Chromosomal evolution in higher plants. London; UK: Arnold Publishing Company, 1971.

[62] Stergianou KK. Habit differentiation and chromosome evolution in *Pleione* (Orchidaceae). Pl Syst Evol 1989; 166: 253-264.

[63] Teppner H. Adventitious embryony in *Nigritella* (Orchidaceae). Folia Geobot Phytotax 1996; 31: 323-331.

[64] Teppner H, Klein E. Etiam atque etiam—*Nigritella* versus *Gymnadenia*: Neukombinationen und *Gymnadenia dolomitensis* spec. nova (Orchidaceae-Orchideae). Phyton 1998; 38: 220-224.

[65] Vosa CG. Heterochromatin and ecological adaptation in southern African *Ornithogalum* (Liliaceae). Caryologia 1997; 50: 97-103.

[66] Vöth W, Greilhuber J. Zur Karyosystematik von *Dactylorhiza maculata* s.l. und ihrer Verbreitung, Insbesondere in Niederösterreich. Linzer Biol. Beitr. 1980; 12: 415-468.

[67] Wood J. Orchideae, Cytogenetics. In: Pridgeon AM, Cribb PJ, Chase M C, Rasmussen FN, eds. Genera Orchidacearum 2: Orchidoideae. Oxford, UK: Oxford University Press, 2001; Part 1: 241-377.

[68] Yamasaki N. Differenzielle Darstellung der Metaphasechromosomen von *Cypripedium* debile mit Chinacrin- und Giemsa-Färbung. Chromosoma (Berl.) 1973; 41: 403-412.

[69] Zoldos V, Papes D, Cerbah M, et al. Molecular-cytogenetic studies of ribosomal genes and heterochromatin reveal conserved genome organization among 11 *Quercus* species. Theor Appl Genet 1999; 99: 969-977.

Heterochromatin and Microevolution in *Phleum*

ANDRZEJ JOACHIMIAK
Department of Plant Cytology and Embryology, Institute of Botany, Jagiellonian University, Cracow, Poland

ABSTRACT

A concise history of systematic studies of the genus *Phleum* L. (timothy) is presented. The results of cytotaxonomical studies and their importance for investigation of phylogeny and interrelationships between diploid and polyploid *Phleum* taxa are discussed.

Karyotypes of ten species belonging to sections *Phleum* Humphries, *Chilochloa* (Beauv.) Griseb., and *Achnodon* Griseb. were studied with conventional and Giemsa C-banding methods and analyzed with regard to nuclear DNA amount. It is suggested that genomic changes in heterochromatin distribution and DNA amount were important factors in the evolution of this genus.

Key Words: *Phleum*, C-banding, heterochromatin, nuclear DNA amount, karyotype, evolution

INTRODUCTION

The genus *Phleum* L. (timothy) comprises about 15 annual and perennial species (Table 4.1) with dense, spikelike, ovoid, or cylindrical panicles. There are two morphologically distinct groups of *Phleum* species: those with panicle branches completely adnate to the rachis (grouped within

Address for correspondence: Department of Plant Cytology & Embryology, Institute of Botany, Jagiellonian University, Grodzka St. 52, PL 31–044 Cracow, Poland. E-mail: a.joachimiak@iphils.uj.edu.pl

Table 4.1: Taxonomic treatment of *Phleum* (from Dogan 1988, supplemented)

1	2	3	4	5	6	7
GRIESEBACH	ASHERSON & GRAEBNER	KOMAROV et al.	BOR	HUMPHRIES	TZVELEV	DOGAN
1853	1902	1963	1970	1980	1984	1988, 1991
Sect. Euphleum	Sect. Euphleum	Sect. Euphleum	Sect. Euphleum	Sect. Phleum	Sect. Phleum	Sect. Phleum
P. pratense	*P. pratense*	*P. pratense*	*P. pratense*	*P. pratense*	*P. pratense*	*P. pratense*
P. alpinum	*P. alpinum*	*P. alpinum*	*P. alpinum*	*P. alpinum*	*P. alpinum*	*P. alpinum*
	P. echinatum	*P. echinatum*	*P. bertolonii*	*P. echinatum*	*P. echinatum*	*P. bertolonii*
			P. montanum			*P. echinatum*
			P. iranicum			
			P. michelii			
			P. phleoides			
Sect. Chilochloa	Sect. Chilochloa	Sect. Chilochloa	Sect. Chilochloa	Sect. Chilochloa	Sect. Chilochloa	Sect. Chilochloa
P. boehmeri	*P. boehmeri*	*P. phleoides*	*P. himalaicum*	*P. phleoides*	*P. phleoides*	*P. phleoides*
P. michelii	*P. michelii*	*P. michelii*	*P. paniculatum*	*P. hirsutum*	*P. hirsutum*	*P. hirsutum*
P. arenarium	*P. arenarium*	*P. arenarium*	*P. boissieri*	*P. arenarium*		*P. arenarium*
P. asperum	*P. montanum*	*P. montanum*	*P. exaratum*	*P. montanum*		*P. montanum*
P. paniculatum	*P. paniculatum*	*P. paniculatum*		*P. paniculatum*		*P. subulatum*
P. graecum		*P. graecum*		*P. himalaicum*		*P. boissieri*
		P. iranicum				

Contd.

Table 4.1 (*Contd.*)

1 GRIESEBACH 1853	2 ASHERSON & GRAEBNER 1902	3 KOMAROV et al. 1963	4 BOR 1970	5 HUMPHRIES 1980	6 TZVELEV 1984	7 DOGAN 1988, 1991
Sect. Achnodon		Sect. Achnodon	Sect. Achnodon	Sect. Achnodon	Sect. Achnodon	Sect. Achnodon
P. tenue		P. tenue	P. subulatum	P. subulatum	P. subulatum	P. subulatum
					P. himalaicum	P. boissieri
					P. arenarium	P. exaratum
					P. paniculatum	
				Sect. Maillea		Sect. Maillea
				P. crypsoides		P. crypsoides

P. subulatum (Savi) Asch. et Gr. = P. tenue (Host) Schrader; P. boehmeri Wib. = P. phleoides (L.) Karst; P. michelii All. = P. hirsutum Honck.; P. paniculatum Huds. = P. asperum Jacq.; P. graecum Boiss. et Heldr. = P. exaratum Hochst.

sect. *Phleum* Humphries) and those with free panicle branches (all other taxa) (Fig. 4.1).

Fig. 4.1: Panicle of *P. phleoides* (sect. *Chilochloa*) (a), and *P. bertolonii* (sect. *Phleum*) (b). (Photograph courtesy of Adam Kula.)

It was speculated that the characteristic spikelike, cylindrical inflorescence originated from loose panicles [73]; thus sect. *Phleum* seems to be the most evolutionarily advanced within the genus.

Based on macromorphological characters, most authors have subdivided genus *Phleum* L. into three sections: *Phleum* Humphries (formerly *Euphleum*), *Chilochloa* (Beauv.) Griseb., and *Achnodon* Griseb.

Humphries [24] recognized a fourth section, Maillea (based on Maillea crypsoides (d'Urv.) Bois.), and proposed a clear subdivision of the group with free branches of panicles in three sections: sect. Chilochloa (3 stamens, grains ovoid-oblong), sect. Achnodon (3 stamens, grains laterally compressed), and sect. Maillea (2 stamens) (Table 4.2).

Table 4.2: Phleum sections (from Humphries 1980)

Sect. Phleum	Sect. Achnodon
* Panicle-branches almost completely adnate to rachis	* Panicle-branches free
* Rachilla not prolonged	* Rachilla prolonged
* Glumes not winged on keel	* Glumes not winged on keel
* Lemma 3- to 7-veined	* Lemma 5- to 7-veined
* Stamens 3	* Stamens 3
* Grain ovoid-oblong, terete	* Grain laterally compressed
Sect. Chilochloa	Sect. Maillea
* Panicle-branches free	* Panicle-branches free
* Rachilla prolonged	* Rachilla prolonged
* Glumes not winged on keel	* Glumes winged on keel
* Lemma 3- to 7-veined	* Lemma 1-veined
* Stamens 3	* Stamens 2
* Grain ovoid-oblonge, terete	* Grain ovoid-oblonge, terete

Despite the relatively few taxa in the classification, the circumscription and subdivision of Phleum L. is still controversial [63]. Attempts have been made to introduce biosystematic methods into the taxonomy of Phleum for investigation of its phylogeny and the interrelationships between and within recognized sections: karyotype analysis (see [33,34], for review), DNA hybridization [7], numerical analysis of selected macro- and micromorphological characters [14], and SEM studies of the lemma surface and hair type [12,63]. Some authors have done biometric studies to identify several species/cytotypes, especially within the group of closely related perennial taxa belonging to sect. Phleum [10,29,30,39,40,45].

All Phleum species are indigenous to the Old World. Most taxa are distributed in the Mediterranean area, the center of diversity of this genus, but some extend to many nontropical countries of both hemispheres [13]. The two most widespread species belong to sect. Phleum: tetraploid P. commutatum Gaud. (= P. alpinum L., P. alpinum var. commutatum /Gaud./ Grisebach, P. alpinum subsp. commutatum /Gaud./

Richter, *P. alpinum* subsp. *alpinum* Humphries), which is widespread in cool and subarctic regions of the Northern Hemisphere and the only member of the genus that is also indigenous to the New World (as *P. alpinum* L. var. *americanum* Fournier); and hexaploid *P. pratense* L., which was introduced intentionally to many countries and occurs commonly in almost all temperate regions of the world.

Hexaploid *P. pratense*, one of the most important cool-season forage grasses, has the longest history of formal breeding activity in both Europe and North America [9]. It has been suggested that the native center of diversity of this timothy is northern Europe (see [21] for review). Most timothy cultivars are synthetics but some have been selected from landraces or local ecotypes. Some American strains developed from European populations were later reintroduced into Europe. American cultivars were also introduced in Japan and many other countries.

Diploid *P. bertolonii* DC (also known as *P. nodosum* L., *P. hubbardii* D. Kov.), is treated by many authors as a chromosomal race or subspecies of *P. pratense*, and seems to have a distribution very similar to that of hexaploid cultivars. The two taxa have often been confused with each other; grains of *P. bertolonii* probably have been mixed with those of *P. pratense*, accounting for the observed overlapping of their current geographic ranges [41].

The agricultural importance of two perennials, *P. pratense* L. and *P. alpinum* L. (a valuable pasture species, especially in the cool and montane regions) has led to many studies of their genetic variation and adaptation (see [9] for review), and also of the taxonomy and evolution of sect. *Phleum*.

This section includes the two abovementioned core species considered in all classifications proposed so far (Table 4.1) but some morphological variants have been recorded within each of them. A variant of *P. pratense* was described under the name *P. nodosum* in 1759 by Linnaeus and as *P. bertolonii* in 1813 by De Candolle (see [40] for review), and a variant of *P. alpinum* was described as *P. commutatum* in 1808 by Gaudin (see [23] for review). For this reason, Schörter [64] divided the group of perennials into two coenospecies: *P. pratense* sp. coll. and *P. alpinum* sp. coll. The existence of two different variants of each core species has not been questioned since then, although they have been variously named and variously treated as species, subspecies or varietes [23,39]. Differences between the cytotypes of *P. pratense* are quantitative rather than qualitative [10,54], while the two forms of *P. alpinum* differ only by the ciliate/glabrous awns on the spikelets (see Fig. 4.6 e and f). As Humphries [23] showed, Linnaeus knew two separate forms of P. *alpinum* but did not distinguish between them, and the phrase name for P. *alpinum* was undoubtelly based on a plant with glabrous awns from Lapland, P. *commutatum Gaud.*

Since the beginning of the twentieth century, the annual P. *echinatum* Host., a Mediterranean element with a narrow range of distribution and rather unknown alliances with other taxa, has been included in sect. *Phleum* (Table 4.1).

Beginning in the early 1930s, intensive and comprehensive experimental and cytotaxonomic studies aimed at determining the relationships between the two basic taxa (P. *pratense* L. and P. *alpinum* L.) were undertaken by British [20] and Swedish researchers [49–53]. It was established that the basic chromosome number in *Phleum* is 7 and that each species has two chromosome races, one diploid (2n = 2x = 14) and one polyploid, i.e., hexaploid (2n = 6x = 42) in P. *pratense* and tetraploid (2n = 4x = 28) in P. *alpinum*. These early cytological studies confirmed the phenotypic splitting of P. *pratense* L. and P. *alpinum* L. into two separate taxa each.

According to Nordenskiöld's [54] oft-cited, exemplary experimental and cytomorphological study, four perennial taxa can be clearly separated genetically, cytologically, and ecologically. In view of the lack of significant morphological differences within P. *pratense* and P. *alpinum*, Nordenskiöld proposed dividing each group into two separate species according to chromosome number: P. *nodosum* L. (= P. *bertolonii* DC.), 2n = 14 ↔ P. *pratense* L., 2n = 42 and P. *alpinum* L. (P. *alpinum* subsp. *rhaeticum* Humphries, P. *rhaeticum* (Humphries) Raushert) 2n = 14 ↔ P. *commutatum* Gaud., 2n = 28.

Nordenskiöld suggested the following phylogenetic relationships between the species:

1. Six genomes of P. *pratense* are closely homologous; P. *pratense* is the autohexaploid of P. *nodosum*.

Experimental studies by Nordenskiöld [55–58] provided further evidence for the autohexaploid origin of P. *pratense*. Other authors have also suggested it [8,76].

2. P. *alpinum*, though separate from the abovementioned species, is genetically much closer to them than to P. *commutatum*.

Nordenskiöld's cytological observations of P. *alpinum* × P. *nodosum* hybrids suggested that the genomes of these two species might be partly homologous. On the other hand, *P. commutatum* produces sterile hybrids both with *P. alpinum* and with hexaploid *P. pratense*. Also, hybrids between *P. commutatum* and *P. nodosum* were apparently intrasterile [20,53]. Nordenskiöld [54] speculated that the two genomes of tetraploid *P. commutatum* are of different origin, but their hypothetical donors (unknown diploid species within *P. pratense* – *P. alpinum* group) were not identified. Another possibility, consistent with the isolated position of this species within the group, is that *P. commutatum* is a highly differentiated polyploid bearing many fixed genetic changes.

In 1948, Litardiere [47] found diploid plants from the *P. alpinum* group in the Pyrenees with glabrous (not cilliate) awns on the spikelets. In accordance with Nordenskiöld's subdivision of the *P. alpinum* group, these plants should be classified by morphology as *P. commutatum* but by chromosome number as *P. alpinum*. The existence of a diploid race of *P. commutatum* has also been confirmed from the Polish Carpathians [18,31,48,62], the Alps and Sweden [32,72]. These findings disturb the clear picture of the *P. alpinum* group established by Nordenskiöld [54]; they mean that it is impossible to identify the taxa recorded here by chromosome counting. They also pose problems for traditional taxonomists – the two cytological variants of *P. commutatum* are practically indistinguishable by morphology. On the other hand, at least one putative parent of tetraploid *P. commutatum* Gaud. has been discovered.

The current nomenclature of the *P. alpinum* group introduces considerable confusion and leads to unnecessary misunderstandings. The Linnean name *P. alpinum* is used by different authors for individuals with ciliate awns [22,42,61], glabrous awns [11], or both [24]. Here I propose the broadly accepted name *P. commutatum* for all plants with glabrous awns (both diploid and tetraploid) and the more unequivocal name *P. rhaeticum* (Humphries) Raushert for plants with ciliate awns.

The validity of the two names (*P. nodosum* L. and *P. bertolonii* DC.) of the diploid form of *P. pratense* s.l. has also been questioned [40]. Kovats suggested that both the Linnaeus and De Candolle holotypes are in fact underdeveloped specimens of the hexaploid *P. pratense*, and proposed a new name for the diploid form (*P. hubbardii* D. Kov.). In Conert [11], however, we find that stomata length in the *P. bertolonii* holotype is typical for the diploid, not the hexaploid form of *P. pratense*. Moreover, in collodium replicas of the culm surface presented by Kovats [40], the stomata of the *P. bertolonii* holotype and of *P. hubbardii* are the same size. Thus, the diploidy of this holotype cannot be questioned and the valid name of the diploid form is still *P. bertolonii* DC.

SECTION *PHLEUM*

Early karyological work showed the usefulness of chromosome studies in *Phleum* taxonomy. In the beginning of the 1990s, Dr. Adam Kula and I began C-banding studies on this interesting genus, first on representatives of sect. *Phleum* from Poland [31] and later from other areas of Europe [32,34]. This review is based on those papers and our other published and unpublished data.

Perennial Taxa

The karyotypes of three diploid taxa (*P. bertolonii*, *P. commutatum* /2x/, and *P. rhaeticum*) are practically undistinguishable by conventional chromosome staining [31,32]. Thus the basic genomes (designated here as B, C, and R) in polyploid forms also cannot be identified using conventional karyotype analysis.

All three diploid species show relatively high intraspecific C-band polymorphism and their karyotypes show similarities if the localization of all (polymorphic and fixed) heterochromatic blocks is analyzed. If only stable heterochromatin segments are considered, however, differences in their genome structure become clear [31,32,35].

When heterochromatin occurring with relatively high frequency (>50% within chromosome pool) was considered, the genome of diploid *P. commutatum* (genome C) showed seven distinct centromeric but only two telomeric blocks of heterochromatin. Chromosomes of *P. bertolonii* (genome B) showed a completely different "banding style" with 12 stable heterochromatin segments gathered at the chromosome termini. In *P. rhaeticum* (genome R), both centromeric and telomeric segments were observed within the majority of chromosomes (Fig. 4.2).

The most variable element of diploid karyotypes is the telomeric heterochromatin. In the taxa analyzed, karyotypes of individual plants vary with respect to the localization, frequency, and size of telomeric segments. For example, in different populations of *P. commutatum* (2x) the amount of this heterochromatin within the karyotype fluctuates between 1.49% and 11.4%, and in *P. rhaeticum* between 16% and 25.47%. The average amount of relatively stable heterochromatin (50% frequencies)

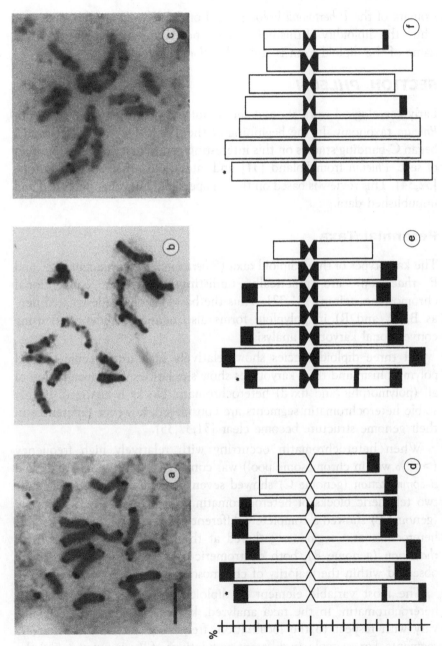

Fig. 4.2: C-banded chromosomes (a–c) and genome structure (d–f) of three perennial diploid taxa of sect. *Phleum*. a,d – *P. bertolonii*; b,e – *P. rhaeticum*; c,f – *P. commutatum* /2x/. * SAT-chromosomes. (Unpublished micrographs [a–c] provided by Adam Kula.)

greatly depends on the frequency and thickness of telomeric blocks; thus the karyotype of *P. commutatum* is poor in heterochromatin while *P. rhaeticum* is heterochromatin-rich and resembles the karyotype of *P. bertolonii* in this respect.

The C-banding karyotype of tetraploid *P. commutatum* (Fig. 4.3a) is very similar to the karyotype of the diploid race of this species. Its four genomes are not recognizable by morphology and greatly resemble the C genome. European (Scottish and Scandinavian) and American (Californian) tetraploids show virtually the same C-banding karyotypes (with only a slight difference in heterochromatin amount), suggesting a common origin for them.

In hexaploid *P. pratense* almost all major heterochromatic segments are terminally located (Fig. 4.3b). With minor differences, the chromosomes of this species resemble those of *P. bertolonii*. The two taxa have similar heterochromatin distribution patterns (similar banding style). On the other hand, the number (2 or 4) of some distinct chromosome markers identified by us in this species suggests either diploidization or an allopolyploid origin for *P. pratense* [31,44]. The genome formula of *P. pratense* deduced from these results is BBBBXX, where X is probably the genome of *P. rhaeticum* or another taxon unidentified as yet [34].

Our interpretation is consistent with the results of C-banding studies made by Cai and Bullen [6]. They also proposed a 4:2 genome formula but suggested four genomes coming from tetraploid *P. alpinum* (i.e. *P. commutatum*) and two from *P. bertolonii*. Further studies of theirs [7] provided the first molecular evidence for close relationships between *P. pratense* and two diploid taxa: species-specific DNA clones of diploid *P. alpinum* (not specified by morphology, but most probably *P. rhaeticum*) and *P. bertolonii* hybridize to genomic probes of hexaploid *P. pratense*. Unfortunately, the number of R and B genomes cannot be deduced from these studies.

In the light of our C-banding studies, it seems clear that the diploid *P. alpinum* analyzed by Cai and Bullen [6,7] was in fact *P. rhaeticum*. The C-banded karyotype they presented for this form is virtually the same as the karyotype of *P. rhaeticum* we found, and very similar to the karyotype of *P. bertolonii*. Moreover, it shows banding patterns that differ from the genome C identified by us in both chromosomal races of *P. commutatum*. In view of the completely different banding styles of their chromosomes, *P. commutatum* should be excluded as the donor of the four *P. pratense* genomes. More likely, not *P. commutatum* but *P. rhaeticum* was the second

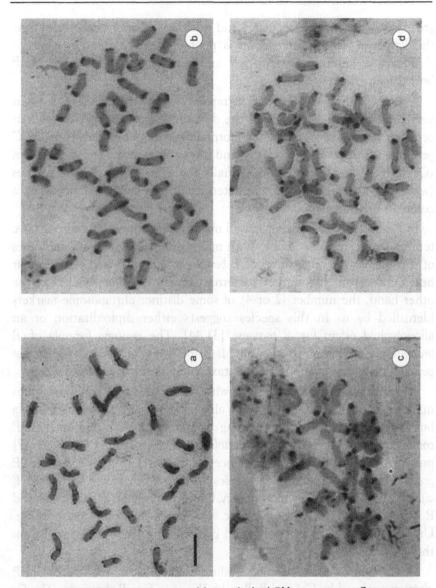

Fig. 4.3: C-banded chromosomes of four polyploid *Phleum* taxa. a – *P. commutatum* /4x/; b – *P. pratense* /6x/; c – *P. pratense* /4x/; d – *P. pratense* /8x/. (Supplied by Barbara Dudziak [a] and Adam Kula [b–d].)

parental species of *P. pratense*. This view accords not only with C-banding studies [6,31,32,34,44], but also with molecular data [7] and cytological observations [54]. All the available data suggest that all *P. pratense* genomes are very similar to the *P. bertolonii* genome and that the genome of *P. rhaeticum* (R) is much closer to the genome of *P. bertolonii* (B) than to that of *P. commutatum* (C).

Another possibility is that *P. pratense* derived from hybridization between two very close relatives with slightly different B genomes. This is very probable because at least two morphological variants of diploid *P. pratense* s.l. can be observed in Europe. The first, Mediterranean (?), is morphologically undistinguishable from hexaploid timothy [10,45]. The second is morphologically distinct and was described by Kovats [40] as *P. hubbardii*. Most probably, this morphological variant of diploid *Phleum* is widespread in central Europe [41,45]. The two forms differ slightly in nuclear DNA amount and heterochromatin content [45]. This view is consistent with the 4:2 genome hypothesis, the Nordenskiöld [54] suggestions about the autohexaploid nature of timothy, and the most recent RAPD and UP-PCR data [8].

All cultivated forage-type timothy is hexaploid [9] and almost all cytologically analyzed, indigenous populations of this species are also hexaploid. For example, in Poland only one of 112 *P. pratense* plants collected from 15 natural populations was tetraploid [31,67]. In some geographic regions, however, plants with nonhexaploid chromosome numbers may not be so rare. Of the 36 natural Italian populations of *P. pratense* s.l. analyzed by Cenci et al. [10], 26 were hexaploid, 8 tetraploid, 1 octoploid, and 1 diploid. All chromosomal races of *P. pratense* show similar distribution of heterochromatin within the karyotype (Figs. 4.2a, and 4.3b – d; Table 4.4). It is difficult to estimate the actual frequency of various forms in nature because morphologically the chromosomal races of *P. pratense* s.l. are practically indistinguishable [10], but it has been shown that these races can be differentiated through analysis of stomatal cell length [29,30] (Table 4.3).

There are some interesting problems concerning the interrelationships among the various chromosomal races of *P. pratense* and these races and other species belonging to sect. *Phleum*. Especially interesting is the genomic constitution of naturally occurring tetraploids. Hexaploid *Phleum pratense* most probably originated from a cross between diploid and tetraploid plants. The C-banded chromosomes of three tetraploid *P. pratense* lines from the Mediterranean (France, Spain, and Italy) resemble chromosomes of *P. bertolonii*, with almost all heterochromatin located terminally (Fig. 4.3c). Thus naturally occurring 4x plants seem to be autotetraploid. Moreover, tetraploid plants have twice as much 2C DNA as *P. bertolonii* [68] (Table 4.4).

Table 4.3: Stomatal cell length in four cytotypes of *P. pratense* s.l. (from Joachimiak et al. 2002)

Cytotype	Average stomatal cell length ± standard deviation (mm)	Duncan test (at P = 0.05)
Diploid	28.34 ± 3.21	d
Tetraploid	32.49 ± 3.33	c
Hexaploid	35.38 ± 2.69	b
Octoploid	42.16 ± 3.62	a

Table 4.4: 2C DNA amount, average size of basal chromosome set (1x) and telomeric and centromeric heterochromatin in perennial taxa of sect. *Phleum*. Dominating type of heterochromatin (over 6% of karyotype length) shaded (from Sliwinska et al. 2003, and unpubl. data)

Species/Cytotype	Nuclear DNA content (pg)		Heterochromatin amount (% of karyotype length)	
	2C	1x	Telomeric	Centromeric
P. bertolonii/2x	3.34	1.67	13.85	1.02
P. pratense/4x	6.78	1.69	7.29	1.56
P. pratense/6x	9.17	1.53	11.09	1.00
P. pratense/8x	12.37	1.53	9.85	1.21
P. rhaeticum/2x	2.93	1.46	11.90	7.47
P. commutatum/2x	2.68	1.34	2.35	6.27
P. commutatum/4x	6.17	1.54	4.10	6.40

Within the *P. pratense* group, the average length of the basal chromosome set at metaphase decreases with increasing ploidy level (Fig. 4.4) [46]. On the other hand, there is no observed decrease in average DNA amount per basic set in tetraploid plants (putative autotetraploids) and only a slight decrease in hexa- and octoploids (putative allopolyploids with slightly different genomes). Thus, as in many other plant genera, allopolyploidization within this group was probably accomplished without substantial DNA loss [59].

The gradual decrease of chromosome length observed here is most likely associated with substantial increases in nuclear 2C DNA amount. Karp and Jewell [36] suggested that the duration of the mitotic cycle is proportional to the nuclear DNA amount and that the duration of prophase affects chromosome length at metaphase. The higher the chromosome number, the longer the time to reach metaphase and the greater the contraction of chromosomes at this stage. Verma and Rees

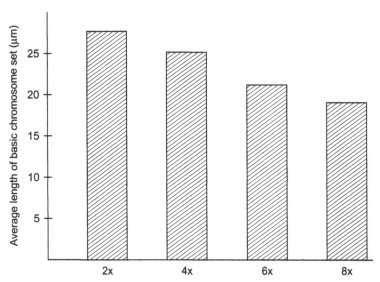

Fig. 4.4:   Length of basic chromosome set in four cytotypes of *P. pratense* s.l.

[74] pointed to relationships between polyploidy, evolution, and chromatin condensation in *Brassica*. In the allotetraploid *Brassica* species, DNA content represents the sum of the DNA content of their diploid ancestors, but the volume of nuclei in long-established tetraploids is much less than in newly synthesized tetraploids.

Despite the lack of substantial DNA loss in 6x and 8x polyploids, some minor evolutionary DNA changes are probable in these plants. For example, the absolute average amount of heterochromatin (calculated in pg) within their karyotypes is obviously lower than that predicted from the average amount of heterochromatin in diploids (Fig. 4.5).

Some other deviations from additivity can be deduced from our PCR-ISSR studies: five apparent ISSR DNA markers of diploid *P. bertolonii* were present in tetraploid timothy but absent in hexaploid [71]. Another deviation was the appearance of new ISSR bands in polyploid plants (Table 4.5). Our ISSR studies also suggest some alliances between *P. rhaeticum* and hexaploid but not tetraploid *P. pratense*.

Deviations from additivity on the DNA level have also been observed in polyploid wheat [17,37,60,65]. It has been suggested that the elimination of some chromosome-specific and genome-specific sequences in newly synthesized allopolyploids is nonrandom and highly reproducible [17,65]. Interestingly, elimination of noncoding DNA sequences observed

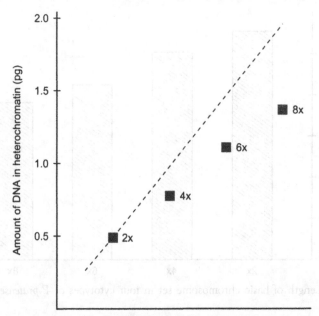

Fig. 4.5: Calculated amount of DNA in heterochromatin in four cytotypes of *P. pratense* s.l.

by the authors cited was rapid and occurred soon after allopolyploidization. Nonrandom and concerted elimination of some noncoding DNA sequences may contribute to cytological diploidization by increasing the divergence of homeologous chromosomes. Genomic changes of this type can facilitate the establishment of newly formed polyploids as successful species [60].

Phleum echinatum

Morphologically, *P. echinatum* is very similar to two perennials, *P. commutatum* and *P. rhaeticum*. Its panicles are short cylindrical or ovoid, spikelets strongly compressed laterally, and glumes relatively long-awned vis-à-vis representatives of the *P. pratense* group (Fig. 4.6). However, unlike all other representatives of sect. *Phleum*, *P. echinatum* is a highly specialized annual species with a relatively narrow range of distribution.

The karyotype of *P. echinatum* was investigated only once, back in 1950 by Ellerström and Tjio [15]; they reported an atypical chromosome number (2n = 10) and atypical chromosome morphology for this species, both of which were confirmed in our recent observations (Fig. 4.7a).

The chromosome complement of *P. echinatum* was highly asymmetric compared to all other taxa of the genus, heterochromatin poor (6.8%),

Table 4.5: Selected ISSR markers in *P. rhaeticum*, *P. commutatum* (4x) and 2x, 4x, and 6x plants from the *P. pratense* group (from Sutkowska et al. 2002)

SSR primer	Marker length (bp)	P. alpinum group		P. pratense group		
		P. rhaeticum	P. commutatum	2x	4x	6x
ISSR 1	530	—	—	+++	+++	—
	350	—	—	+++	+++	—
	290	+++	+++	—	—	—
ISSR 2	360	+++	+++	—	—	—
ISSR 3	420	—	—	+++	—	+++
	240	—	—	—	+++	+++
ISSR 6	430	—	—	+++	+++	—
	370	—	—	+++	—	+++
	310	—	—	+++	+++	—
	270	—	(+)	—	+++	+++
	220	(+)	(+)	+++	+++	—
ISSR 7	220	+++	—	—	—	++

+++: in all analyzed plants

(+): only in some plants

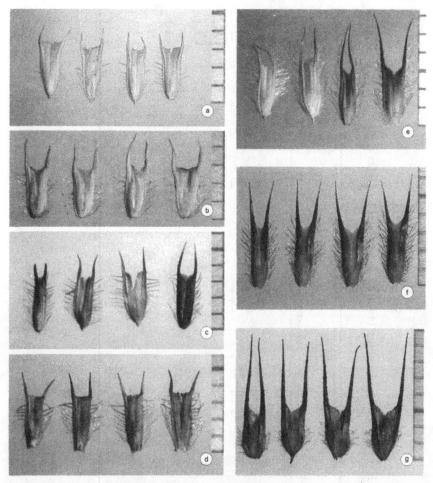

Fig. 4.6: Shape of spikelets in six taxa of *Phleum* section. a – *P. bertolonii*; b – *P. pratense* /4x/; c,d – *P. pratense* /6x/ (from two different populations); e – *P. rhaeticum*; f – *P. commutatum* /4x/; g – *P. echinatum*. (Supplied by Adam Kula).

and showed a telomeric C-band distribution (Fig. 4.7b). Its banding style was very similar to that of representatives of *P. pratense*. Surprisingly, the nuclear 2C DNA amount of this diploid species (7.28 pg) was twice that of *P. bertolonii* (3.35 pg).

Thus this puzzling species shows great morphological similarity to representatives of the *P. commutatum* – *P. rhaeticum* group, chromosomes with a telomeric heterochromatin distribution characteristic for the *P. pratense* group, and a DNA amount similar to that of polyploid species within sect. *Phleum*.

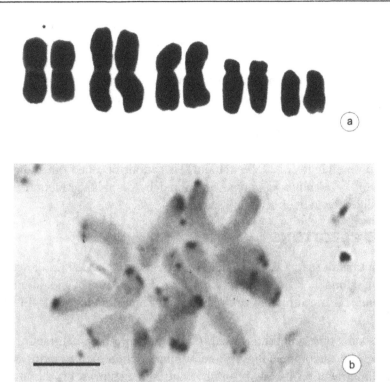

Fig. 4.7: Chromosomes of *P. echinatum*. a – morphology of chromosomes; b – C-banded metaphase plate. * SAT-chromosomes. (Supplied by Adam Kula.)

The unusual basic chromosome number (x = 5) and asymmetric karyotype of *P. echinatum* have long intrigued scientists. It seems unlikely that a genome of such shape could be primitive; asymmetrical karyotypes are generally regarded as more specialized in higher plants [70]. Moreover, the basic chromosome number x = 7 is considered primary in this group of grasses [38,73] and to all taxa of the genus investigated so far.

The calculated basic genome size (3.64 pg) in *P. echinatum* is also very unusual. In other taxa belonging to *Phleum* sections it does not exceed 2 pg [68]. All these observations suggest that *P. echinatum* possesses a derived genome of totally unknown origin, reduced in chromosome number and highly differentiated.

Secondary reductions in chromosome number have been observed in many Mediterranean or Australian plants [1,2,66,69,75]. Such reductions have often been observed in plants growing in semidesert or seasonally xeric regions and are associated with the evolution of an annual habit

[75]. Probably the observed reduction of chromosome number in *P. echinatum* was also associated with its ecological specialization and evolution of an annual habit.

Reduced chromosome number associated with a more asymmetric karyotype is usually achieved through dysploid reduction, i.e., unequal translocation of large chromosome segments followed by loss of a centromere [2,66,75]. Dysploid reductions would usually increase average chromosome length but would not increase genome size (total chromosome length or DNA amount). It cannot be ruled out, however, that the *P. echinatum* karyotype originated by dysploid reduction in an originally polyploid (tetraploid) species.

OTHER SECTIONS

All the taxa of sect. *Chilochloa* analyzed by us (*P. phleoides* (L.) Karst., *P. hirsutum* Honck., *P. montanum* Koch, *P. graecum* Boiss. et Heldr., and *P. arenarium* L.) are diploids with the chromosome number 2n = 14 (Fig. 4.8a – e).

The karyotypes of these species are generally heterochromatin poor (under 7% karyotype length) and telomeric heterochromatin is the prevailing fraction (Table 4.6). Only *P. arenarium* shows a higher amount of heterochromatin (9.2%) and a different C-band distribution. Hence two different chromosome banding styles are also evidenced (Fig. 4.9).

P. subulatum, the only species of sect. *Achnodon* [24], is also diploid (Fig. 4.8f) with a telomeric distribution of heterochromatin. Unfortunately, there is still no statistical data from different populations for a careful karyological examination of all these taxa.

The nuclear 2C DNA amount in all species mentioned above ranges between 2.88 and 3.72 pg ([68] and unpubl. data). Average size of the basic chromosome set (1.64 pg) of the analyzed species of sect. *Chilochloa* is almost identical to that of representatives of the *P. pratense* group (1.63 pg) but the set is distinctly larger in the two widespread perennial taxa, *P. phleoides* (1.78 pg), the section type, and *P. hirsutum* (1.92 pg) [68]. It may be that the perennial habit is the ancestral one in this section (as in section *Phleum*); hence the larger genome seems more primitive here.

The most interesting species within sect. *Chilochloa* is *P. arenarium*. It is a winter annual, photosynthesizing from autumn to early summer. It is typical of sand poor in nitrogen and phosphorus but rich in calcium. This highly specialized species colonizes coastal dunes from S. Sweden down to

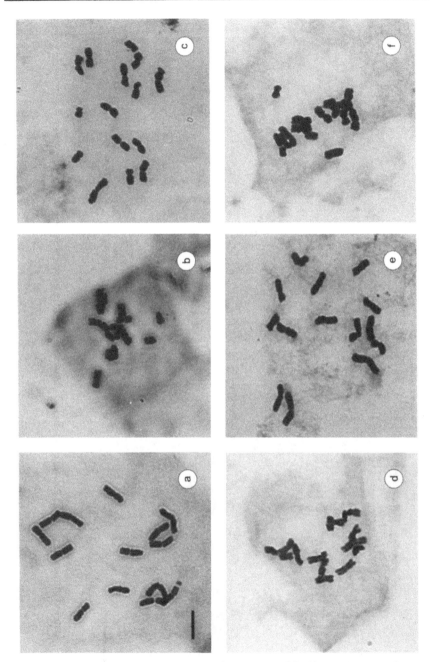

Fig. 4.8: Metaphase plates of six species from sect. *Chilochloa* and sect. *Achnodon*. a – *P. phleoides*; b – *P. hirsutum*; c – *P. montanum*; d – *P. arenarium*; e – *P. graecum*; f – *P. subulatum*. (Supplied by Adam Kula.)

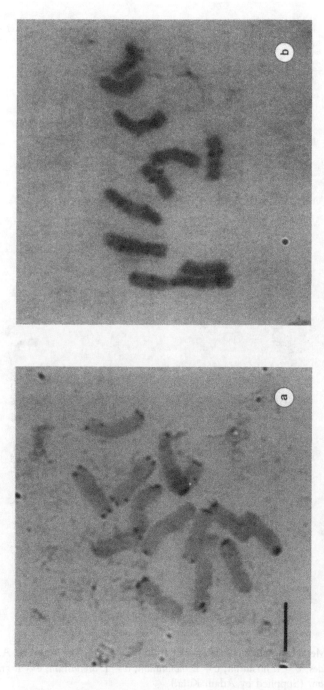

Fig. 4.9: C-banded chromosomes of *P. phleoides* (a) and *P. arenarium* (b). (Supplied by Adam Kula.)

Table 4.6: Distribution of heterochromatin in taxa belonging to sect. *Chilochloa* and sect. *Achnodon*

Species	Heterochromatin distribution	
	Telomeric	Centromeric
P. phleoides (L.) Karst.	+++	
P. hirsutum Honck.	+++	
P. montanum Koch	+++	
P. graecum Boiss. et Heldr.	+++	
P. arenarium L.		+++
P. subulatum (Savi) Asch. et Gr.	+++	

the Iberian Pennisula and the Mediterranean region, along the Danish, German, Dutch, Belgian, and French coasts [16]. *P. arenarium* differs from the other representatives of the section in both its heterochromatin and its DNA amount. The plants analyzed by us from two different sites (Kent, UK and Castricum, The Netherlands) have very similar 2C DNA values (2.86 pg and 2.91 pg on average, respectively) and karyotype structure [68]. This species probably is the most advanced one within the section, and the characteristic organization of its genome is associated with its ecological specialization and evolution of an annual habit.

The phylogenetic relationships within and between the two sections analyzed here are not clear. Most of the species are diploids with a uniform 2n = 14 chromosome number and only sligtly differentiated karyotypes. Two perennial species, *P. phleoides* and *P. hirsutum*, are the core species of sect. *Chilochloa*, just as the annual *P. subulatum* is for sect. *Achnodon* (Table 4.1). Polyploid chromosome numbers have been reported for three species: *P. himalaicum* Mez. (2n = 8x = 58), *P. paniculatum* Huds. (2n = 4x = 28) and *P. montanum* (2n = 6x = 42) (http:\mobot.mobot.org\W3T\Search\ipcn.html). Unfortunately, the polyploid karyotypes of these species were not analyzed by C-banding.

Perennial *P. montanum* probably has different chromosomal races; 2n = 42 was reported from Bulgaria [43] but the plants we analyzed from Makhedonia showed a diploid chromosome number (Fig. 4.8c). Moreover, Humphries [24] reported that *P. montanum* is a highly variable species, intermediate between *P. phleoides* and *P. hirsutum*, and that different morphological variants of this species have been described (*P. serrulatum* Boiss., *P. ambiguum* Ten.). The nuclear DNA amount of plants from Makhedonia is 2.87 pg (Joachimiak et al., unpubl.); thus the calculated size of the basic chromosome set (1.43 pg) in the diploid race of this species is much smaller than in its two closest relatives, *P. phleoides* and *P. hirsutum* [68].

CONCLUDING REMARKS

Phleum is a diverse Old World genus with four sections and a small number of taxa. The basic chromosome number in *Phleum* is x = 7 and there are both diploid (2n = 14) and polyploid taxa (2n = 28, 2n = 42, 2n = 56). Within the two main sections, sect. *Phleum* and sect. *Chilochloa*, a few perennial core species have been included in all classifications proposed so far (Table 4.1), but the circumscription and taxonomical range of other forms, both perennial and annual, remain controversial to this day. Almost all proposed infrageneric *Phleum* classifications have been based on a few selected morphological characters and are not natural classifications [14]. Introduction of cytogenetic methods into *Phleum* taxonomy for investigation of phylogeny and the interrelationships between species has been attempted, in particular to explain the origin of polyploid taxa.

A group of taxa especially well suited to this type of analysis is sect. *Phleum*, the most evolutionally advanced section within the genus. Two of its perennial taxa, *P. pratense* and *P. commutatum*, are now widespread worldwide. The nuclear genomes of diploid taxa are well recognized and characterized by their different DNA amount (1.34 pg – 1.69 pg per basic chromosome set) and different heterochromatin distribution (telomeric vs. centromeric). The diagnostic characters of the basic genome of diploid *P. commutatum*, *P. rhaeticum*, and *P. bertolonii*, designated here as C, R and B, enable study of the chromosomal constitution of their closest polyploid relatives: tetraploid *P. commutatum* and different chromosomal races (4x, 6x, and 8x) of *P. pratense*.

Cytological evidence indicates that both European and American tetraploid *P. commutatum* have four genomes of type C (or very similar to C and to each other). Tetraploid *P. pratense* has four B genomes (genome formula BBBB) and the most widespread *P. pratense* hexaploids are themselves allopolyploids (genome formula 4:2) derived from hybridization involving two different species (*P. bertolonii* and most probably *P. rhaeticum* or an unknown diploid similar to *P. bertolonii*). Thus, its deduced genome formula is BBBBRR or BBBBB'B'.

The fourth diploid of this section, *P. echinatum*, has a basic chromosome number of x = 5, a highly asymmetric karyotype and an unusually large genome (1C DNA = 3.64 pg). The genome of this species is heterochromatin poor, with only a few small terminal blocks of heterochromatin. All these observations suggest that *P. echinatum* possesses a derived genome of totally unknown origin, reduced in chromosome number and highly differentiated.

The three basic genomes (B, C, and R) of perennial taxa in sect. *Phleum* are symmetric and relatively rich in heterochromatin, distributed mainly terminally and/or proximally. The heterochromatin distributions of genome B (majority of heterochromatin located terminally) and genome C (majority of heterochromatin located proximally) are very different. Genome R, possessing both telomeric and centromeric blocks of heterochromatin, is intermediate between these two genomes. In terms of heterochromatin amount, genome R is especially rich in heterochromatin (19.37% of the genome length), roughly the same as the sum of telomeric heterochromatin in the B genome (13.85%) and centromeric heterochromatin in the C genome (6.27%).

The main question is the evolution of such shaped genomes within sect. *Phleum* and the structure of the ancestral genome. The simple hypothesis that the genome of type R was ancestral and that the other two were derived should be rejected. All available experimental and molecular data suggest that the R and B genomes might be partly homologous but not R and C. Thus, either genome of type B originated by elimination of heterochromatin from the genome of type R, or R originated from B by the addition of heterochromatin. The latter scenario is more plausible because, with one exception, genomes of type B (with telomeric distribution of heterochromatin) have been observed in all other *Phleum* taxa analyzed to date.

The origin of genome C, rich in centromeric heterochromatin, is still unresolved. This genome is the smallest within the genus (1.34 pg of DNA). Presumably its evolution was accomplished by amplification of centromeric heterochromatin and massive elimination of some other DNA fractions. The only species within the genus with a very similar genome structure and size is the specialized annual species *P. arenarium* (sect. *Chilochloa*). The structure of its genome, obviously derivative, seems associated with ecological specialization and/or evolution of an annual habit.

B chromosomes are an additive, dispensable element of the *Phleum* karyotype. They have been observed in many species: *P. bertolonii* [5, 19],

P. rhaeticum [32], diploid *Phleum commutatum* (Joachimiak et al., unpubl.), tetraploid *P. commutatum* [32], *P. phleoides* [3,4,25–28] and *P. arenarium* [16]. Interestingly, this dispensable element has been absent in all *Phleum* species with larger nuclear genomes (> 4 pg DNA per diploid nucleus) analyzed so far.

Acknowledgments

I thank Dr. A. Kula (Department of Plant Breeding and Seed Science, Agricultural University, Cracow) for his support and collaboration, especially for producing the Figures, and Dr. Alau Stewart (PEE Seeds, New Zealand) for Phleum seeds and stimulating comments and suggestions.

[1] Babcock EB. The genus *Crepis*. Univ Calif Publ Bot 21, 22. Berkeley, CA: University Press, 1947.

[2] Bigazzi M, Selvi F. Karyotype morphology and cytogeography in *Brunnera* and *Cynoglottis* (Boraginaceae). Bot J Linn Soc 2001; 136: 365-378.

[3] Böcher T. Chromosome behaviour and syncyte formation in *Phleum phleoides* (L.) Karst. Bot Not 1950: 353-368.

[4] Bosemark NO. Cytogenetics of accessory chromosomes in *Phleum phleoides*. Hereditas 1956; 42: 443-466.

[5] Bosemark NO. Further studies on accessory chromosomes in grasses. Hereditas 1957; 43: 236-297.

[6] Cai Q, Bullen MR. Characterization of genomes of timothy (*Phleum pratense* L.). I. Karyotypes and C-banding patterns in cultivated timothy and two wild relatives. Genome 1991; 34: 52-58.

[7] Cai Q, Bullen MR. Analysis of genome-specific sequences in *Phleum* species: identification and use for study of genomic relationships. Theor Appl Genet 1994; 88: 831-837.

[8] Cai H-W, Yuyama N, Tamaki H, Yoshizawa A. Isolation and characterization of simple sequence repeat markers in the hexaploid forage grass timothy (*Phleum pratense* L.). Theor Appl Genet 2003; 107: 1337-1349.

[9] Casler MD. Patterns of variation in a collection of timothy accessions. Crop Sci 2001; 41: 1616-1624.

[10] Cenci CA, Pegiati MT, Falistocco E. *Phleum pratense* (Gramineae): chromosomal and biometrical analysis of Italian populations. Willdenowia 1984; 14: 343-353.

[11] Conert HJ. *Phleum*. In: Hegi, G. ed. Ilustrierte Flora von Mitteleuropa I/3. Berlin: Parey Buchverlag, 1998, 190-206.

[12] Dogan M. A scanning electron microscope survey of the lemma in *Phleum*, *Pseudophleum* and *Rhizocephalus* (Gramineae). Notes RBG Edinb 1988; 45: 117-124.

[13] Dogan M. A taxonomical revision of the genus *Phleum* L. (Gramineae). Karaca Arb Mag 1991; 1: 53-70.

[14] Dogan M, Us J. Infrageneric classification of the genus *Phleum* L. (Gramineae) estimated by numerical taxonomy. In: Ozturk M, Secmen O, Gork G, eds. Plant life in southwest and central Asia. Izmir, 1996, pp. 160-165.

[15] Ellerström S, Tjio JH. Note on the chromosomes of *Phleum echinatum*. Bot Not 1950: 463-465.

[16] Ernst WHO, Malloch AJC. Biological flora of the British Isles. *Phleum arenarium* L. (*Phalaris arenaria* Willd.; *Chilochloa arenaria* P. Beauv.). J Ecol 1994; 82: 403-413.

[17] Feldman M, Liu B, Segal G, Abbo S, Levy AA, Vega JM. Rapid elimination of low-copy DNA sequences in polyploid wheat: A possible mechanism for differentiation of homeologous chromosomes. Genetics 1997; 147: 1381-1387.

[18] Frey L, Mirek Z, Mizianty M. Contribution to the chromosome numbers of Polish vascular plants. Fragm Flor Geobot 1977; 23: 317-325.

[19] Fröst S. The inheritance of accessory chromosomes in plants, especially in *Ranunculus acris* and *Phleum nodosum*. Hereditas 1969; 61: 317-326.

[20] Gregor JW, Sansome F. Experiments on the genetics of wild populations. II. *Phleum pratense* L. and the hybrid *P. pratense* x *P. alpinum* L. J Genet 1930; 22: 373-387.

[21] Guo Y-D, Yli-Matilla T, Pulli S. Assessment of genetic variation in timothy (*Phleum pratense* L.) using RAPD and UP-PCR. Hereditas 2003; 138: 101-113.

[22] Hess H, Landolt E, Hirzel R. Flora der Schweiz und angrenzender Gebiete. 1. Basel: Birkhauser, 1967.

[23] Humphries CJ. Notes on the genus *Phleum* L. Bot J Linn Soc 1978; 76: 337-340.

[24] Humphries CJ. *Phleum* L. In: Tutin TG, Heywood VM, Burges NA, et al, eds. Flora Europea. 5. Cambridge; UK, Cambridge University Press, 1980: 239-241.

[25] Joachimiak A. Karyological studies in *Phleum boehmeri* Wib. from Poland. Acta Biol Cracov Ser Bot 1978; 21: 65-73.

[26] Joachimiak A. *Phleum boehmeri* Wib. Hy-banding karyotype analysis. Acta Biol Cracov Ser Bot 1981: 147-152.

[27] Joachimiak A. Cytogenetics of standard B-chromosomes in *Phleum boehmeri* from Poland. Acta Biol Cracov Ser Bot 1982; 24: 63-77.

[28] Joachimiak A. B-chromosome condensation in *Phleum* pollen grains. Genetica 1986; 68: 169-174.

[29] Joachimiak A, Grabowska-Joachimiak A. Stomatal cell length and ploidy level in four taxa belonging to the *Phleum* sect. *Phleum*. Acta Biol Cracov Ser Bot 2000; 42: 103-107.

[30] Joachimiak A, Klos J, Kula A, Stewart A. The length of stomatal cells in four cytotypes of *Phleum* from New Zealand Collection (in Polish). Zesz Probl Post Nauk Rol 2002; 488: 267-272.

[31] Joachimiak A, Kula A. Cytotaxonomy and karyotype evolution in *Phleum* sect. *Phleum* (Poaceae) in Poland. Pl Syst Evol 1993; 188: 11-25.

[32] Joachimiak A, Kula A. Karyosystematics of the *Phleum alpinum* polyploid complex (Poaceae). Pl Syst Evol 1996; 203: 11-25.

[33] Joachimiak A, Kula A. Systematics and karyology of *Phleum* section in *Phleum* genus (in Polish). Zesz Probl Post Nauk Rol 1997; 451: 13-19.

[34] Joachimiak A, Kula A. Systematics and karyology of section *Phleum* in the genus *Phleum*. J Appl Genet 1997; 38: 463-470.

[35] Joachimiak A, Kula A, Grabowska-Joachimiak A. On heterochromatin in karyosystematic studies. Acta Biol Cracov Ser Bot 1997; 39: 69-77.

[36] Karp A, Jewell AW. The effects of nucleotype and genotype upon pollen grain development in Hyacinth and *Scilla*. Heredity 1982; 48: 251-261.

[37] Kashkush K, Feldman M, Levy AA. Gene loss, silencing and activation in a newly synthesized wheat allotetraploid. Genetics 2002; 160: 1651-1659.

[38] Kellog EA. Relationships of cereal crops and other grasses. Proc Natl Acad Sci USA 1998; 95: 2005-2010.

[39] Kovats D. *Phleum* studies I. Data on the taxonomy and morphology of *Phleum bertolonii* DC. and *Phleum pratense* L. Acta Bot Acad Sci Hung 1976; 22: 107-126.

[40] Kovats D. *Phleum* studies II. *Phleum hubbardii* a new species of Poaceae (Gramineae). Acta Bot Acad Sci Hung 1977; 23: 119-142.

[41] Kovats D. Distribution and diversity of *Phleum hubbardii* and *Phleum pratense* (Poaceae) in the Carpathian Basin. Stud Bot Hung 1980; 14: 107-116.

[42] Kovats D. Distribution and diversity of *Phleum alpinum* L. and *Phleum commutatum* Gaud. (Poaceae) in the Carpathians. Stud Bot Hung 1981; 15: 65-76.

[43] Kozuharov SI, Petrova AV. Chromosome numbers of Bulgarian angiosperms. Fitologija 1991; 39: 72-77.

[44] Kula A. The investigations of the karyotype of *Phleum pratense* L. with the C-banding method and the problem of the origin of this species (in Polish). Fragm Flor Geobot 1988; 33: 257-265.

[45] Kula A. Morphology and cytogenetics of *Phleum hubbardii*. In: L. Frey, ed. Problems of grass biology. Kraków: W. Szafer Institute of Botany, Acad. Sci. 2003: 293-306.

[46] Kula A, Joachimiak A, Klos J, Kras M, Stewart A. Efekty nukleotypowe u czterech różniących się stopniem ploidalności linii *Phleum*. V Ogólnopolskie Spotkanie Naukowe "Taksonomia, Kariologia i Rozmieszczenie Traw w Polsce". Kraków: Instytut Botaniki im. W. Szafera, PAN; 14-15 listopada 2002.

[47] Litardiere R. Sur l'existence dans les Pyrénées d'une nouvelle race chromosomique du groupe du *Phleum alpinum* L C R Acad Sci Paris 1948; 226: 1337-1339.

[48] Michalski T. Cytomorphological study in *Phleum commutatum* from the Tatra Mts. Acta Soc Bot Pol 1955; 24: 181-188.

[49] Müntzing A. Cytogenetic studies on hybrids between two *Phleum* species. Hereditas 1935; 20: 103-136.

[50] Müntzing A. Note on heteroploid twin plants from eleven genera. Hereditas 1938; 24: 487-491.

[51] Müntzing A, Prakken R. The mode of chromosome pairing in *Phleum* twins with 63 chromosomes and its cytogenetic consequences. Hereditas 1940; 26: 463-501.

[52] Nordenskiöld H. Intra- and interspecific hybrids of *Phleum pratense* and *P. alpinum*. Hereditas 1937; 23: 304-316.

[53] Nordenskiöld H. Cytological studies in triploid *Phleum*. Bot Not 1941: 12-32.

[54] Nordenskiöld H. Cytogenetic studies in the genus *Phleum*. Acta Agr Suecana 1945; 1: 1-137.

[55] Nordenskiöld H. Synthesis of *Phleum pratense* L. from *P. nodosum* L. Hereditas 1949; 35: 190-202.

[56] Nordenskiöld H. A genetical study of the mode of segregation in hexaploid *Phleum pratense*. Hereditas 1953; 39: 469-488.

[57] Nordenskiöld H. Segregation ratios in progenies of hybrids between natural and synthesized *Phleum pratense*. Hereditas 1957; 43: 525-540.

[58] Nordenskiöld H. The mode of segregation in a family of hexaploid *Phleum pratense*. Hereditas 1960; 46: 504-510.

[59] Ohri D. Genome size variation and plant systematics. Ann Bot 1998; 82: 75-83.

[60] Ozkan H, Levy AA, Feldman M. Allopolyploidy-induced rapid genome evolution in the wheat (*Aegilops-Triticum*) group. Plant Cell 2001; 13: 1735-1747.

[61] Pignatti S. 1982. Flora d'Italia 3. Bologna: Edagricole, 1982.

[62] Pogan E, Wcislo H, Jankun A, et al. Further studies in chromosome numbers of Polish angiosperms. Acta Biol Cracov Ser Bot 1980; 22: 36-49.

[63] Scholz H. Short notes on *Phleum* sect. *Achnodon* (Gramineae). Willdenowia 1999; 29: 45-49.

[64] Schörter C. Beitrage zur Kenntnis schweizerischer Blüthenpflanzen. Ber St Gallischen Naturwiss Ges 1889; 1887/88: 223-245.

[65] Shaked H, Kashkush K, Ozkan H, Feldman M, Levy AA. Sequence elimination and cytosine methylation are rapid and reproducible responses of the genome to wide hybridization and allopolyploidy in wheat. Plant Cell 2001; 13: 1749-1759.

[66] Shan F, Yan G, Plummer JA. Karyotype evolution in the genus *Boronia* (Rutaceae). Bot J Linn Soc 2003; 142: 309-320.

[67] Skalinska M, Pogan E, Czapik R, et al. Further studies in chromosome numbers of Polish angiosperms. Acta Biol Cracov Ser Bot 1978; 21: 31-63.

[68] Sliwinska E, Kula A, Joachimiak A, Stewart A. Genome size in the genus *Phleum*. In: Application of Novel Cytogenetic and Molecular Techniques in Genetics and Breeding of the Grasses, International Workshop, Poznan: Poland, 1-2 April 2003.

[69] Stace HM. Cytoevolution in the genus *Calotis* R. Br. (Compositeae: Asterae). Austr J Bot 1978; 26: 287-307.

[70] Stebbins GL. Chromosomal evolution in higher plants. London, UK: Edward Arnold, 1971.

[71] Sutkowska A, Joachimiak A, Idzik I. Analiza DNA trzech gatunków *Phleum* z sekcji *Phleum* metod[1] ISSR. V Ogólnopolskie Spotkanie Naukowe "Taksonomia, Kariologia i Rozmieszczenie Traw w Polsce". Kraków: Instytut Botaniki im. W. Szafera, PAN; 14-15 listopada 2002.

[72] Teppner H. Karyologie und Systematik einiger Gefasspflanzen der Ostalpen. Botanische Studien im Gebiet der Planneralm (Niedere Tauern, Steirmark), VII. Phyton 1980; 20: 73-94.

[73] Tzvelev NN. The system of grasses (Poaceae) and their evolution. Bot Rev 1989; 55: 141-204.

[74] Verma SC, Rees H. Nuclear DNA and the evolution of allotetraploid *Brassicae*. Heredity 1974; 33: 61-68.

[75] Watanabe K, Yahara T, Denda T, Kosuge K. Chromosomal evolution in the genus *Brachyscome* (Asteraceae, Asterae): Statistical tests regarding correlation between changes in karyotype and habit using phylogenetic information. J Plant Res 1999; 112: 145-161.

[76] Wilton AC, Klebesadel LJ. Karyology and phylogenetic relationships of *Phleum pratense*, *P. commutatum* and *P. bertolonii*. Crop Sci 1973; 13: 663-665.

Molecular Marker Technology for the Study of Genetic Variation and Comparative Genetics in Pasture Grasses

JOHN W. FORSTER[1], ELIZABETH S. JONES[1], KEVIN F. SMITH[2], KATHRYN M. GUTHRIDGE[1], MARK P. DUPAL[1], SHARON HOWLETT[1], LEONIE J. HUGHES[1], SALLY GARVIE[3], and *CHRISTOPHER PRESTON[3]*

[1]Primary Industries Research Victoria, Plant Biotechnology Centre, La Trobe University, Bundoora, Victoria 3086, and Molecular Plant Breeding Australia.

[2]Primary Industries Research Victoria, Hamilton Centre, Hamilton, Victoria 3300, Cooperative Research Centre Australia.

[3]Department of Applied and Molecular Ecology, University of Adelaide, Glen Osmond, South Australia 5064, and Cooperative Research Centre for Weed Management Systems, Australia.

ABSTRACT

Molecular genetic marker systems and associated linkage maps were developed for the most important temperate pasture grasses. SSR and AFLP systems are of particularly high value for genetic analysis of perennial ryegrass (*Lolium perenne* L.). Due to the obligate outbreeding reproductive system of this

Address for correspondence: Prof. John W. Forster, Primary Industries Research Victoria, Plant Biotechnology Centre, La Trobe University, Bundoora, Victoria 3086, Australia. Tel: +61-3-9479-5645, Fax: +61-3-9479-3618, E-mail: john.forster@dpi.vic.gov.au

species, natural and synthetic populations are genetically heterogeneous. AFLP-based DNA profiling has been used to relate genetic variation within and between populations to their breeding histories, and parallel analysis with SSRs has produced largely congruent results. The merits of different analytical techniques for SSR-based diversity analysis are discussed. SSRs provide a more readily implementable high throughput method for large-scale characterization of perennial ryegrass germplasm collections, with high potential value for population-based marker-trait linkage detection. AFLPs have also been used to detect genetic variation between different taxa in the *Lolium* genus, permitting identification of species-specific diagnostic markers. Molecular markers are also suitable for comparative genetic analysis. This is generally achieved with RFLPs, but the close phylogenetic relationship between perennial ryegrass and annual ryegrass (*L. rigidum* Gaud.) suggests that genomic DNA-derived SSRs may also prove useful. This marker system has been used to study the genetic control of resistance to the herbicide glyphosate in crosses based on resistant and susceptible Australian ecotypes, providing support for the hypothesis of simple genetic control, but also identifying possible divergences of genome structure between *Lolium* species.

Key Words: pasture grass, perennial ryegrass, molecular marker, simple sequence repeat polymorphism, genetic diversity, comparative map, annual ryegrass, herbicide resistance, glyphosate

Abbreviations: AFLP: amplified fragment length polymorphism, AMOVA: analysis of molecular variance, BSA: bulked segregant analysis, CDS: coding sequence, EPSP: 5-enolpyruvylshikamate 3-phosphate synthase, EST: expressed sequence tag, FAM: 6-carboxyfluorescein, GD: genetic distance, HEX: hexachloro-6-carboxyfluorescein, LD: linkage disequilibrium, LG: linkage group, NED: 7',8'-benzo-5'-fluoro-2,4,7-trichloro-5-carboxyfluorescein, PIC: polymorphism information content, QTL: quantitative trait locus, RFLP: restriction fragment length polymorphism, SI: self-incompatibility, SNP: single nucleotide polymorphism, SSR: simple sequence repeat, STS: sequence tagged site, UPGMA: unweighted pair group method using arithmetic averages, UTR: untranslated region

INTRODUCTION

The most important pasture species on a global basis are members of the Poaceae (grass and cereal) and Leguminosae (legume) families. Among the agricultural grasses cultivated in temperate regions of the world for dairy, beef, and wool production, members of *Lolium* and *Festuca* genera show the widest distribution. The two genera are closely related, located within tribe Poeae of super tribe Poodae within subfamily Pooideae [75]. Individual cultivated members include perennial ryegrass (*Lolium perenne* L.), Italian ryegrass (*Lolium multiflorum* Lam.), tall fescue (*Festuca arundinacea* Schreb.), and meadow fescue (*Festuca pratensis* Huds.). The different pasture grass species show variation for favorable agronomic traits such as herbage yield and quality, palatability, persistence and ease of establishment, disease resistance, and tolerance to abiotic stresses such as drought and cold [85].

Development and implementation of molecular genetic marker systems for improvement of the key temperate pasture grass species occurred later than for many other major crop plant species, but comprehensive marker systems and associated genetic maps are now available for perennial ryegrass [42-44], meadow fescue [2, 3], and tall fescue [67, 68, 86]. Simple sequence repeat (SSR) markers are of especially high value for genetic map construction, marker-assisted selection, and DNA profiling [62]. A set of ca. 400 unique SSR markers have been developed for perennial ryegrass [42] and used in concert with restriction fragment length polymorphism (RFLP) and amplified fragment length polymorphism (AFLP: [82]) systems for genetic map construction [43, 44] and detection of genes and quantitative trait loci (QTLs) for morphological and developmental traits [87], disease resistance [24], and other characters [29].

Specific applications of molecular marker systems, especially genomic DNA-derived SSRs, to two aspects of the genetics of ryegrasses are described here: assessment of molecular variation within and between perennial ryegrass populations and comparative genetic analysis of herbicide resistance in annual ryegrass. These examples exemplify the power and flexibility of molecular marker technology as applied to pasture grasses.

MOLECULAR ANALYSIS OF GENETIC VARIATION IN PASTURE GRASSES

Background

Pasture grasses exhibit a large range of variation in the breeding system, including inbreeding (autogamous), outbreeding (allogamous), vegetatively propagated and apomictic species. The most important of the temperate forages are cross-pollinated. As a consequence of the outbreeding habit, both natural ecotypes and synthetic populations of these species are likely to be highly genetically heterogeneous. Strategies for genetic diversity analysis based on DNA profiling must address this issue and permit quantification of variation within and between populations. The optimal sampling size to determine variation within a cross-pollinated population depends on the frequency of the least common allele or genotype [18] and is highly influenced by the number of founder genotypes and the genetic distances between these genotypes. For comparison of different cultivars and populations, the genetic structure of each population and the degree of relatedness between populations are also important. Close-bred varieties that share a higher proportion of common alleles are more difficult to discriminate than those based on genetically divergent parents.

Molecular Diversity in Perennial Ryegrass

Perennial ryegrass is an obligate outbreeding grass species with a gametophytic self-incompatibility (SI) system controlled by two loci (S and Z). Incompatible matings occur when the alleles at both loci in the male gametophyte (pollen grain) match one of the two alleles at each locus in the female sporophyte [17]. The S and Z loci have been genetically mapped and located to linkage groups (LGs) 1 and 2 respectively [79] in the equivalent genomic regions defined by conserved synteny with the locations of their ortholoci in other Self Incompatibility Poaceae species [31, 47, 83]. A high level of allelic diversity has been detected at both S and Z, with up to 20 different alleles observed at each locus [20, 28]. This multiallelic character of the SI loci limits suppression of fertility in large, complex populations. However, varietal development in perennial ryegrass and other SI grass species such as meadow fescue, tall fescue, and phalaris is dependent on outcrossing between multiple parental genotypes, rather than inbreeding strategies used for varietal development in cereal species such as wheat, barley, maize, and rice.

Both natural and synthetic populations of perennial ryegrass are genetically heterogeneous. Synthetic populations form the basis for cultivar development and are based on the selection of suitable parental genotypes, followed by complex crossing (either polycrosses or combined diallel crosses involving all parental genotypes). Varietal development schemes may include pair-crosses between parental genotypes to define combining ability or recurrent selection among progeny generations. The level of genetic diversity in a population is a function of two variables [30, 37]. The first variable is the genetic distance between the parental genotypes, defining a population as having a wide or narrow genetic base. A wide genetic base is obtained from parental plants with highly differentiated genotypes, and is characterized by a low proportion of common alleles at a given genetic locus, while the converse applies to a narrow genetic base. The second variable is the number of parental genotypes used to found the population. A large number of foundation clones (>6) is characteristic of a nonrestricted base population, while a small number of foundation clones (3-6) is characteristic of a restricted base population. Even when the foundation clones are of a similar genetic provenance, a nonrestricted base variety will generally show greater genetic heterogeneity than a corresponding restricted base variety, due to the potential for sampling a large number of alleles. Nonetheless, it is possible for a restricted base variety to exhibit a high level of intrapopulation diversity if the parental genotypes were genetically divergent, and for a nonrestricted base variety to be relatively invariable if the parental genotypes were closely related.

Forage breeders have traditionally been conservative in the number of parental clones selected for varietal development [9] but a number of perennial ryegrass varieties with restricted genetic bases (from 4 parental clones) have been highly successful in the market-place, including Yatsyn 1 [4], Embassy [11], and Banks [12].

The levels of genetic diversity within and between populations of perennial ryegrass will vary, with differing breeding histories reflecting the relative contributions of genetic distance between founder genotypes and number of parental clones. The extent of such variation is a key determinant of the ability to discriminate between different populations. For current varieties and natural ecotypes, it is generally only possible to have *a priori* knowledge of the extent of the restricted base, while the genetic distance between the parents must be deduced *post hoc*. However, DNA profiling methods allow routine evaluation of genetic distance between parents of newly synthesized varieties.

Analysis of Genetic Diversity within and between Populations of Perennial Ryegrass with AFLP

For a preliminary evaluation of genetic diversity, six contrasting populations were selected. Kangaroo Valley is a long-established ecotype renowned for winter growth [45, 78], which has undergone local adaptation to Australian conditions and is believed to be derived from several independent introductions from England during the mid-nineteenth century. Grasslands Nui and Ellett are widely used cultivars obtained from the Mangere district of New Zealand. Grasslands Nui has a partially restricted genetic base, as it was derived from nine parental clones [5], of which four were closely related by descent. Ellett has a highly nonrestricted genetic base, as it was derived from several hundred plants from the same original population as Grasslands Nui [69]. The three restricted base varieties described above were also used in this study: Yatsyn 1 was also derived from Mangere genotypes, while Embassy was derived from two New Zealand and two Uruguayan parents, while Banks was derived from two parents from Embassy and two parents from a breeding line designated "C3".

The AFLP technique provides a simple means for individual genotyping over a large number of target loci and was earlier demonstrated to be capable of discrimination between different genotypes with some indication of genetic affinity based on geographic groups [30]. The first phase of analysis was performed with the Kangaroo Valley, Grasslands Nui, Ellett and Yatsyn 1 populations, that differ in breeding histories, especially with respect to the restricted/nonrestricted nature of the genetic base [37]. The genetic relationships between the groups correlated closely with respective breeding history and partial discrimination between populations was achieved. The three restricted base cultivars were then studied and shown to be almost completely discriminated by the AFLP analysis, even in the case of the two varieties (Banks and Embassy) with a common genetic heritage [37]. DNA profiling of bulked samples of >15 individuals with AFLPs revealed similar relationships to those seen with the individual phylogenetic trees, suggesting that this technique may be suitable for high throughput analysis of germplasm collections.

Analysis of Genetic Diversity within and between Populations of Perennial Ryegrass with SSR Polymorphism

Although the AFLP technique provides results with high information content due to the reproducible nature and high multiplex ratio of the profiles, allowing an evaluation of polymorphism at multiple loci, the technical requirements for the method limit its applicability for high throughput genotyping. The AFLP technique requires relatively large amounts of high quality genomic DNA and multiple manipulations to generate the linker-adapted template, primary amplification product, and secondary amplification product. By contrast, the SSR technique requires small quantities (5-50 ng) of relatively lower quality genomic DNA and the template requires only sequence-specific amplification prior to fragment analysis. The multiallelic nature of SSR markers, associated with high levels of genetic polymorphism, also make them ideal for genetic diversity analysis.

To validate use of perennial ryegrass genomic DNA-derived SSR (LPSSR) markers [42] for genetic diversity analysis, and to compare the data obtained with SSRs with that obtained with AFLPs, the contrasting populations that had been used for AFLP analysis were used for further study. Eighteen distinct genotypes from each of the populations Yatsyn 1, Grasslands Nui, Embassy, Banks, and Kangaroo Valley were selected. In parallel, 12 perennial ryegrass SSR primer pairs were selected on the initial basis of the allelic complexity (as represented by polymorphism information content [PIC] values) in a small panel of diverse perennial ryegrass genotypes [42]. These primer pairs were used individually to produce amplification profiles from the 90 different genotypes, and a range of allele complexity from a minimum of two to a maximum of eight observed (Table 5.1). In total, 53 different alleles were scored across the 12 primer pairs.

Table 5.1: Perennial ryegrass SSR (LPSSR) primer pairs selected for analysis of genetic diversity

Primer pair name	SSR motif	Optimal annealing temperature (°C)	Expected fragment size (bp)	Number of alleles detected[a]
LPSSRH01A10	$[CCT]_{20}$	54	152	5
LPSSRH01H06	$[CA]_9$	50.6	150	4
LPSSRH02D01	$[GT]_3..[GT]_6$	47	204	4
LSSRH02D12	$[CA]_{12}$	50	219	4
LPSSRH02E12	$[TG]_{12}$	54	210	3
LPSSRH02F01	$[CA]_{13}$	52	122	4
LPSSRH05D11	$[CA]_8$	51	145	5
LPSSRH05F02	$[CA]_4..[CA]_7$	53	168	4
LPSSRH05G07	$[CA]_{11}$	50	186	3
LPSSRK02E08	$[CTT]_7$	57	174	2
LPSSRK03A02	$[GA]_{27}$	52	227	7
LPSSRK10G04	$[GAT]_9..[CAT]_4$	52	299	8

[a]Assessment of allelic complexity was based on evaluation of a small panel of diverse perennial ryegrass genotypes [42].

In the first instance, the SSR alleles were scored as dominant markers, as were AFLPs, according to presence and absence. Analysis of genetic diversity within populations using the NT-SYSPc software package (Version 2.02 [65]) revealed values similar to those derived from AFLP analysis, with a range of genetic distance (GD) values from 0.42 to 0.54.

In general, the GD value for a given population correlated inversely with the presumptive number of parents, with lower values for the restricted base cultivars. The only exception was variety Banks.

A phenogram based on the unweighted pair group method using arithmetic averages (UPGMA) application was constructed based on average taxonomic distance [74]. Partial separation of the populations was observed, with some overlap between the respective groups (Fig. 5.1). Members of the restricted base cultivars seem to show more distinct clustering than nonrestricted base varieties and ecotypes, a result consistent with that observed for AFLPs. The main difference observed with the corresponding AFLP analysis was the limited clustering of Grasslands Nui genotypes. This variety has a relatively restricted genetic base and showed tight clustering in the dendrograms derived from AFLP analysis. In addition, individual genotypes from the restricted base varieties Yatsyn 1, Banks, and Embassy are separated from their main clusters, but showed a lower tendency for this effect to occur in the AFLP analysis.

Analyses of molecular variance (AMOVA: [27]) based on squared Euclidean distance, as described for RAPD markers by Huff [40], was used to calculate variance components within and between populations. Most of the variation was detected within populations (85.5%) compared to variation between populations (14.5%). Estimates were highly comparable to those obtained from AFLP analysis [37]. There were significant differences in variance within individual populations based on AMOVA but the relative values were small.

Since SSRs, unlike AFLPs, generally detect codominant loci, other methods of marker-based genetic diversity analysis may be more appropriate. The analytical technique based on characteristics of individual SSR loci using Arlequin ver. 2000 [70] employs such variables as variation in repeat length and observed number of alleles to determine the amount of variation within and between populations. Five LPSSR primer pairs were selected on the basis of the simple allelic segregation patterns (of the form AB x BB) detected in the one-way pseudotestcross [35, 64] p150/112 reference genetic mapping population [43]. Data were generated for the sample genotypic set and analyzed for the estimated number of repeats per allele (based on the repeat array structure in the source clone and the fragment size expected from PCR primer design) and the number of different alleles per locus (Table 5.2). A total of 19 different alleles were amplified by the five primer pairs, with a range from

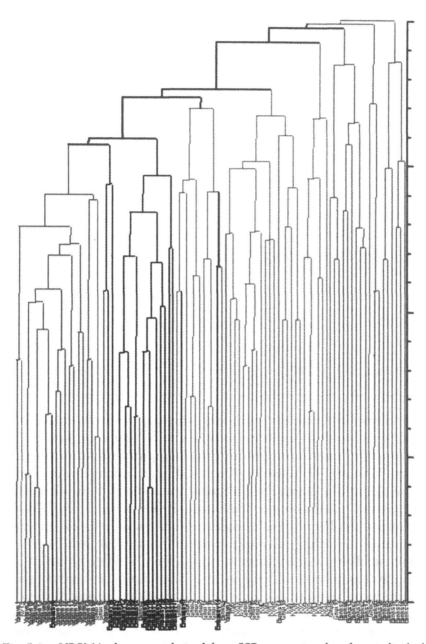

Fig. 5.1: UPGMA phenogram derived from SSR genotyping data from individual plants from the perennial ryegrass populations Yatsyn 1 (red); Banks (green), Embassy (blue), Kangaroo Valley (ochre), and Grasslands Nui (pink), calculated using measurements of average taxonomic distance.

Table 5.2: Allelic complexity and variation of array repeat number in the genetic diversity sample set based on analysis of variation at 5 LPSSR loci

Primer pair	SSR motif and expected fragment length		Allele	Estimated array repeat number
LPSSRK03A02	$[GA]_{27}$	227bp	A	51
			B	37
			C	30
			D	27
			E	44
			F	59
			G	87
LPSSRH02E12	$[TG]_{12}$	210bp	A	12
			B	10
			C	8
LPSSRH05F02	$[CA]_4...[CA]_7$	168bp	A	13
			B	11
			C	12
			D	10
LPSSRH05G07	$[CA]_{11}$	186bp	A	10
			B	11
			C	12
LPSSRK02E08	$[CTT]_7$	174bp	A	7
			B	6

two to seven. The data were then analyzed using two different methods (F_{ST} for the number of different alleles and R_{ST} for the squared size differences). These two analytical methods produced very different results. The AMOVA result based on the F_{ST} was similar to that based on the scoring of presence and absence of bands (83.5% variance within cultivars and 16.5% between cultivars). However, when a UPGMA phenogram based on squared Euclidean distance was produced from the R_{ST} analysis data, the relationships observed from the analysis based on dominant alleles was not preserved, and only a limited amount of variation was detected. Despite these differences, analysis of gene diversity at the intrapopulation level based on allelic complexity at each locus revealed molecular variation consistent with the known breeding histories of the populations. Restricted base cultivars generally have a low gene diversity (17% for Yatsyn 1, 18% for Embassy), except, as noted above, for Banks (42%). Conversely, nonrestricted base populations showed relatively high gene diversity values (34% for Grasslands Nui, 37% for Kangaroo Valley).

Analysis based on R_{ST} values suggested that all the observed variation was present within each population and consequently population discrimination was not possible using this method. An increase in number of loci tested, or selection of a subset of highly informative and multiallelic loci, could provide the basis for effective population discrimination.

Lastly, the method of locus-by-locus AMOVA can be used to identify those loci that contribute the largest proportion of the variation present within and between populations. This is a useful tool for the selection of the most informative SSR loci for genetic diversity analysis. For example, primer pair LPSSRK03A02 detects a highly polymorphic locus with seven distinct alleles in the genotypic sample set, with 25% of the variance detected between populations. Highly significant genetic variation between populations ($P < 0.001$) was only detected by two primer pairs (LPSSRK03A02 and LPSSRH05G07). On the other hand, primer pairs LPSSRK03E08, LPSSRH02E12 and LPSSRH05F02 detected limited allelic variation ($P < 0.05$) between populations, with most of the variance detected within populations. This result suggests that loci should be preselected on the criterion of multiple allele detection across different germplasm sources in order to provide maximum information for population discrimination.

To summarize, perennial ryegrass SSRs can provide useful information for detection of genetic diversity within and between populations and are capable of discrimination between populations, but the analytical technique is used will influence the outcomes. Although the scoring of SSR alleles as dominant (presence/absence) features may seem inappropriate for generally codominant markers, the chief advantage of this method is presumably the large data set obtainable with relatively few primer pairs. Although allelic variants at the same genetic locus are not fully independent in this analysis, as would generally be the case for AFLP loci, similar results may be obtained. It is possible that dissimilarities between the two marker systems may be related to the difference between genetic codominance and dominance, as well as the differential distribution of marker types within the genome. Conversely, analytical techniques based on individual locus characteristics are vulnerable to differences in allelic complexity between different loci. In the study described here, SSR loci with high allelic complexity provide the most efficient means for detection of within and between population diversity. As ca. 450 unique SSR-containing clones have been identified for perennial ryegrass [42, 43], it should be straightforward to select a subset of highly polymorphic multiallelic loci for such analysis. However, there are potential complicating factors in the detailed analysis of single locus variation. Estimates of genetic variation based on estimated repeat number may be confounded by the incidence of size homoplasy [26, 36] in which two different alleles share a common fragment size but have arisen by independent convergent molecular evolutionary events. Size homoplasy underestimates levels of genetic variation and consequently biases measurements of genetic diversity. Direct sequence analysis of different SSR alleles is the only reliable method for identifying such effects.

Molecular Genetic Variation in Genus *Lolium*

As described above, molecular genetic marker systems can be used to assess diversity within species and also to determine variation between species within a genus or at higher taxonomic levels. The AFLP technique provides a particularly useful method for interspecific analysis compared to other genetic marker systems such as RFLPs and SSRs. This is because AFLP is a sequence-independent marker system and hence expected to function with equal efficiency for a diverse range of taxa. RFLPs and even more so SSRs on the other hand, may show a species-specific character that results in variable detection among taxa and complications in subsequent analysis.

Genetic diversity was analyzed in genus *Lolium* with particular emphasis on the outbreeding taxa, using accessions derived from a number of germplasm storage centers. Accessions were obtained from the following outbreeding taxa: *L. perenne* L. (perennial ryegrass); *L. multiflorum* Lam. var. *italicum* (biennial Italian ryegrass); *L. multiflorum* Lam. var. *westerwoldicum* (Westerwolds-type [annual] Italian ryegrass); *L. rigidum* Gaud. var. *rigidum* (annual ryegrass); *L. rigidum* Gaud. var. *rottobolliodes* (annual ryegrass); and *L. boucheanum*, the F_1 hybrid between *L. perenne* and *L. multiflorum*. The following inbreeding taxa were additionally represented: *L. temulentum* L. (darnel), *L. persicum* L. (Persian darnel), *L. remotum* L., and *L. loliaceum* L.

AFLP analysis was performed to evaluate the degree of relatedness of the taxa and comparison with preexisting data obtained using other molecular marker systems [13, 77, 84]. Analysis was performed using 40 genotypes representing the eight different species. Seven different genotypes of *Lolium multiflorum* from different accessions were represented, along with a single *L. boucheanum* genotype, four genotypes of *L. temulentum*, one genotype of *L. persicum*, two genotypes of *L. loliaceum*, ten genotypes of *L. rigidum*, two genotypes of *L. remotum*, and fourteen genotypes of *L. perenne*. The outbreeding *Lolium* taxa accessions included populations from across the Northern Hemisphere and the *L. rigidum* sample was augmented by the addition of a single genotype from the Western Australian WLR population, which typically shows multiple herbicide resistance (see below).

Eight AFLP primer pair combinations were used to generate profiles across this genotypic panel. Four *EcoRI/MseI* combinations were employed, three with seven (+3/+4) selective nucleotides (*EcoRI* + ACA/*MseI* + ACGT; *EcoRI* + ACT/*MseI* + GGGT; *EcoRI* + AGG/*MseI* + GGGT) and one with six (+3/ +3) (*EcoRI* + CCA/*MseI* + AGG), while four *PstI/MseI* combinations were employed, three with seven (+3/ +4) selective nucleotides (*PstI* + ACC/*MseI* + ACGT; *PstI* + ACC/*MseI* + GGGT; *PstI* + AAC/*MseI* + GGGT) and one with six (+3/ +3) (*PstI* + CCA/*MseI* + AGG). All the primer pairs produced profiles with a high degree of complexity. The results obtained with the *EcoRI* + ACA/*MseI* + ACGT primer combination are shown in Figure 5.2a and those with the *PstI* + ACC/*MseI* + ACGT primer combination in Figure 5.2b. As indicated, these and a number of other primer combinations detected prominent banding features that appeared to discriminate between the three outbreeding taxa. For the *EcoRI* + ACA/*MseI* + ACGT profiles, a strong band at a relatively high molecular weight was present in all genotypes of *L. multiflorum* (and also *L. boucheanum*) but apparently not in the *L. rigidum* or *L. perenne* genotypes. For this and other apparently discriminatory bands, a larger range of germplasm must be surveyed to determine whether the bands correspond to conserved monomorphic species-specific features. It is possible that such AFLP loci may be derived from species-specific (or highly prevalent) repetitive DNA sequence family structures [41], in which case the basic sequence structure is likely to be conserved between the three species, but homogenization events following speciation may have fixed certain structural variants defined by restriction site variation.

For further analysis, the putative discriminatory bands may be excised from the gel, cloned and sequenced in order to design sequence tagged site (STS) markers to test the specificity of the sequence [14]. If such sequences are derived from dispersed repetitive DNA families, they are potentially of high value not only for population-based diagnostics, but as

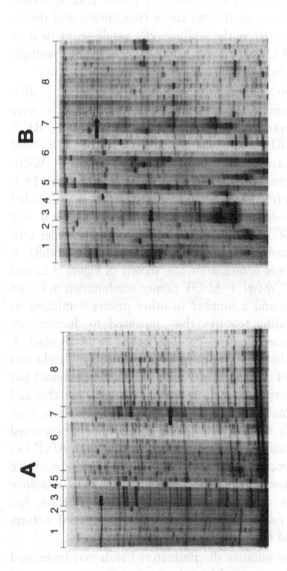

Fig. 5.2: Amplified fragment length polymorphism (AFLP) profiles generated from genomic DNA templates derived from a selection in outbreeding and inbreeding *Lolium* taxa. PCR amplification products were generated using rare cutter-specific primers end-labelled with ³³P, and the products separated by vertical polyacrylamide gel electrophoresis followed by autoradiography. The samples were loaded as follows: 1: *L. multiflorum*; 2: *L. boucheanum* (F₁ hybrid between *L. perenne* and *L. multiflorum*); 3: *L. temulentum*; 4: *L. persicum*; 5: *L. loliaceum*; 6: *L. rigidum*; 7: *L. remotum*; 8: *L. perenne*. (A) primer pair combination EcoRI + ACA/MseI + ACGT and (B) primer pair combination PstI + ACC/MseI + ACGT. In both instances, two prominent banding features that apparently distinguish between the outbreeding taxa are shown: the arrowed band in the group 1 samples is apparently present in *L. multiflorum* (and *L. boucheanum*) but is not apparently present in *L. rigidum* or *L. perenne*, while the arrowed band in the group 8 samples is present in *L. perenne* but is not apparently present in *L. multiflorum* or *L. rigidum*.

possible tools to monitor the introgression of genetic material in advanced hybrid germplasm. The origin and applications of such germplasm were described by Breese and Lewis [10].

Significant differences were also observed between the profiles of the inbreeding and outbreeding *Lolium* taxa, consistent with earlier studies and accepted taxonomic relationships [8], while strong similarities were observed between the profiles generated by the two inbreeding taxa *L. temulentum* and *L. remotum*, especially with the *Eco*RI + ACA/*Mse*I + ACGT primer combination.

Further studies of AFLP-based interspecific genetic variation in the *Lolium* genus could involve an extension of the study with a larger number of genotypes from present taxa, the addition of other *Lolium* species such as *L. canariensis* and members of the closely related *Festuca* genus. Rapid analysis with a large number of primer pair combinations could be obtained through use of fluorochrome-labeled primers [71] and separation using an automated capillary electrophoresis platform. A large genotypic data set of this kind is accessible to phenetic analysis in the form of a dendrogram representing genetic affinity between different taxa, as performed in plant groups such as genus *Oryza* (rice) [1], *Cichorium* (endive and chicory) [46], and *Daucus* (carrot) [56], and fungal species such as *Neotyphodium* endophytes of pasture grasses [81]. Such data could then be exploited for taxonomic classification of unknown or equivocal accessions.

COMPARATIVE GENETICS IN PASTURE GRASSES

Background

A number of studies have indicated that related plant species may have genomes closely related in terms of genetic map segmental structure (conserved synteny) and gene order (conserved colinearity). These properties were discovered through use of conserved sequence RFLP probes (anchor probes) which detect equivalent loci (ortholoci). Resultant genetic maps could then be aligned to reveal structural relationships. The most striking results have been found among Solanaceae crop plants (potato, tomato, tobacco, aubergine; [7]) and especially among comparisons of various species of cereals. Poaceae is thought to have evolved as a major taxonomic group relatively recently, about 60 mya. Genetic maps of Triticeae cereals (wheat, rye, and barley) have long been deduced to be similar on the basis of chromosome studies

[21]. Other studies have revealed striking colinearities between wheat and more distantly related species such as maize and rice [48]. Relationships among a number of Poaceae species (rice, maize, pearl millet, foxtail millet, maize, sorghum, sugarcane, wheat, barley, and rye) may be represented in the form of concentric alignments of circularly permuted genetic maps [22,23,55].

Comparative genetic mapping between perennial ryegrass and other Poaceae species has been done through construction of a comprehensive reference genetic map based on a mapping population with a simplified structure. The p150/112 F_1 progeny set is a one-way pseudotestcross population based on the pair-cross of a multiple heterozygous genotype of complex descent with a doubled haploid genotype [6,44]. Clonal ramets from each progeny plant were transferred to a number of laboratories affiliated to the International *Lolium* Genome Initiative (ILGI) and genotypic data for nonproprietary marker systems collated from four main sources. The group at Institut National de la Recherche Agronomique (INRA), Clermont-Ferrand, France contributed *Eco*RI/*Mse*I amplified fragment length polymorphism data; the Institute of Grassland and Environmental Research (IGER) Primary Industries Research Victoria-Plant Biotechnology Centre (PIRVic-PBC), Melbourne, Australia contributed wheat, barley, and oat cDNA probe-detected RFLP data, while groups at the Yamanashi Prefectural Dairy Experiment Station (YPDES), Yamanashi, Japan and the National Agricultural Research Centre for Hokkaido Region (NARCH), Sapporo, Japan, contributed rice cDNA probe-detected RFLP data.

Polymorphism data were obtained for 343 markers, of which 192 were AFLP loci and the remainder primarily heterologous RFLP loci, with the addition of a small number of *L. perenne Pst*I genomic probe RFLPs, EST, and isoenzyme markers. A total of 322 loci were assigned to the genetic map, with seven LGs and a total map distance of 811 cM [44]. One hundred and twenty-four loci were codominant, of which 109 were heterologous RFLP loci. Mapping of heterologous RFLP loci detected by the cDNAs from wheat, barley, rice, and oat, allowed evaluation of conserved syntenic relationships with other species of Poaceae [32]. Syntenic chromosomal regions may be represented either in the form of linkage block idiographs, or as a concentric circle alignment based on circular permutation of chromosome arm orders. At the macrosyntenic level, each of the 7 linkage groups of perennial ryegrass chiefly corresponds to one of the seven basic homeologous chromosome groups of Triticeae cereals and they have been numbered accordingly. For instance, LG1 is the syntenic counterpart of wheat chromosomes 1A, 1B, and 1D, barley 1H and rye 1R. Perennial ryegrass LGs 1, 3, and 5 showed uninterrupted synteny with their Triticeae counterparts, while LGs 2, 4, 6 and 7 contained nonsyntenic regions [44]. Overall, 80% of common markers between perennial ryegrass and Triticeae maps were syntenic and colinearity was well conserved. The observed relationships between the perennial ryegrass map and the oat and rice maps were also consistent with previous comparisons of these species to the Triticeae [22, 80]. For instance, LG3 corresponds to rice chromosome 1 and oat groups C and G.

Comparative genetic mapping between distantly related species is dependent on conserved genetic markers, such as the heterologous RFLPs described above. In general, SSR loci are expected to show lower levels of conservation, especially when derived from noncoding DNA sequences. EST-SSRs, located within either 5'- or 3'-untranslated regions (UTRs) as well as the coding sequence (CDS) [15, 72], are expected to show higher levels of cross-species amplification [16], but for genomic DNA-derived SSR loci obtained by enrichment library technology [25], marker transfer is often restricted to members of the same genus [57, 66]. In genus *Lolium*, genomic DNA-derived SSR (LPSSR) primer pairs derived from perennial ryegrass show high levels of cross-amplification in the other outbreeding *Lolium* taxa *L. rigidum* (ca. 80%) and *L. multiflorum* (ca. 70%) [42]. This result implies that LPSSR markers may be highly suitable for comparative genetic mapping studies between the outbreeding ryegrasses.

Molecular Marker-based Analysis of Herbicide Resistance in Annual Ryegrass

Outbreeding Lolium *species*

There are three well-recognized outbreeding species within the *Lolium* genus: perennial ryegrass (*Lolium perenne* L.); Italian ryegrass (*Lolium multiflorum* Lam.), and annual ryegrass (*Lolium rigidum* Gaud.). The three species show a very high degree of morphological similarity and chiefly differ in reproductive development and growth habit, with a continuous range of variation from short-lived annual ecotypes of annual ryegrass at one extreme to long-lived perennial ecotypes of perennial ryegrass at the other extreme. Italian ryegrass has two major subspecies or varietal types.

The Westerwolds–type Italian ryegrasses (*L. multiflorum* var. *westerwoldicum*) show an annual growth habit, while var. *italicum*-type Italian ryegrasses show a biennial growth habit. The different species show a high degree of interfertility and are reproductively isolated from one another in nature through separate geographic locations and mean flowering time. The annual-type ryegrasses are more characteristic of Mediterranean environments and are early flowering, while the perennial-type ryegrasses are characteristic of cooler temperate environments and are generally later flowering. The F_1 hybrids formed between the species are highly fertile, suggesting regular chromosome pairing at meiotic metaphase and a high degree of genome structure conservation. For this reason, and from studies of SSR cross-amplification described above, genetic mapping tools generated for perennial ryegrass are expected to have a high degree of applicability for other outbreeding taxa.

Herbicide resistance in annual ryegrass

Herbicide resistance in weeds, especially in annual ryegrass, is widespread across much of the cropping zones of Australia and is a major problem for grain growers [38]. Herbicides may be classified into different groups based on their biochemical modes of action. Weeds are at high risk of developing resistance to herbicides from group A (inhibitors of lipid biosynthetic genes such as ACCase) and B (inhibitors of the enzyme acetolactate synthase [ALS]), but at lower risk for other groups such as group M (inhibitors of EPSP synthase) [19, 34]. Resistance to class M herbicide glyphosate, commonly known under the (Monsanto) trademark Roundup^{TM}, is a novel feature of annual ryegrass populations in Australia [58] and is of significant economic importance due to its widespread global use for crop protection and associated commercialization of transgenic resistant crop varieties.

Annual ryegrass populations may also display cross and multiple resistance [60]. Some populations, such as the South Australian ecotype SLR31, are resistant to many different herbicides with different modes of action. This complex behavior may be due to the accumulation of multiple genes in single genotypes selected by low rates of herbicide application in Australian agriculture.

Determination of genetic control mechanisms for herbicide resistance is important for strategies of control. A weed management strategy based on occasional high rates of herbicide application will only prove effective if resistance is under multigenic control. If resistance is due to a single major gene, such a strategy will be counterproductive and will lead to more rapid selection for resistance [33]. Use of molecular marker mapping can determine whether herbicide resistance is controlled by a single major gene, major and minor genes, or many genes of minor effect. Determination of the number of genes, their linkage relationships, pleiotropic effects and interactions will provide vital information for management strategies as well as insights into the evolution of such resistance.

Design and phenotypic analysis of mapping families for glyphosate resistance

Segregating families for genetic mapping analysis of glyphosate resistance in annual ryegrass were generated in a three-generation structure involving the initial pair-cross of single plants selected from the resistant ecotype NLR70 (collected from New South Wales, Australia) with single plants from the susceptible ecotype VLR1 (collected from Victoria, Australia). Single F_1 individuals were then selected and crossed to other distinct genotypes from VLR1. This cross-structure resembles a backcross, but due to the annual reproductive habit of the species, it is not possible to produce a true backcross and, indeed, such a cross-structure would not be favored due to the effects of the annual ryegrass gametophytic self-incompatibility mechanism on crosses between closely related genotypes. The structure of the families may be represented as [NLR70 x VLR1-1] F_1 x VLR1-2 and described as a pseudobackcross (ψ-BC). Phenotypic analysis of a number of such families has suggested the presence of a major semidominant nuclear resistance gene [50] which is fixed in a homozygous state in the NLR70 parents.

The mechanism of glyphosate resistance was investigated. No substantial metabolism of glyphosate was evident, but accumulation of the metabolite shikamate (an intermediate in the pathway to aromatic amino acid biosynthesis inhibited by glyphosate) following herbicide application is higher in susceptible plants, suggesting that there may be some barrier to penetration into the chloroplast [51, 59]. Studies of the biochemical properties of the target enzyme for glyphosate action (the shikamate pathway enzyme 5-enolpyruvylshikamate 3-phosphate [EPSP] synthase) in resistant and susceptible ecotypes suggest that no major gene effect is mediated through target site resistance or overexpression.

Preliminary phenotypic analysis of two pseudobackcross populations designated ([NLR70 x VLR1-1]F_1 x VLR1-2)#2 and ([NLR70 x VLR1-

1]F_1 x VLR1-2)#10 was performed with replication over time at three different levels of dose application (0.75, 1, and 1.25 L ha^{-1} of a glyphosate isopropylamine solution at 450g acid equivalent [a.e.] (a.e. = theoretical yield of parent acid, glyphosate, from the formulated salt) L^{-1}. A single clone of each progeny plant was represented at each dose application. Twofold replication was performed for ψ-BC#2 and two separate rounds of twofold replication were performed for ψ-BC#10, with 150 progeny plants used for each experiment. Resistance was assessed qualitatively, with susceptible plants scored as 1 and resistant plants as 0. For each dose, progeny were divided into three classes: survival in both replicates was scored as 0, survival in one replicate but not the other as 0.5, and death in both replicates as 1. The average frequency distribution data over the three doses for each family are shown in Figure 5.3. Variance analysis (ANOVA) showed that the population means for ψ-BC#2 across dose replicates were similar but the average broad sense heritability value across doses showed an intermediate value (H^2 = 0.67). This indicates that individual progeny showed inconsistent phenotypes between replicates, with individuals altering resistance in both directions. Distribution of the resistance phenotype was not fully discontinuous, as would be expected were a major gene segregating in the population, and suggests either the effect of genotype x environment interaction, or the action of modifier genes. Contrarily, the average broad sense heritability value for ψ-BC#10 was higher (H^2 = 0.86) and the distribution more discontinuous, with a small proportion of the progeny individuals falling into the intermediate classes.

Genotypic analysis of mapping families for glyphosate resistance based on bulked segregant analysis

As the two pseudobackcross families were generated from parents selected from the same base populations, the same major glyphosate resistance gene should presumably segregate in both crosses. Bulked segregant analysis (BSA: [54]) is suitable for analysis of such populations. Of the major molecular marker systems, AFLP is highly suitable for BSA due to the feasibility of rapidly surveying multiple genomic loci. However, it is not straightforward to then assign the linked markers to positions on reference genetic maps, due to the largely pedigree-specific nature of AFLPs. BSA with SSR markers, on the other hand, is more labor intensive due to the lower multiplex ratio but provides access to robust map

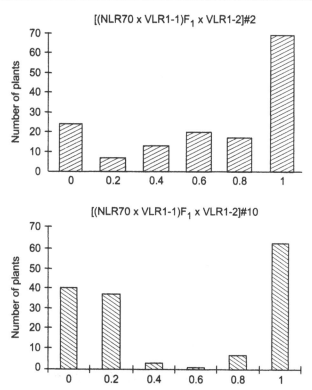

Fig. 5.3: Frequency distribution bar-charts for glyphosate resistance in the ψ-BC populations ([NLR70 x VLR1-1] x VLR1-2) #2 (averaged over 3 doses with twofold replication) and ([NLR70 x VLR1-1] x VLR1-2) #10 (averaged over 3 doses and 4 replicates). Resistance was scored qualitatively as 1 for susceptibility and 0 for resistance.

location data. Perennial ryegrass SSR (LPSSR) markers provide an ideal technology for genotyping of annual ryegrass segregating populations. A subset (54) of LPSSR primer pairs was screened earlier [42] for efficiency in cross-amplification annual ryegrass with a success rate of 80% or greater, and amplification of genuine annual ryegrass SSR loci confirmed by DNA sequence analysis. As a framework set of 93 LPSSR loci was superimposed on the heterologous RFLP and AFLP-based perennial ryegrass reference map [43, 44], genomic analysis of new mapping populations could be performed on a map-structured basis. Hence, both AFLP and SSR technologies were employed in this study.

Eight AFLP primer pair combinations were used to generate profiles from bulk samples comprising up to 20 resistant (R) and 20 susceptible (S) progeny genotypes from each cross. Six of the primer combinations that detected distinct differences between bulked samples were then screened using genomic DNA from the individual genotypes used in composing these samples. No band showing complete cosegregation with the glyphosate resistance phenotype was found but several showed a frequency shift between individuals of the R and S groups; none of these putatively associated bands showed a similar pattern of association in the two crosses, however.

As the initial AFLP analysis was of limited value, the bulked samples were screened with 69 primer pairs designed for LPSSR loci that had priorly been assigned to the perennial ryegrass reference genetic map [43]. Eleven of these primer pairs detected polymorphism between the R and S bulks from ψ-BC#2, of which four also detected polymorphism between the bulks of ψ-BC#10. The 11 primer pairs were then screened on the debulked individuals, with seven showing significant association with the glyphosate resistance character (Table 5.3). The seven primer pairs were then used to genotype the full progeny set of ψ-BC#2. ANOVA showed that six of the seven primer pairs detected loci in significant association with glyphosate resistance, corresponding to loci mapped to perennial ryegrass LGs 2, 4, and 5.

Genotypic analyses of mapping families for glyphosate resistance based on progeny genotyping

To enhance this study, other LPSSR primer pairs which did detect polymorphism in ψ-BC#2 but had not in every instance detected differences between bulked samples, were used to genotype the parents and full progeny set. The overall level of polymorphism was very high (83%), even compared to the generally high values (60-65%) seen in perennial ryegrass pseudotestcross populations [29]. The total set of 42 primer pairs, including those previously demonstrated to detect association with the R character, were assembled into triplexed panels based on the three fluorochromes FAM (6-carboxyfluorescein), HEX (hexachloro-6-carboxyfluorescein) and NED (=7',8'-benzo-5'-fluoro-2,4,7-trichloro-5-carboxyfluorescein). The aim was to map the resistance gene through coverage of the L. rigidum genome with at least two markers from each LG of perennial ryegrass. This was fulfilled with a minimum of three markers per linkage group, allowing for good preliminary genomic coverage.

As several of the primers revealed complex banding patterns, segregating alleles were scored as dominant markers in the test cross

Table 5.3: *Lolium perenne* SSR loci polymorphic between glyphosate resistant and susceptible bulks of the pseudo-backcross (ψ-BC) populations ([NLR70 x VLR1-1] F_1 x VLR1-2)#2 and ([NLR70 x VLR1-2] F_1 x VLR1-2)#10

Locus	LG^a	Polymorphic in bulks of:		Associated in de bulks of ψ-BC #2	F value and significance^c		
		ψ-BC #2	ψ-BC #10		0.75	1	1.25 l ha^{-1}
LPSSRH02F10	2	Y	N	N	ns	ns	ns
LPSSRH02C11	3	Y	N	N	ns	ns	ns
LPSSRK01G06	4	Y	Y	Y	30***	32***	11**
LPSSRK05A11	4	Y	N	Y	ns	ns	ns
LPSSRK02F04	4	Y	Y	Y	14***	13***	ns
LPSSRH03F03	2	Y	Y	N	ns	ns	ns
LPSSRKXX104	4	Y	N	Y	10*	ns	ns
LPSSRH11G05	5	Y	Y	Y	7*	9*	ns
LPSSRK09C10	5	Y	nd^b	Y	15***	25***	11**
LPSSRK14C04	1	Y	nd^b	N	ns	ns	ns
LPSSRH09E12	2	Y	N	Y	10**	ns	ns

^{a}Linkage group assignment on the p150/112 perennial ryegrass reference map.

^{b}nd = no data

^{c}0.75, 1 and 1.25 l ha^{-1} are the dose rates for glyphosate application. ns = $P > 0.01$, * = $P £ 0.01$, ** = $P £ 0.001$, *** = $P £ 0.0001$

progeny (i.e., presence/absence). Data were analyzed for association of each marker with resistance using Proc GLM (general linear modeling) in SAS 8.02. Although in some cases the relevant allele could be identified in the NLR70 parent, the "alternative" (null) allele could likewise be evaluated in this analysis. Table 5.4 shows the results for 18 primer pairs that detected loci located on LGs 2, 4, and 5 of perennial ryegrass. The most distinctly associated markers in this analysis are those detecting the perennial ryegrass LG4 markers xlpssrk01g06, xlpssrk02f04, and xlpssrk03c05, and that detecting LG5 marker xlpssrk09c10 [43]. The three LG4 loci are located in the middle of the linkage group, in the putative pericentromeric cluster, with xlpssrk01g06 and xlpssrk03c05 in particularly close linkage and xlpssrk02f04 approximately 20 cM distant from the other markers (Fig. 5.4). The strongest effects are associated with the primer pair detecting the LG5 marker xlpssrk09c10.

Table 5.4: Summary of association data for selected LPSSR primer pairs detecting genetic polymorphism in the ([NLR70 x VLR1-1] x VLR1-2) ψ-BC#2 showing segregation for glyphosate resistance and susceptibility

LPSSR primer pair	Linkage group of detected locus in perennial ryegrass	Molecular size of scored allele (bp)[a]	Significance probability value[b]
LPSSRH02D10	2	139	0.433
LPSSRH02F10	2	169	0.236
LPSSRH03F03	2	103	0.535
LPSSRH09E12	2	116	0.0503
LPSSRHXX285	2	84	0.127
LPSSRK09F06	2	103	0.206
LPSSRH01H06	4	142	0.228
LPSSRK01G06	4	143	2.76×10^{-5}
LPSSRK02D08	4	264	0.404
LPSSRK02F04	4	197	8.02×10^{-6}
LPSSRK03C05	4	319	6.09×10^{-4}
LPSSRK05A11	4	114	0.23
LPSSRK15F05	4	213	0.482
LPSSRKXX104	4	100	0.0457
LPSSRH11G05	5	243	0.0292
LPSSRK03B03	5	297	0.472
LPSSRK03F09	5	288	0.0578
LPSSRK09C10	5	297	2.37×10^{-8}

[a]Molecular size of clearest allele representing a dominant marker tested for association with the resistance character.

[b]Higher values indicate low levels of association, and lower indicate high levels of association.

The four parental genotypes and 81 test cross progeny from the ψ-BC#10 population were also genotyped using a set of 12 LPSSR primer pairs which either identified polymorphism between the bulked samples or detected polymorphic loci in close linkage to the bulk-discriminatory markers. Results are shown in Table 5.5. In this cross, the locus detected by primer pair LPSSRK02F04 is again significantly associated at a high level. Segregation of the 195 bp allele inherited in the pseudobackcross progeny from the NLR70 parent is shown in Figure 5.5. The effect of this LG4 marker is hence consistent between the two crosses. However, the nearby xlpssrk03c05 marker does not show significant association in the ψ-BC#10 cross. The other significant association is with the LG5-located marker detected by primer pair LPSSRH11G05, which is located in the upper part of LG5 in perennial ryegrass close (c. 5 cM) to xlpssrk09c10. The strongest effects are again associated with an LG5 marker, although differing from those detected in the ψ-BC#2 population.

Table 5.5: Summary of association data for selected LPSSR primer pairs detecting genetic polymorphism in the ([NLR70 x VLR1-1] x VLR1-2) ψ-BC#10 showing segregation for glyphosate resistance and susceptibility

LPSSR primer pair	Linkage group of detected locus in perennial ryegrass	Molecular size of scored allele (bp)[a]	Significance probability value[b]
LPSSRH02D10	2	108	0.161
LPSSRH02F10	2	171	0.794
LPSSRH03F03	2	100	0.191
LPSSRH09E12	2	108	0.632
LPSSRK02D08	4	260	0.218
LPSSRK02F04	4	195	1.69×10^{-7}
LPSSRK03C05	4	309	0.907
LPSSRK05A11	4	142	0.526
LPSSRKXX104	4	102	0.411
LPSSRH11G05	5	242	3.6×10^{-15}
LPSSRK03F09	5	341	0.192
LPSSRK09C10	5	265	0.14

[a]Molecular size of clearest allele representing a dominant marker tested for association with the resistance character.

[b]Higher values indicate low levels of association and lower indicate high levels of association.

Fig. 5.4 SSR-based framework genetic map of perennial ryegrass, indicating the location of loci detected by LPSSR primer pairs evaluated for association with the glyphosate resistance character in the pseudobackcross (ψ-BC) populations ([NLR70 x VLR1-1] x VLR1-2) #2 and #10. Loci that do not show a high level of association are indicated in green Loci showing a high level of association in cross #2 are shown in red type and those in cross #10 in pink. Loci showing a high level of association in both crosses are shown in blue. Condt. on next page

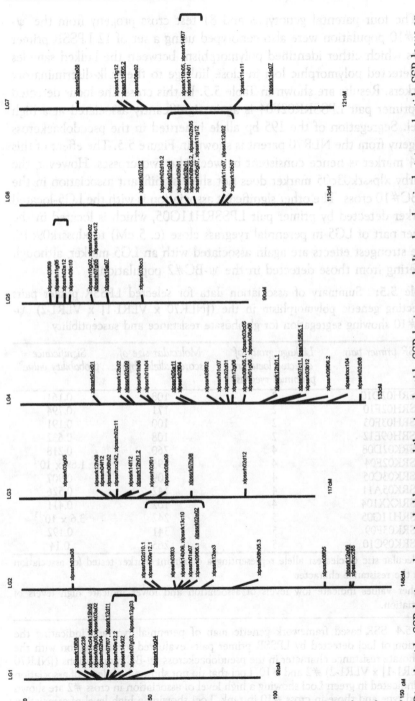

Dinucleotide SSR loci are shown in bold type, while trinucleotide SSR loci are underlined and a single tetranucleotide SSR locus on LG4 in Romant type.

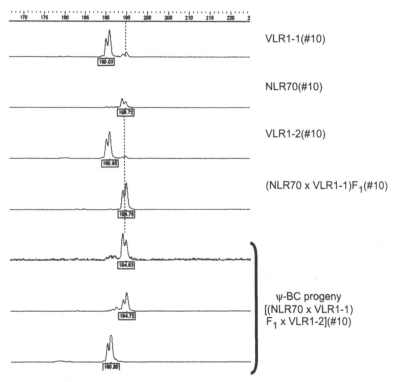

Fig. 5.5: Segregating genetic polymorphism detected in the progeny of the ([NLR70 x VLR1-1] x VLR1-2) cross #10, detected by the primer pair designed to locus LPSSRK02F04. Segregation of the NLR70-derived 195 bp allele associated with glyphosate resistance is indicated.

The regions of effect are consistent between the two crosses, supporting the view that similar genetic factors are segregating in both progeny sets, and consistent with a putative common origin of the resistance gene or genes in the NLR70 population. However, although the more complex mode of inheritance in ψ-BC#2 is compatible with the effects of several genomic regions, this is not the case for ψ-BC#10. Linkage of the markers associated with the resistance character in ψ-BC#2 was evaluated using the mapping program MAPMAKER 3.0 [49], and the four associated loci were demonstrated to all be in linkage, despite their locations on different linkage groups in perennial ryegrass. A similar analysis was performed for the two markers associated with the resistance character in ψ-BC#10, but they did not show linkage to one another.

Interpretation of data

Linkage analysis suggests that at least in ψ-BC#2, a single region of the annual ryegrass genome may be responsible for most of the genetic variation for the herbicide resistance character. Despite failure to detect linkage between the associated markers in ψ-BC#10, the same situation may also apply based on the discontinuous mode of inheritance in this population. The relevant primer pairs have detected loci on different linkage groups in perennial ryegrass, as demonstrated by the construction of the SSR-based framework map in combination with heterologous RFLP and AFLP markers in the p150/112 population [44]. One possible explanation for this anomaly is that a genomic rearrangement differentiates the genomes of *L. perenne* and *L. rigidum*, with the appropriate regions of perennial ryegrass LG4 and LG5 belonging to a single linkage group in annual ryegrass. However, the high level of fertility in F_1 interspecific hybrids between the outbreeding *Lolium* taxa suggests that chromosome pairing at meiosis is highly regular, implying a very high level of conserved colinearity between the chromosomes of these species. It is conceivable that a small subcytological region of perennial ryegrass LG5 was reduplicated or translocated to the LG4 orthologous chromosome in annual ryegrass but is of insufficient size to disrupt chromosome pairing. Comparative genetic mapping analysis between perennial ryegrass (as well as other members of the Poaceae family) and meadow fescue (*Festuca pratensis* Huds.), a member of the closely related *Festuca* genus, has revealed some evidence for evolutionary translocations involving regions syntenic with the Triticeae homeologous group 4 and 5 chromosomes [2, 3, 44]. Comparative genetic mapping of perennial ryegrass compared to wheat, rice and oats has revealed the presence of a Triticeae group 5 marker, xpsr580, on the upper part of perennial ryegrass LG4, while a number of rice cDNA RFLP markers on the upper part of LG4 define a region syntenic with rice chromosome 3, which is the syntenic counterpart of Triticeae group 5L. The same order is observed in meadow fescue [3].

Another possible explanation for the anomalous mapping data is the effect of paralogous sequences. Several of the LPSSR primer pairs detecting loci associated with the glyphosate resistance character amplified more than one segregating locus in the relevant crosses. Detection of multiple loci by a small proportion of LPSSR primer pairs (ca. 10%) was observed during construction of the perennial SSR-based framework genetic map [43]. It is possible that due to limited sequence duplication, an LPSSR primer pair may amplify sequences located on both LG4 and LG5, but that the polymorphic locus mapped in the perennial ryegrass reference population may be derived from one linkage group (e.g. LG4) and the polymorphic locus mapped in annual ryegrass may be derived from the other linkage group (e.g. LG5). All of the significantly associated markers in ψ-BC#2 were scored from multiple amplification products, supporting this possibility.

Future prospects

The evidence from biochemical studies suggests that the resistance gene or genes in the NLR70 parents do not correspond to the structural gene for the target enzyme for glyphosate action (EPSP synthase). This conclusion may be confirmed through a molecular marker-based approach by genetic mapping of the EPSP synthase structural gene in the segregating families, with the expectation of a location in another region of the genome. The EPSP synthase gene may be mapped either as an RFLP or as a sequence tagged site (STS) or single nucleotide polymorphism (SNP) marker. The complete cDNA sequence for the *L. rigidum* EPSP synthase is available as a GenBank accession (AF349754) for such analysis.

Comparative genetic analysis based on in silico data [76] may contribute to the molecular characterization of herbicide resistance in *L. rigidum*. Genetic location of the rice EPSP synthase gene may be determined through the current public domain sequencing efforts, and the comprehensive comparative genetic map constructed for perennial ryegrass may be used to predict the syntenic locations between species. The comparative map locations of other possible candidate genes involved in herbicide metabolism (e.g. cytochrome P450 genes for detoxification) from other species (rice, barley etc.) may also be inferred from comparative genetics and tested for colocation with the genomic regions identified as of interest in this study.

CONCLUSIONS

The empirical data presented demonstrate the value of molecular genetic marker systems, especially those based on PCR-mediated detection such as AFLP and SSR, for analysis of important aspects of pasture grass biology. For genetic variation analysis, demonstration of consistency between breeding history and genetic diversity within and between perennial ryegrass populations with both marker systems is highly significant for future studies. Linkage disequilibrium (LD) analysis [52] has been proposed as a suitable method for marker-trait discovery in obligate outbreeding species [29,73]. LD mapping is a method for statistical analysis of nonrandom associations between genetic loci using population samples rather than full-sib mapping populations. This approach is well established for genetic studies of humans [63] and some livestock species [53] but is still relatively novel for plant species. Levels of LD between markers and target genes are variable between species, populations and regions of the genome, and dependent on average rates of recombination per generation, number of elapsed generations, and effective population size [39]. In general, we would expect an obligate outbreeding species such as perennial ryegrass to show LD over relatively short molecular distances [52]. This is especially true for long-established populations derived from a large number of parental individuals, as would be expected for ecotypes and many older varieties, in which many rounds of recombination have occurred.

Selection of populations for LD analysis is an important theoretical and empirical issue. Key resources for the design of such populations are extensive germplasm collections. A global perennial ryegrass germplasm resource has been established (N. Bannan and K.F. Smith, unpubl. data), consisting of 477 accessions from most of the areas of natural distribution and cultivation, with similar numbers of domesticated and nondomesticated sources. Hierarchical measurement of genetic variation between and within these accessions may be performed using genomic DNA-derived SSR markers and may provide a comprehensive view of the genetic architecture of the species. Such data are used in determining the degree of genetic affinity among different components of the gene pool and hence in designing optimized populations for LD studies in perennial ryegrass. Sophisticated analytical techniques are required for the most appropriate analysis of large data sets based on SSR polymorphism. Other systems such as candidate gene-based SNP markers [29] may also be used for supplementary analysis.

Comparative genetic mapping between different grass and cereal species has provided a valuable tool for structured gene location prediction, and this approach has been validated in perennial ryegrass for the SI genes [79], major genes for crown rust resistance [24], and flowering time control genes [87]. Comparative genetic analysis among members of the same genus, such as perennial and annual ryegrass, would be expected to reveal very high levels of conserved synteny and colinearity. On this basis, the results obtained from genetic analysis of glyphosate resistance in L. *rigidum* permit predictions of ortholocus location in perennial ryegrass. Large-scale mapping of EST-derived candidate genes in perennial ryegrass, especially as EST-RFLPs and EST-SNPs [29], will then allow positional evaluation of candidate genes for herbicide resistance in annual ryegrass; such studies will be further reinforced by comparative genomics with model species such as rice. However, observed interruptions of microsynteny between rice and wheat [76] demonstrate limitations of comparative genetic analysis. Similarly, in this study, detection of genetic markers attributed to LGs 4 and 5 in perennial ryegrass with an apparent single locus effect in annual ryegrass suggests that small-scale genetic map rearrangements may also differentiate even closely related taxa. Nonetheless, the present data do support the hypothesis of simple genetic control for glyphosate resistance in L. *rigidum*, along with the appropriate control strategy.

The genomic DNA-derived SSR marker system described here will remain a valuable tool for genetic analysis of *Lolium* species. In future, greater emphasis on functionally associated genetic markers is likely, especially EST-SSRs and EST-SNPs [29]. The anticipated low level of LD in such outcrossing species is especially suitable for the application of candidate gene-based markers [61].

ACKNOWLEDGMENTS

The original research work described here was funded by Dairy Australia, the Molecular Plant Breeding Co-operative Research Centre, and the Victorian Department of Primary Industries. The authors thank Dr. Noel Cogan, Prof. Michael Hayward, and Prof. German Spangenberg for careful critical reading of the manuscript.

[1] Aggarwal RK, Brar DS, Nandi S, Huang N, Khush GS. Phylogenetic relationships among *Oryza* species revealed by AFLP markers. Theor Appl Genet 1999; 98: 1320-1328.

[2] Alm V. Comparative genome analyses of meadow fescue (*Festuca pratensis* Huds.): Genetic linkage mapping and QTL analyses of frost and drought tolerance. PhD thesis, Agricultural University of Norway, Ås, 2001.

[3] Alm V, Cheng F, Busso C, et al. A linkage map of meadow fescue (*Festuca pratensis* Huds.) and comparative mapping with the Triticeae species, *Lolium*, oat, rice, maize and sorghum. Theor Appl Genet 2003; 108: 25-40.

[4] Anonymous. Perennial ryegrass (*Lolium perenne*) variety 'Yatsyn 1'. Pl Var J 1988; 1: 5-7.

[5] Armstrong CS. 'Grasslands Nui' perennial ryegrass. NZ J Exper Agr 1977; 5: 381-384.

[6] Bert PF, Charmet G, Sourdille P, Hayward MD, Balfourier F. A high-density molecular map for ryegrass (*Lolium perenne*) using AFLP markers. Theor Appl Genet 1999; 99: 445-452.

[7] Bonierbale MW, Plaisted RL, Tanksley SD. RFLP maps based on a common set of clones reveal modes of chromosomal evolution in potato and tomato. Genetics 1988; 120: 1095-1103.

[8] Borrill M. Temperate grasses. In: Simmons NW, ed. Evolution of crop plants: Longman, London; UK: Longman, 1976: 137 – 142.

[9] Bray RA, Irwin JAG. *Medicago sativa* L. (lucerne) c. Hallmark. Austr J Exper Agric 1999; 39: 643-644.

[10] Breese EL, Lewis EJ. Breeding versatile hybrid grasses. Span 1984; 27: 3-5.

[11] Cameron N. Perennial ryegrass (*Lolium perenne* L.) variety 'Embassy'. Pl Var J 1994a; 7: 10-11.

[12] Cameron N. Perennial ryegrass (*Lolium perenne*) variety 'Banks'. Pl Var J 1994b; 7: 14-15.

[13] Charmet G, Ravel C, Balfourier F. Phylogenetic analysis in the *Festuca-Lolium* complex using molecular markers and ITS rDNA. Theor Appl Genet 1997; 94: 1038-1046.

[14] Cho YG, Blair MW, Panaud O, McCouch SR. Cloning and mapping of variety-specific rice genomic DNA sequences: amplified fragment length polymorphisms (AFLP) from silver-stained polyacrylamide gels. Genome 1996; 39: 373-378.

[15] Cho YG, Ishii T, Temnykh S, et al. Diversity of microsatellites derived from genomic libraries and GenBank sequences in rice (*Oryza sativa* L.). Theor Appl Genet 2000; 100: 713-722.

[16] Cordeiro GM, Casu R, McIntyre CL, Manners JM, Henry RJ. Microsatellite markers from sugarcane (*Saccharum* spp.) ESTs cross transferable to erianthus and sorghum. Pl Sci 2001; 160: 1115-1123.

[17] Cornish MA, Hayward MD, Lawrence MJ. Self-incompatibility in ryegrass. I. Genetic control in diploid *Lolium perenne* L. Heredity 1979; 43: 95-106.

[18] Crossa J. Methodologies for estimating the sample size required for genetic conservation of outbreeding crops. Theor Appl Genet 1989; 77: 153-161.

[19] Darmency H. Genetics of herbicide resistance in weeds and crops. In: Powles SB, Holtum JAM, eds. Herbicide resistance in plants: biology and biochemistry. Boca Raton, FL: Lewis Publishers, 1994: 263-267.

[20] Devey F, Fearon CH, Hayward MD, Lawrence MJ. Self-incompatibility in ryegrass. XI. Number and frequency of alleles in a cultivar of *Lolium perenne* L. Heredity 1994; 73: 262-264.

[21] Devos KM, Gale MD. The genetic maps of wheat and their potential in plant breeding. Outlook Agric 1993; 22: 93-99.

[22] Devos KM, Gale MD. Comparative genetics in the grasses. Pl Mol Biol 1997; 35: 3-15.

[23] Devos KM, Gale MD. Genome relationships: the grass model in current research. Pl Cell 2000; 12: 637-645.

[24] Dumsday JL, Smith KF, Forster JW, Jones ES. SSR-based genetic linkage analysis of resistance to crown rust (*Puccinia coronata* Corda f. sp. *lolii*) in perennial ryegrass (*Lolium perenne* L.). Pl Path 2003; 52: 628-637.

[25] Edwards KJ, Barker JHA, Daly A, Jones C, Karp A. Microsatellite libraries enriched for several microsatellite sequences in plants. Biotechniques 1996; 20: 758-759.

[26] Estoup A, Jarne P, Cornuet J-M. Homoplasy and mutation model at microsatellite loci and their consequences for population genetics analyses. Mol Ecol 2002; 11: 1591-1604.

[27] Excoffier L, Smouse PE, Quattro JM. Analysis of molecular variance inferred from metric distances among DNA haplotypes: application to human mitochondrial restriction data. Genetics 1992; 131: 479-491.

[28] Fearon CH, Cornish MA, Hayward MD, Lawrence MJ. Self-incompatibility in ryegrass. X. Number and frequency of alleles in a natural population of *Lolium perenne* L. Heredity 1994; 73: 254-261.

[29] Forster JW, Jones ES, Batley J, Smith KF. Molecular marker-based genetic analysis of pasture and turf grasses. In: Wang Z-Y et al., eds. Molecular breeding of forage and turf, Dallas, Texas. Dordrecht, Netherlands: Kluwer Academic Press pp. 197-239.

[30] Forster JW, Jones ES, Kölliker R, et al. DNA profiling in outbreeding forage species. In: Henry RJ, ed. Plant genotyping: the DNA fingerprinting of plants. Wallingford, UK: CAB International, 2001; 299-320.

[31] Fuong FT, Voylokov AV, Smirnov VG. Genetic studies of self-fertility in rye (*Secale cereale* L.). 2. The search for molecular marker genes linked to self-incompatibility loci. Theor Appl Genet 1993; 87: 619-623.

[32] Gale MD, Devos KM. Plant comparative genetics after 10 years. Science 1998; 282: 656-659.

[33] Gardner SN, Gressel J, Mangel M. The revolving dose strategy to delay the evolution of both quantitative vs. major monogene resistances to pesticides and drugs. Intl J Pest Manag 1998; 44: 161-180.

[34] Gasquez J. Genetics of herbicide resistance within weeds. Factors of evolution, inheritance and fitness. In: De Prado JJR, Garcia-Torres L, eds. Weed and crop resistance to herbicides. Dordrecht, Netherlands; Kluwer Academic Press, 1997; 181-189.

[35] Grattapaglia D, Sederoff R. Genetic linkage maps of *Eucalyptus grandis* and *Eucalyptus urophylla* using a pseudo-testcross: mapping strategy and RAPD markers. Genetics 1994; 137: 1121-1137.

[36] Grimaldi M-C, Cronau-Roy B. Microsatellite allelic homoplasy due to variable flanking sequences. J Mol Evol 1997; 44: 336-340.

[37] Guthridge KM, Dupal MD, Kölliker R, Jones ES, Smith KF, Forster JW. AFLP analysis of genetic diversity within and between populations of perennial ryegrass (*Lolium perenne* L.). Euphytica 2001; 122: 191-201.

[38] Heap I. International survey of herbicide-resistant weeds. Online from August 24th 2000 at http://www.weedscience.com.

[39] Hill WG, Robertson A. Linkage disequilibrium in finite populations. Theor Appl Genet 1968; 38: 226-231.

[40] Huff DR. RAPD characterization of heterogeneous perennial ryegrass cultivars. Crop Sci 1997; 37: 557-564.

[41] Jenkins G, Head J, Forster JW. Probing meiosis in hybrids of *Lolium* (Poaceae) with a discriminatory repetitive genomic sequence. Chromosoma 2000; 109: 280-286.

[42] Jones ES, Dupal MP, Kölliker R, Drayton MC, Forster JW. Development and characterisation of simple sequence repeat (SSR) markers for perennial ryegrass (*Lolium perenne* L.). Theor Appl Genet 2001; 102: 405-415.

[43] Jones ES, Dupal MD, Dumsday JL, Hughes LJ, Forster JW. An SSR-based genetic linkage map for perennial ryegrass (*Lolium perenne* L.). Theor Appl Genet 2002; 105: 577-584.

[44] Jones ES, Mahoney NL, Hayward MD, et al. An enhanced molecular marker-based map of perennial ryegrass (*Lolium perenne* L.) reveals comparative relationships with other Poaceae species. Genome 2002; 45: 282-295.

[45] Kemp DR. The effects of flowering time and leaf area on sward growth in winter of temperate pasture grasses. Austr J Agric Res 1988; 39: 597-604.

[46] Kiers AM, Mes THM, van der Meijden R, Bachmann K. A search for diagnostic AFLP markers in *Chicorium* species with an emphasis on endive and chicory groups. Genome 2000; 470-476.

[47] Korzun V, Malyshev S, Voylokov AV, Börner A. A genetic map of rye (*Secale cereale* L.) combining RFLP, isozyme, protein, microsatellite and gene loci. Theor Appl Genet 2001; 102: 709-717.

[48] Kurata N, Moore G, Nagamura Y, Foote T, Yano M, Minobe Y, Gale MD. Conservation of genome structure between rice and wheat. Biotech 1993; 12: 276-278.

[49] Lander ES, Green P, Abrahamson J, Barlow A, Daly MJ, Lincoln SE, Newburg L. MAPMAKER: an interactive computer package for constructing primary linkage maps of experimental and natural populations. Genomics 1987; 1: 174-81.

[50] Lorraine-Colwill DF, Powles SB, Hawkes TR, Preston C. Inheritance of evolved glyphosate resistance in *Lolium rigidum* (Gaud.). Theor Appl Genet 2001; 102: 545-550.

[51] Lorraine-Colwill DF, Hawkes TR, Williams PH, et al. Resistance to glyphosate in *Lolium rigidum*. Pest Sci 1999; 55: 486-503.

[52] Mackay TFC. The genetic architecture of quantitative traits. Ann Rev Genet 2001; 35: 303-309.

[53] Meuwissen THE, Karlsen A, Lien S, Olsaker I, Goddard ME. Fine mapping of a quantitative trait locus for twinning rate using combined linkage and linkage disequilibrium mapping. Genetics 2002; 161: 373-379.

[54] Michelmore RW, Paran I, Kesseli R.V. Identification of markers linked to disease-resistance genes by bulked segregant analysis: a rapid method to detect markers in specific genomic regions by using segregating populations. Proc Natl Acad Sci USA 1991; 88: 9828-9832.

[55] Moore G, Devos KM, Wang Z, Gale MD. Grasses, line up and form a circle. Curr Biol 1995; 5: 737-739.

[56] Nakajima Y, Oeda K, Yamamoto T. Characterisation of genetic diversity of nuclear and mitochondrial genomes in *Daucus* varieties by RAPD and AFLP. Pl Cell Repts 1998; 17: 848-853.

[57] Peakall R, Gilmore S, Keys W, Morgante M, Rafalksi A. Cross species amplification of soybean (*Glycine max*) simple sequence repeats (SSRs) within the genus and other legume genera: implications for the transferability of SSRs in plants. Mol Biol Evol 1998; 15: 1275-1287.

[58] Powles SB, Lorraine-Colwill DF, Dellow JJ, Preston C. Evolved resistance to glyphosate in rigid ryegrass (*Lolium rigidum*) in Australia. Weed Sci 1998; 46: 405-411.

[59] Preston C, Powles SB. Mechanisms of multiple herbicide resistance in *Lolium rigidum*. In: Clark JM, ed. Pesticide science: pesticide resistance. Washington DC: American Chemical Society, 2000.

[60] Preston C, Tardif FJ, Powles SB. Multiple mechanisms and multiple herbicide resistance in *Lolium rigidum*. In: Brown TM, ed. Molecular genetics and evolution of pesticide resistance. Washington DC: American Chemical Society, 1996; 117-129.

[61] Rafalski A. Applications of single nucleotide polymorphisms in crop genetics. Curr Op Pl Biol 2002; 5: 94-1000.

[62] Rafalski JA, Vogel JM, Morgante M, et al. Generating and using DNA markers in plants. In: Birren B, Lai E, eds. Non-mammalian genomic analysis: a practical guide. San Diego, CA: Academic Press Inc., 1996; 75-135.

[63] Risch N, Merikangas K. The future of genetic studies of complex human diseases. Science 1996; 273: 1516-1517.

[64] Ritter E, Gebhardt C, Salamini F. Estimation of recombination frequencies and construction of linkage maps from crosses between heterozygous parents. Genetics 1990; 125: 645-654.

[65] Rohlf, F.J. NTSYS-pc: Numerical taxonomy and multivariate analysis system. Setauket, NY: Exeter Publishers, 1993.

[66] Rossetto M. Sourcing of SSR markers from related plant species. In: Henry RJ, ed. Plant genotyping: the DNA fingerprinting of plants. Wallingford, UK: CAB International, 2001; 211-255.

[67] Saha MC, Zwonitzer JC, Hopkins AA, Mian MAR. A molecular linkage map of a tall fescue population segregating for forage quality traits. Development and application of molecular technologies in forage and turf improvement. In: Wang Z-Y et al., eds. Molecular Breeding of Forage and Turf, Dallas, Texas 2003; P3.

[68] Saha MC, Chekhovksi K, Zwonitzer JC, Eujayl I, Mian MAR. Development of microsatellite markers for forage grass and cereal species. Pl Anim Genome XI. San Diego, California 2003: 201.

[69] Sanders PM, Barker DJ, Wewala GS. Phosphoglucoisomerase-2 allozymes for distinguishing perennial ryegrass cultivars in binary mixtures. J Agrc Sci Camb 1989; 112: 179-184.

[70] Schneider S. Excoffier L. Estimation of demographic parameters from the distribution of pairwise differences when the mutation rates vary among sites: applications to human mitochondrial DNA. Genetics 1999; 152: 1079-1089.

[71] Schwarz G, Herz M, Huang X-Q, et al. Application of fluorescence-based semi-automated AFLP analysis in barley and wheat. Theor Appl Genet 2000; 100: 545-551.

[72] Scott KD, Eggler P, Seaton G, et al. Analysis of SSRs derived from grape ESTs. Theor Appl Genet 2000; 100: 723-726

[73] Skøt L, Sackville-Hamilton NR, Mizen S, Chorlton KH, Thomas ID. Molecular genecology of temperature response in *Lolium perenne*: 2. Association of AFLP markers with ecogeography. Mol Ecol 2002 11: 1865-1876.

[74] Sneath PHA, Sokal RR. Numerical taxonomy. San Francisco, CA: Freeman, 1973, 573 pp.

[75] Soreng RJ, Davis JI. Phylogenetics and character evolution in the grass family (Poaceae): Simultaneous analysis of morphological and chloroplast DNA restriction site character sets. Bot Rev 1998; 64: 1-85.

[76] Sorrells ME, La Rota M, Bermundez-Kandianis CE, et al. Comparative DNA sequence analysis of wheat and rice genomes. Genome Res 2003; 13: 1818-1827.

[77] Stammers M, Harris J, Evans GM, Hayward MD, Forster JW. Use of random PCR (RAPD) technology to analyse phylogenetic relationships in the *Lolium/Festuca* complex. Heredity 1995; 74: 19-27.

[78] Strang J. Kangaroo Valley perennial ryegrass. Agric Gazette NSW 1961; 72: 75-133.

[79] Thorogood D, Kaiser WJ, Jones JG, Armstead I. Self-incompatibility in ryegrass: 12. Genotyping and mapping the S and Z loci of *Lolium perenne* L. Heredity 2002; 88: 385-390.

[80] Van Deynze AE, Nelson JC, O'Donoghue LS, et al. Comparative mapping in grasses. Oat relationships. Mol Gen Genet 1995; 249: 349-356.

[81] van Zijll de Jong E, Guthridge KM, Spangenberg GC, Forster JW. Development and characterization of EST-derived simple sequence repeat (SSR) markers for pasture grass endophytes. Genome 2003; 46: 277-290

[82] Vos P, Hogers R, Bleeker M, et al. AFLP-a new technique for DNA fingerprinting. Nucl Acids Res 1995; 23: 4407-4414.

[83] Voylokov AV, Korzun V, Börner A. Mapping of three self-fertility mutations in rye (*Secale cereale* L.) using RFLP, isozyme and morphological markers. Theor Appl Genet 1998; 97: 147-153.

[84] Warpeha KMF, Gilliland TJ, Capesius I. An evaluation of rDNA variation in *Lolium* species (ryegrass). Genome 1998; 41: 307-311.

[85] Wilkins PW. Breeding perennial ryegrass for agriculture. Euphytica 1991; 52: 201-214.

[86] Xu WW, Sleper DA, Chao S. Genome mapping of polyploid tall fescue (*Festuca arundinacea* Schreb.) with RFLP markers. Theor Appl Genet 1995; 91: 947-955.

[87] Yamada T, Jones ES, Cogan NOI, et al. QTL analysis of morphological, developmental and winter hardiness-associated traits in perennial ryegrass (*Lolium perenne* L.). Crop Sci 2004; 44: 925-935.

Genetic Diversity and Relationships of Coconut (*Cocos nucifera* L) Revealed by Microsatellite Markers

LALITH PERERA

Genetics and Plant Breeding Division, Coconut Research Institute, Lunuwila, Sri Lanka

ABSTRACT

A recent advance in the field of molecular biology has been the development and application of molecular markers in plant and animal species for establishing genetic relationships and phylogeny, accurate estimation of amount and distribution of genetic diversity, and identification of marker-trait associations/linkages useful in marker-assisted selection. Among the many molecular marker techniques available, microsatellites (Simple Sequence Repeats or SSRs) are attractive, since they are reproducible, easy to perform through PCR, highly informative, codominant, and multiallelic in nature.

This chapter reviews the application of microsatellites in studying genetic diversity and relationships of coconut. Overall a very high level of genetic diversity has been found in coconut, more so in outbreeding Tall coconuts than in inbreeding Dwarf coconuts. Most genetic variation in Dwarfs was found between populations but was more or less equally distributed within and between populations in Talls. The number of alleles detected in Talls was nearly double the number of alleles detected in Dwarfs, indicating a loss of allelic richness in the latter. Genetic relationship studies revealed two major groups of genetically different Tall coconuts, one from Southeast Asia and the Pacific,

Address for Correspondence: Dr. Lalith Perera, Genetics and Plant Breeding Division, Coconut Research Institute, Lunuwila, Sri Lanka. Tel: 94 31 2257391, Fax: 94 31 2257391. E-mail: gpb@cri.lk, lalithperera2004@hotmail.com

the other from southern Asia and Africa. Dwarfs, on the other hand, formed a genetically tight single cluster and moreover fell in with the Southeast Asia and Pacific Tall cluster. This observation and loss of allelic richness in Dwarfs tend to suggest that Dwarfs originated from Talls from Southeast Asia and the Pacific, probably as a result of domestication.

Key Words: Coconut, microsatellites, SSR, genetic diversity, genetic relationships

INTRODUCTION

The coconut palm (*Cocos nucifera* L., Arecaceae) is the most widely cultivated/occurring tropical plantation crop. The only species of genus *Cocos*, it has a diploid genome with 32 chromosomes (2n = 2x = 32) and mainly comprises two morphotypes, the mostly outbreeding Talls and the mostly inbreeding Dwarfs. The Talls are late bearing with medium to large nuts and in general are hardy and thrive in various environmental conditions. Capable of cross-pollination, they form highly heterogeneous populations both naturally and in cultivation. The Dwarfs are less heterogeneous and an individual Dwarf palm can be logically considered a homozygous line. In the natural state, Dwarfs occur only very sparingly, are early bearing, and generally produce small bright colored nuts in large numbers.

Harries [15] proposed two main types of Tall coconuts based on morphology of the nut and nut component analysis. The one which evolved naturally and was disseminated by ocean currents he termed Niu kafa (the wild type); it produces large, long, angular, thick-husked and slow-germinating nuts with less free-water content. The second type purportedly evolved as a result of selection under cultivation for increased nut water content from the Niu kafa type and disseminated by human intervention, he termed Niu vai; it produces spherical nuts with an increased proportion of endosperm, reduced husk thickness, and early germinating nuts with resistance to disease. Harries [15] suggested further that introgression of these two types and further selection and dissemination by man gave the wide range of varieties and pantropical distribution of coconut seen today.

Conflicting theories regarding the origin and domestication of coconut were published in the last century [4, 5, 6, 7, 9, 12, 13, 15, 17, 34, 36]. The most likely region today for coconut domestication is considered to be Melanesia, on the coasts and islands between Southeast Asia and the western Pacific, approximately between New Guinea and Fiji [5, 16, 34].

Swaminathan and Nambiar [40] suggested that Dwarf coconuts might have originated as a result of inbreeding among Tall coconuts which show limited self-pollination. Purseglove [34] stated that Dwarfs are probably mutations of Tall coconuts. However Harries [15, 17] suggested that some characters of Dwarfs (i.e. early germination, precocity, nut shape, variously bright-colored fruit, proportion of husk, and for some dwarf forms high resistance to lethal yellowing disease) indicate domestication as the more likely cause of Dwarf origination.

MORPHOLOGICAL DIVERSITY OF COCONUT

Within both Tall and Dwarf groups, there are other numerous types of coconuts loosely categorized as varieties, forms, populations or ecotypes based on some morphological differences, breeding habit, and place of origin [12a, 20, 22, 23, 26]. To date some 900 accessions have been entered in the Coconut Genetic Resources Database (IPGRI/COGENT Version 4). However, many of the coconut names are either vernacular labels given by local people or labels arising from slight morphological differences. Some duplication is also possible as a result of the same variety or ecotype being differently named in different locales.

Within coconuts there is a great morphological diversity in nut size, shape, and color, weight, and proportion of husk, shell, and endosperm, volume of nut water, etc. [3]. Variation is also high in yield-related characters such as number of bunches, number of female flowers, number of nuts, etc. Today coconut is mainly evaluated according to morphological variations, which are highly dependent on the environment and therefore very little is known about the accurate genetic variation of coconut in terms of its magnitude and structure.

APPLICATION OF MOLECULAR MARKERS IN COCONUT

Molecular markers are independent of environment, reflecting variations at the DNA sequence level. They are also unlimited in number and can detect diversity at a very high level of resolution. Therefore, use of molecular markers for assessing the amount and distribution of genetic diversity in coconut is of paramount importance. Among the molecular marker systems currently in use, Restriction Fragment Length Polymorphism (RFLP) [19], Amplified Fragment Length Polymorphism (AFLP) [32, 41], Randomly Amplified Polymorhpic DNA (RAPD) [2, 8, 11], and Microsatellites or Simple Sequence Repeats (SSR) [8, 21, 27, 28, 29, 30, 31, 37, 41] are employed more often.

Microsatellite Markers

Of the scant reports available on application of molecular markers in coconut, microsatellite markers hold a prominent place. Microsatellites are tandem SSRs that occur abundantly and at random throughout most eukaryote genomes and are composed of monotonous motifs, consisting of repeats of one to six nucleotides in length. They occur in perfect tandem repetition, imperfect (interrupted) repetition or compound repetition. Visualization of the hypervariability of the SSR array is achieved by PCR. PCR primers flank the repeat region and allow amplification via the PCR. The resultant products are discriminated on the basis of length, using denaturing polyacrylamide gel electrophoresis. The relative abundance of microsatellites, their good genome coverage, ease of visualization and highly informative content facilitate their use as molecular markers. Their multiallelic and codominant nature (ability to detect heterozygotes from homozygotes) suggest that microsatellites may be the most powerful Mendelian marker available.

Although microsatellites are one of the favored candidate markers for population analysis, they are developmentally expensive since sequencing and isolation are required for each plant, though some examples of cross-amplification of microsatellites between closely related species have been published [42]. The high cost involved, notwithstanding, a substantial number of coconut microsatellite primers have been developed [28, 29, 37, 41]. A set of 39 coconut-specific microsatellite primer pairs was developed by Perera and colleagues [28] from a small insert genomic library enriched for CA repeats using genomic DNA from the coconut variety Sri Lankan Tall following the method described by White and Powell [43] (contact author for primer sequences) while a set of 38 polymorphic microsatellite primer pairs has been isolated by Rivera et al. [37] from a small insert genomic library enriched for several types of microsatellite repeats following the method described by Edwards and coworkers [10], using genomic DNA from the Philippine coconut variety Tagnanan Tall (primer sequences available in [41]).

AMOUNT AND DISTRIBUTION OF GENETIC DIVERSITY OF COCONUT

A high level of genetic diversity was observed by Perera et al. [28, 29, 31] using 12 pairs of highly polymorphic coconut-specific microsatellite primers in a collection of 130 coconut palms representing 51 Tall coconut varieties and 49 Dwarf coconut varieties sampled across the entire geographic range (Table 6.1). The total diversity values based on Nei's unbiased statistics [25] varied from 0.386 ± 0.035 to 0.837 ± 0.009 among 12 microsatellite primer pairs with a mean gene diversity of 0.647 ± 0.139. A total of 84 alleles were observed for Talls ranging from 3 alleles for primer pair CAC13 to 11 alleles for primer pair CAC56 with an average of 7 alleles per locus versus only 42 alleles for Dwarfs ranging from 2 for primer pairs CAC3, CAC13 and CAC68 to 6 for primer pair CAC56 with an average of 3.5 alleles per locus. The mean gene diversity of Talls was 0.703 ± 0.125 ranging from 0.499 ± 0.015 for primer pair CAC13 to 0.867 ± 0.008 for primer pair CAC56. The mean diversity index of Dwarfs, on the other hand, was much lower (0.374 ± 0.204) ranging from 0.072 ± 0.033 for primer pair CAC3 to 0.693 ± 0.024 for primer pair CAC56, comparable to the observed reduction in number of alleles detected.

A similar level of genetic diversity based on Nei's diversity index [24] was observed by Rivera et al. [37], in 20 coconut varieties, mainly from Southeast Asia and the Pacific, using 38 microsatellite primer pairs developed by them. They generated a total of 198 alleles with an average of 5.2 alleles per microsatellite locus and noted genetic diversity ranging between 0.141 for primer pair CNZ03 to 0.809 for primer pair CNZ43 among 38 primers. In a continuation of the previous work, they generated 64 alleles per locus with an average of 8 alleles per locus, using only highly polymorphic 8 primer pairs of the 38 pairs previously applied, in a large collection of germplasm from 40 coconut varieties representing the entire geographic range. Genetic diversity of coconut in their collection ranged from 0.579 for primer pair CN2A5 to 0.790 for pair CN11E10 [37], which is very much comparable with Perera et al. results.

Teulat and colleagues [41] studied 31 individuals of coconut comprising 14 coconut varieties across the geographic range, including 10 Tall coconut varieties [West African Tall (Mensah), Mozambique Tall, Malayan Tall, Baybay Tall, Sri Lankan Tall, Andaman Tall, Kar Kar Tall, Tonga Tall, Rennel Island Tall, Panama Tall (Monagre)] and 4 Dwarf coconut varieties [Malayan Yellow, Kiribati Green, Madang Brown, Niu Leka] using 37 microsatellite primers developed by Rivera et al. [37]. Once again a very high level of genetic diversity in coconut (based on Nei's diversity index [24]) was found, ranging from 0.47 ± 0.062 for CNZ03 to 0.9 ± 0.009 for CNZ11A10 among primer pairs with an average of 9.4 alleles per locus, confirming the results of the two earlier researches (Table 6.2). The number of alleles ranged between 2 and 16 per locus and produced a total of 339 alleles for 14 varieties.

Table 6.1: Number of alleles detected, diversity statistics and F_{ST} for 12 coconut microsatellite primers [28]

Locus	Number of alleles			Observed product size range (bp)	Gene index			F_{ST} (Tall/Dwarf)
	All	Tall	Dwarf		All (n = 130)	Tall (n = 75)	Dwarf (n = 55)	
CAC2	9	9	5	210 - 154	0.713 ± 0.024	0.830 ± 0.016	0.418 ± 0.016	0.184 ***
CAC3	5	5	2	187 - 203	0.386 ± 0.035	0.549 ± 0.036	0.072 ± 0.033	0.187 ***
CAC4	8	8	8	182 - 216	0.762 ± 0.015	0.792 ± 0.012	0.573 ± 0.02	0.154 ***
CAC6	9	9	5	150 - 168	0.704 ± 0.019	0.793 ± 0.016	0.476 ± 0.041	0.125 ***
CAC8	9	8	5	188 - 210	0.722 ± 0.018	0.74 ± 0.025	0.303 ± 0.055	0.381***
CAC10	5	5	3	195 - 205	0.468 ± 0.032	0.616 ± 0.028	0.156 ± 0.045	0.191***
CAC13	5	3	3	158 - 172	0.488 ± 0.013	0.499 ± 0.015	0.348 ± 0.176	0.285 ***
CAC52	7	7	7	148 - 162	0.568 ± 0.031	0.731 ± 0.02	0.124 ± 0.042	0.292 ***
CAC68	6	6	6	136 - 152	0.630 ± 0.014	0.534 ± 0.036	0.203 ± 0.047	0.549 ***
CAC65	7	7	4	154 - 180	0.765 ± 0.013	0.800 ± 0.011	0.575 ± 0.022	0.148 ***
CAC20	6	6	4	122 - 138	0.723 ± 0.012	0.684 ± 0.026	0.545 ± 0.043	0.243 ***
CAC56	11	11	6	144 - 168	0.837 ± 0.009	0.867 ± 0.008	0.693 ± 0.024	0.100 ***
Mean	7.4	7	3.5		0.647 ± 0.139	0.703 ± 0.125	0.374 ± 0.204	0.233***
Total	85	84	42		0.995 ± 0.001	0.999 ± 0.001	0.971 ± 0.006	

Table 6.2: Microsatellite loci, number of alleles per locus allele size range and gene diversity for the microsatellites tested on 2-3 samples of 14 coconut populations (31 individuals) (adopted from [14])

Name of the SSR locus	Number of alleles	Diversity value	Standard Error (D)
CNZ01	7	A	A
CNZ02	10	0.7133	0.0546
CNZ03	3	0.4688	0.0623
CNZ04	11	0.8033	0.0358
CNZ05	7	0.5765	0.0616
CNZ06	7	0.8033	0.0173
CNZ09	11	0.7820	0.0395
CNZ10	13	0.8096	0.0328
CNZ12	7	0.7237	0.0264
CNZ13	A	A	A
CNZ16	5	0.6597	0.0333
CNZ17	8	0.6108	0.0644
CNZ18	13	0.8923	0.0132
CNZ19	9	0.8137	0.0259
CNZ20	5	0.6660	0.0348
CNZ21	14	0.9037	0.0100
CNZ22	A	A	A
CNZ23	5	0.6805	0.0285
CNZ24	10	A	A
CNZ26	12	0.8642	0.0158
CNZ29	13	0.7773	0.0368
CNZ31	5	0.7560	0.0201
CNZ32	8	0.5375	0.0707
CNZ33	2	0.4870	0.0202
CNZ34	11	0.8424	0.0183
CNZ37	14	0.8845	0.0171
CNZ40	8	0.8189	0.0193
CNZ42	8	0.6873	0.0352
CNZ43	10	0.8028	0.0291
CNZ44	9	0.8215	0.0761
CNZ46	9	0.8304	0.0163
CNIH2	9	0.7680	0.0299
CNIC6	6	0.6993	0.0346
CN2A5	12	0.7674	0.0457
CNIG4	7	0.7854	0.0212
CNIIE6	16	0.7711	0.0511
CN2A4	13	0.8392	0.0239
CNIIE10	9	0.7305	0.0401
CNIIA10	12	0.9027	0.0090

A: Number of alleles and diversity values not estimated for various reasons.

An allele frequency analysis carried out by Perera et al. [28, 29] for each locus and each group of coconuts (Tall and Dwarf) in their samples (Table 6.3) showed 42 unique alleles for the Tall coconut group, which are not present in Dwarfs. A loss of allelic richness was therefore apparent in Dwarf palms sampled. Frequency distribution of alleles was greater in Talls compared to Dwarfs. At five of the 12 loci, the most frequent allele was the same for both Tall and Dwarf (e.g. CAC2: 254bp, CAC3: 197bp, CAC10: 201bp, CAC52: 160bp, and CAC65: 154bp). In contrast to locus CAC8 the most frequent allele in the Tall group was not found within the Dwarf genotypes sampled. However in some cases there were two equally common alleles in the dwarfs (e.g. CAC4: 212bp and 186bp, CAC65: 176bp and 154). The 210bp allele for Tall and 198bp allele for Dwarf in Sri Lanka appeared to be fixed at locus CAC8.

Perera et al. [28] reported heterozygous loci in both Talls and Dwarfs, but a higher frequency in Talls than in Dwarfs. Heterozygote loci in Dwarfs were as low as 2.5% vs. more than 30% observed in Talls. Almost all the Talls were heterozygous at at least one locus while only 13 Dwarf varieties out of 43 were heterozygous at one or more loci, the 30 other Dwarf varieties being completely homozygous for all 12 loci studied. Variety Niu Leka, known to exhibit cross-pollination [33], was heterozygous for six loci. Variety Banigan was heterozygous for five loci while varieties Malayan Green Dwarf and San Isidro were heterozygous for three. The other nine varieties—Tacunan Green Dwarf, Kapatagan Green Dwarf, Salak, Nok Khum, Malayan Yellow Dwarf, Sri Lanka Yellow Dwarf, and Cameroon Red Dwarf, Ghana Yellow Dwarf, and Brazilian Green Dwarf—were heterozygous for only one or two loci. Rivera et al. [37] also noted higher heterozygosity values for Talls than Dwarfs in their microsatellite study. Of the 38 microsatellite loci analyzed in a collection of 10 Dwarf varieties, only 5 Dwarf varieties were heterozygous at any one locus. Variety San Isidro was heterozygous for 11 loci while varieties Santo. Nino Green Dwarf, Kapatagan Green Dwarf, Tacunan Green Dwarf, and Banga Green Dwarf were heterozygous for 4 loci. Teulat et al. [41] studied four dwarf coconuts using 37 microsatellite primers and also found that Niu Leka and Malayan Yellow Dwarf were heterozygous for several loci and a single locus respectively, while Kiribati Green Dwarf and Madang Brown Dwarf were wholly homozygous for all 37 microsatellite loci. Teulat et al. [41] reported a higher heterozygosity in east African Tall coconuts than in west African.

Allelic distribution, estimates of genetic diversity, and associated population differentiation statistics (F_{ST}) for coconuts from Sri Lanka,

Table 6.3: Allele frequencies at 12 microsatellite loci in the worldwide collection of coconut genotypes studied. The most common allele at each locus is shown in bold [28]

Locus	Allele (bp)	Frequency		
		Tall	*Dwarf*	*Both*
CAC2	210	0.007	—	0.004
	220	0.007	—	0.004
	232	0.061	—	0.035
	234	0.142	—	0.083
	240	0.270	0.038	0.173
	246	0.061	0.094	0.075
	248	0.128	0.057	0.098
	252	0.034	0.057	0.043
	254	**0.291**	**0.755**	**0.484**
CAC3	187	0.027	—	0.015
	197	**0.623**	**0.963**	**0.768**
	199	0.226	0.037	0.146
	201	0.116	—	0.067
	203	0.007	—	0.004
CAC4	182	0.007	—	0.004
	186	**0.314**	**0.462**	**0.377**
	188	0.193	0.077	0.143
	200	0.036	—	0.020
	204	0.186	—	0.107
	208	0.193	—	0.111
	212	0.064	**0.462**	0.234
	216	0.007	—	0.004
CAC6	150	0.033	—	0.019
	152	0.155	—	0.088
	154	0.101	0.009	0.062
	156	0.020	—	0.012
	158	**0.304**	0.273	0.291
	160	0.277	**0.672**	**0.446**
	162	0.081	0.009	0.050
	164	0.014	0.036	0.023
	168	0.014	—	0.008
CAC8	188	0.050	0.020	0.040
	196	0.014	—	0.008
	198	0.130	**0.830**	**0.430**
	200	0.210	0.080	0.160
	202	—	0.020	0.008
	204	0.113	0.047	0.085
	206	0.010	—	0.010
	208	0.040	—	0.020
	210	**0.430**	—	0.250
CAC10	195	0.007	—	0.004

Contd.

Table 6.3 (Contd.)

Locus	Allele (bp)	Frequency		
		Tall	Dwarf	Both
	197	0.188	—	0.107
	201	**0.542**	**0.917**	**0.702**
	203	0.243	0.074	0.171
	205	0.021	0.009	0.016
CAC13	158	0.423	**0.833**	**0.594**
	162	**0.570**	0.167	0.402
	172	0.007	—	0.004
CAC52	148	0.097	—	0.056
	150	0.014	0.028	0.019
	154	0.021	—	0.012
	156	0.313	0.037	**0.194**
	158	0.069	—	0.039
	160	**0.389**	**0.935**	**0.623**
	162	0.097	—	0.056
CAC68	136	0.089	**0.887**	**0.425**
	140	0.007	—	0.004
	146	**0.630**	0.113	0.413
	148	0.007	—	0.004
	150	0.253	—	0.147
	152	0.014	—	0.008
CAC65	154	**0.264**	**0.500**	**0.363**
	156	0.223	0.083	0.164
	164	0.034	—	0.019
	168	0.142	—	0.082
	170	0.237	—	0.137
	176	0.095	0.417	0.234
	180	0.007	—	0.004
CAC20	122	0.007	—	0.004
	130	0.021	—	0.012
	132	**0.479**	0.204	**0.360**
	134	0.211	0.019	0.128
	136	0.197	0.148	0.176
	138	0.085	**0.629**	0.320
CAC56	144	0.169	—	0.097
	146	0.014	—	0.008
	150	0.007	0.009	0.008
	152	0.155	0.066	0.117
	154	0.007	—	0.004
	156	0.134	—	0.077
	158	0.121	0.189	0.149
	160	**0.197**	0.293	**0.238**
	164	0.070	0.009	0.044
	166	0.092	**0.434**	**0.238**
	168	0.035	—	0.020

Indonesia, and the Philippines are presented in Figure 6.1 [28]. The allelic distribution of Dwarfs highlights the unique position of these palms from the Philippines that exhibit as much variation as observed in the Tall group (0.442 ± 0.19 for Dwarf vs. 0.512 ± 0.22 for Talls). The lowest population differentiation statistic was observed between Philippine Tall and Philippine Dwarf compared to those of other countries (F_{ST} = 0.21 vs. 0.51 for Sri Lanka and 0.36 for Indonesia). Dwarf palms sampled from Sri Lanka appear to show a low level of allelic diversity with the highest population differentiation statistic for Sri Lankan Talls and Sri Lankan Dwarfs (F_{ST} = 51%). This was further evident with only 10 of 12 microsatellites detecting any polymorphism within Sri Lankan Dwarfs (Table 6.4). The diversity value for Indonesian Dwarfs (0.223 ± 0.23) is comparable to that of Sri Lankan Dwarfs (0.223 ± 0.29). Of the 12 microsatellites described, Indonesian Dwarfs showed diversity at seven, while Sri Lankan Dwarfs were polymorphic at only five loci. All eight unique alleles observed in Sri Lankan Dwarf coconuts were observed in Talls from Indonesia, the Philippines or Pacific Islands. Pacific Island Tall coconuts recorded the highest level of diversity with the highest number of alleles (51). Teulat et al. [41] reported that within and between populations variability was highest in the Tall varieties of the Pacific and of Southeast Asia. Therefore it is likely that higher levels of genetic diversity exist in the Pacific regions compared to the rest of the regions. As Perera et al. [28] had sampled only one individual each from only two Dwarf varieties from Pacific Islands, an accurate analysis of the status of dwarfs in the Pacific Islands is not possible. However, Meerow et al. [21], using microsatellites, analyzed a population of Niu Leka Dwarf introduced in Florida (popularly known there as Fiji Dwarf), found it to be genetically very diverse (0.436) compared to other dwarfs (i.e., 0.202 for Malayan dwarfs) and observed a large number of unique alleles. Tonga Tall and Fiji Tall likewise had unique alleles that were not observed in other South Pacific coconut populations.

The distribution of total genetic variation between Talls and Dwarfs for each primer pair was analyzed by Perera et al. [28] and is given in Table 6.1. Overall, there was a high level of population differentiation between Tall and Dwarf (23.3%). The F_{ST} statistics for Talls and Dwarfs for the Philippines, Indonesia and Sri Lanka were 0.21, 0.36 and 0.51 respectively (Fig. 6.1). A high level of population differentiation between Tall and Dwarf can be observed in Indonesian coconut and Sri Lankan coconut populations (36% and 51% respectively) vs. the low level of population differentiation observed in the Philippine coconuts (21%).

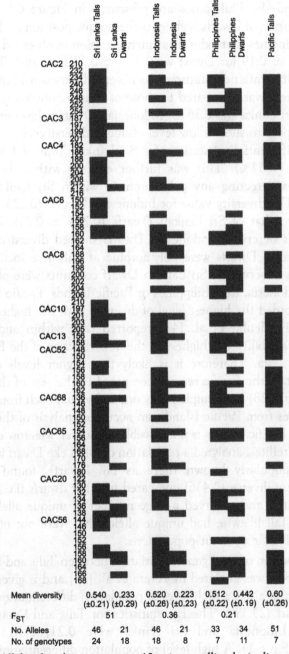

Fig. 6.1: Allelic distribution among 12 microsatellite loci, diversity index, F_{ST} among coconuts from Sri Lanka, Indonesia, Philippines, and Pacific Islands (Perera et al., 1999).

Table 6.4: Primer information, number of alleles detected, diversity statistics, and F_{ST} values for 12 coconut microsatellite primers [28]

Locus	Size Range (bp)	Number of alleles			Gene diversity index ± sd			F_{ST}
		All	Tall	Dwarf	All	Tall	Dwarf	
CAC2	210-254	6	6	1	0.64 ± 0.04	0.75 ± 0.04	—	0.49***
CAC3	187-203	4	4	1	0.61 ± 0.04	0.69 ± 0.03	—	0.52***
CAC4	182-216	6	4	2	0.77 ± 0.02	0.62 ± 0.04	0.42 ± 0.07	0.47***
CAC6	150-168	6	5	3	0.76 ± 0.02	0.69 ± 0.05	0.55 ± 0.05	0.29***
CAC8	188-210	2	1	1	0.50 ± 0.02	—	—	1.00***
CAC10	195-205	3	3	1	0.48 ± 0.05	0.63 ± 0.03	—	0.41***
CAC13	158-170	2	2	1	0.55 ± 0.02	0.33 ± 0.07	—	0.16***
CAC52	154-160	3	3	1	0.49 ± 0.03	0.54 ± 0.05	—	0.56***
CAC68	136-152	5	5	1	0.65 ± 0.02	0.52 ± 0.06	—	0.71***
CAC65	154-176	6	5	3	0.77 ± 0.02	0.59 ± 0.05	0.50 ± 0.05	0.44***
CAC20	132-138	5	4	3	0.63 ± 0.05	0.44 ± 0.08	0.68 ± 0.02	0.23***
CAC56	144-166	6	4	3	0.81 ± 0.01	0.68 ± 0.03	0.64 ± 0.03	0.31***
CAC72	131-139	4	4	2	0.62 ± 0.03	0.59 ± 0.06	0.30 ± 0.08	0.41***
CAC84	158-162	3	3	2	0.60 ± 0.04	0.66 ± 0.02	0.10 ± 0.06	0.46***
CAC71	182-186	3	3	2	0.55 ± 0.04	0.52 ± 0.03	0.29 ± 0.08	0.36***
CAC39	140-154	5	5	2	0.70 ± 0.02	0.55 ± 0.07	0.45 ± 0.05	0.43***
CAC21	154-156	2	2	2	0.50 ± 0.01	0.51 ± 0.01	0.51 ± 0.02	-0.02(ns)
CAC23	170-182	3	3	1	0.59 ± 0.02	0.41 ± 0.07	—	0.75***
Mean		4.1	3.6	1.7	0.62 ± 0.10	0.54 ± 0.17	0.25 ± 0.26	0.49***
Total		74	66	32	0.98 ± 0.00	0.99 ± 0.00	0.94 ± 0.01	0.49***

DNA profiling of almost all the coconut varieties and forms indigenous to Sri Lanka were published by Perera and his group [28, 30] using 18 highly polymorphic microsatellite primers (Table 6.4). A high level of genetic diversity was reported with overall genetic diversity exceeding 0.6. Most variation was detected in the outcrossing Tall group of coconuts rather than the inbreeding Dwarf group. There are eight forms of Tall coconut within the main Sri Lankan Tall group and four forms of coconut within the Dwarf coconuts in Sri Lanka [20]. In addition, according to Liyanage's [20] classification of Sri Lankan coconuts, there is an intermediate group termed Aurantiaca in Sri Lanka, which resembles Talls in many agronomic features but also resembles Dwarf coconuts by its inbreeding nature, early and seasonal bearing habit, and production of large number of small nuts. Most variation was observed between rather than within these forms in inbreeding Dwarfs and Aurantiaca. In contrast, the outbreeding Tall forms exhibited as much variation within as between forms. These observations have important implications, revealing that for Tall coconut forms emphasis should be given to both number of individuals and number of populations whereas for inbreeding Dwarfs and intermediates the number of populations is the important factor. The Aurantiaca group of coconut in Sri Lanka, although considered to be intermediate between the Tall and Dwarf varieties based on agronomic features, is closer to the Dwarf group than to the Tall. Microsatellites also showed a closer relationship between Dwarf varieties and variety Bodiri, which was grouped under Tall forms of coconut based on phenotype, also confirmed by Dassanayake ([8] and unpubl. data) in another independent study on Sri Lankan coconut germplasm. Dassanayake ([8] and unpubl. data) analyzed 43 coconut accessions (19 distinct coconut varieties comprising 7 Tall varieties, 9 Dwarf varieties and 3 intermediate varieties, 7 San-Ramon Tall-like populations and 17 Tall populations within the Sri Lankan Ordinary Tall variety) using 17 pairs of coconut-specific microsatellite primers developed by both Perera [28] and Rivera [37] and detected 82 alleles with an average of 4.8 alleles per locus ranging from 2 to 10 alleles per locus. Genetic distances among the accessions ranged from 0.13 to 1.0 with an average of 0.63, indicating that all coconut germplasm accessions within Sri Lankan Ordinary Tall coconuts shared a similar genome with a narrow genetic base within them.

Meerow et al. [21] reported genetic variation within coconut germplasm collections at two locations in southern Florida, representing eight cultivars comprising 110 palms using 15 microsatellite loci from Perera and colleagues [28]. They detected a total of 67 alleles, with eight alleles as the highest number (CAC56) among 15 loci and gene diversity ranging from 0.778 for CAC56 to 0.223 for CAC84 with a mean of 0.574 ± 0.140. They found Niu Leka (Fiji Dwarf), Malayan Dwarf, Green Niño, and Red Spicata coconut varieties to be distinct clusters in a neighbor-joining tree using modified Rogers distance. Closer examination of the neighbor-joining tree showed that Niu Leka, Green Niño, and Red Spicata individual clusters joined to form a main cluster before joining with the Malayan Dwarf cluster. The highest gene diversity was found in the Tall cultivars in the Floridan collection (0.583 cumulatively and 0.679 for Panama Tall) compared to the Dwarf collection, although genetic identity was highest among the Tall varieties. The lower genetic diversity in Floridan coconut compared to the diversity of Tall varieties from other regions may be due to a narrow genetic base in Floridan coconut as a result of the bottleneck effect. The lowest gene diversity was found in the Malayan Dwarf (0.202) indicating that genetic identity among Malayan Dwarfs was high compared to the other dwarfs studied. Niu Leka (Fiji Dwarf) was the most genetically diverse (0.436) after Talls and had the largest number of unique alleles. Meerow et al. [21] stated a very important line of reasoning, namely that although observed heterozygosity levels are mostly commensurate with the gene diversity values, both Niu Leka and Green Niño were more allelically diverse than heterozygous. Among the three Malayan Dwarf forms, the Malayan Red Dwarf was genetically distinct from the Green and Yellow forms of Malayan Dwarfs, and which as a matter of fact could not be differentiated with the 15 microsatellite loci used.

Perera et al. [30] also analyzed 33 Sri Lankan coconut populations within the Sri Lankan Ordinary Tall coconut variety using 8 microsatellite primers (Table 6.5). They found a very high overall genetic diversity index (0.9994) for Sri Lankan Ordinary Tall coconuts but a very low level of population variation (even less than 1%) accounting for about 99% of the variation within, thus indicating that there were no distinct populations within the native range of Sri Lankan Ordinary Tall coconuts in Sri Lanka. High level within population variation is a common observation in outbreeding crops, for example in *Theobroma cacao* [1, 38] and tropical tree species *Calycophyllum spruceanum* [39] in which more than 90% of the variation observed was within population. It appears likely that there is a common history and a narrow genetic base

Table 6.5: Primer information, gene diversity, number of alleles detected, and population differentiation statistics (F_{ST}) for eight coconut microsatellite primers [28]

Locus	Repeat	Primer (5' – 3')	Size range (bp)	Gene diversity	Number of alleles	F_{ST}	$*F_{ST}$
CAC2	$(CA)_{12}(AG)_{14}$	AGCTTTTTCATTGCTGGAAT CCCCTCCAATACATTTTTCC	210-254	0.8461 ± 0.00053	9	0.051 ***	0.031 ***
CAC3	$(CA)_{13}$	GGCTCTCCAGCAGAGGCTTAC GGGACACCAGAAAAAGCC	187-203	0.6882 ± 0.0132	5	0.078***	0.013 ns
CAC4	$(CA)_{19}(AG)_{17}$	CCCTATGCATCAAAACAAG CTCAGTGTCCGTCTTTGTCC	182-216	0.7486 ± 0.0105	9	0.049 ***	−0.005 ns
CAC6	$(AG)_{14}(CA)_{9}$	TGTACATGTTTTTGCCCAA CGATGTAGCTACCTTCCC	150-168	0.7315 ± 0.0095	7	0.061 ***	0.029 ***
CAC8	$(AG)_{10}(CA)_{9}$	ATCACCCAATACAAGGACA AATTCTATGGTCCACCCACA	188-210	0.6768 ± 0.0129	9	0.024 ***	0.001 ***
CAC10	$(TA)_{6}CATA(CA)_{11}(TA)_{8}$	GGAACCTCTTTTGGGTCATT GATGGAAGGTGGTAATGCTG	195-205	0.4260 ± 0.0221	4	0.014 ***	0.014 ***
CAC13	$(CA)_{9}(TA)_{5}A(TA)_{4}(CA)_{6}$	GGGTTTTTAGATCTTCGGC CTCAACAATCTGAAGCATCG	158-172	0.5167 ± 0.0082	3	0.056 ***	0.0176 ns
CAC56	$(CA)_{14}$	ATTCTTTGGCTTAAAACATG TGATTTTACAGTTACAAGTTTGG	144-168	0.8221 ± 0.0021	10	0.052***	0.028 ***
			Mean:	0.6820 ± 0.1446	7		
			Overall	0.9994 ± 0.0001	56	0.054 ***	0.015 ***

*F_{ST} values calculated without introduced populations

for commercially grown Sri Lankan Ordinary Tall coconuts. These results emphasize the importance of prior knowledge about the amount and distribution of genetic diversity for a successful collection and conservation program.

Perera et al. [27] employed chloroplast microsatellite primers in detecting variability in coconuts of different geographic origin. They earlier reported the use of chloroplast microsatellite primers developed for monocotyledons such as rice, barley, and wheat to amplify coconut chloroplast DNA [27, 28] but later used three coconut-specific chloroplast microsatellite primers developed by sequencing and isolating mononucleotide microsatellite motifs observed in the PCR amplified chloroplast regions of intergenic spacers between *trn*T, and *trn*L, *trn*S and *trn*T and *trn*C and *trn*D, on a set of 130 coconut genotypes. The cross-amplification of several chloroplast microsatellite primers from other plants was able in coconut to generate PCR products in the expected size but no length polymorphisms were observed. The three coconut-specific chloroplast microsatellite primers too have shown no length polymorphisms indicating that coconut from all the tested geographic regions shares a common origin of chloroplast.

GENETIC RELATIONSHIPS AMONG COCONUT VARIETIES/POPULATIONS

Perera et al. [28, 31] constructed a phenetic tree diagram showing the genetic relationships among the same 130 individual coconuts that comprised 51 Tall varieties and 49 Dwarf varieties, using the same data set generated for the previously mentioned genetic diversity study. The genetic distances were calculated based on the proportion of shared microsatellite alleles using the program MICROSAT (Version 1.5; Eric Minch, Stanford University, USA, the NEIGHBOR and DRAWGRAM options in the PHYLIP software package (V3.57c; Joe Felsenstein, University of Washington, USA), and the unweighted pair-group method using arithmetic averages (UPGMA). These 130 genotypes were sampled from the Philippines, Indonesia, Malaysia, Thailand, Vietnam, Sri Lanka, Andaman Island, Comoro Island, Mozambique, Ghana, Ivory Coast, Cameroon, Panama, Brazil, Papua New Guinea, and from some Pacific Islands (Vanuatu, Rennell, Tahiti, Fiji, Tonga, Rotuma and Solomon Island), representing southern Asia, Southeast Asia, eastern and western Africa, the Pacific, and America.

This phenetic tree (Fig. 6.2) divided all Tall coconuts into two main groups. The first group comprised three subgroups; two main Tall subgroups and one main Dwarf group. All the Tall varieties/populations in the two Tall subgroups were from Southeast Asia, except variety Kalok from Thailand. All Tall varieties sampled from Southeast Asia, the Pacific and Papua New Guinea were intermingled among them and between the two Tall subgroups and showed no distinct geographic clusters within them. Coconut varieties/populations from the west coast of Panama (varieties Panama Aquadulce and Panama Monagre) and one east African coconut, Comoro Tall, were also tightly grouped within these Tall subgroups.

Interestingly, the other subgroup consisted of only dwarf varieties, except a single Tall variety, Niu Damu, with a greater genetic distance, and comprised 42 of the 49 Dwarf varieties analyzed. This dwarf subgroup also included Sri Lanka varieties King Coconut and Rathran Thembili, which were earlier classified as intermediate varieties between Tall and Dwarf coconuts. However, Dwarf varieties Banigan, Banga, Kinabalan, Aromatic Green Dwarf, Santo Nino, Malayan Green Dwarf, and Niu Leka Dwarf from Southeast Asia and the Pacific Islands were grouped with Talls from the same regions in either of the two Tall subgroups. Contrarily, Rivera et al. [37] who studied the genetic relationships among ten Philippine Dwarf varieties and ten Philippine and Pacific Tall varieties using data generated from 38 microsatellites and analyzed to construct a dendrogram by UPGMA cluster analysis using a similarity matrix based on Simple matching coefficient (i.e., allele sharing) found that all the samples from Dwarf varieties including varieties Banga, Santo Nino, and Kinabalan grouped separately from those from the Tall cultivars. Teulat et al. [41] who analyzed four Dwarf varieties along with 10 other Tall varieties likewise observed that varieties Malayan Yellow, Kiribati Green, Madang Brown, although collected from different regions, clustered together. Similar to Perera et al. [31], Teulat et al. [41] also observed variety Niu Leka clustered with Tall coconuts and moreover with Tonga Tall from the South Pacific. Further, Perera et al. [28, 29] found that Dwarf varieties Raja Brown, Tebing Tinggi, Madang Brown, Malayan Red, and Sri Lanka Red had identical microsatellite allelic profiles and therefore they could not be discriminated among themselves using their 12 pairs of microsatellite primers, highlighting the fact that these five Dwarfs were genetically very similar. Dwarf varieties Nias Yellow, Nias Green, Thailand Red, Nali Kei, and Sri Lanka Yellow among themselves and Dwarf varieties Tam Quan Yellow, Nam Horm, and Thung Keld likewise among themselves had identical microsatellite banding patterns, making them genetically indistinguishable [28, 29].

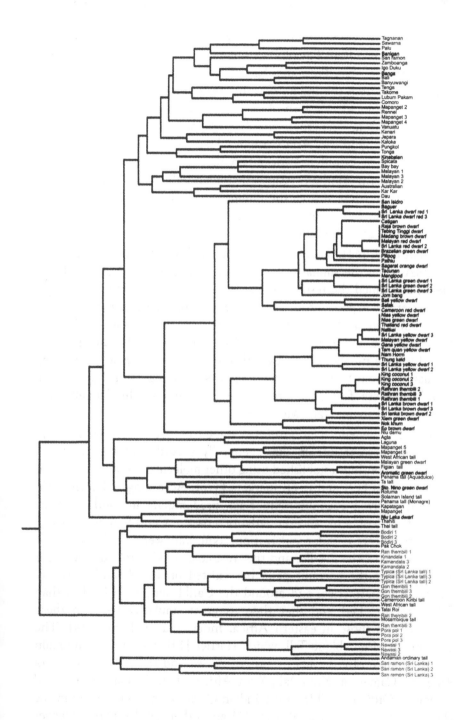

Tagnanan
Sawarna
Palu
Banigan
San Ramon
Zamboanga
Igo Duku
Banga
Bali
Banyuwangi
Tenga
Takome
Lubum Pakam
Comoro
Mapanget 2
Rennel
Mapanget 3
Mapanget 4
Vanuatu
Kenari
Jepara
Kaloke
Pungkol
Tonga
Kinabalan
Spicata
Bay bay
Malayan 1
Malayan 3
Malayan 2
Australian
Kar Kar
Dau
San Isidro
Baguer
Sri Lanka dwarf red 1
Sri Lanka dwarf red 3
Catigan
Raja brown dwarf
Tebing Tinggi dwarf
Madang brown dwarf
Malayan red dwarf
Sri Lanka red dwarf 2
Brazzilian green dwarf
Pilipog
Pahhu
Bagerat orange dwarf
Tacunan
Mangipod
Sri Lanka green dwarf 1
Sri Lanka green dwarf 2
Sri Lanka green dwarf 3
Jom bang
Bali yellow dwarf
Salak
Cameroon red dwarf
Nias yellow dwarf
Nias green dwarf
Thailand red dwarf
Nathai
Sri Lanka yellow dwarf 3
Malayan yellow dwarf
Gana yellow dwarf
Tam quan yellow dwarf
Niam Horm
Thung keld
Sri Lanka yellow dwarf 1
Sri Lanka yellow dwarf 2
King coconut 1
King coconut 2
King coconut 3
Rathran thembili 2
Rathran thembili 3
Rathran thembili 1
Sri Lanka brown dwarf 1
Sri Lanka brown dwarf 3
Sri lanka brown dwarf 2
Xiem green dwarf
Nok khum
Ep brown dwarf
Niu damu
Agta
Laguna
Mapanget 5
Mapanget 6
West African tall
Malayan green dwarf
Fijian tall
Aromatic green dwarf
Panama tall (Aquadulce)
Ta tall
Sto. Nino green dwarf
Rotuma
Solaman Island tall
Panama tall (Monagre)
Kapatagan
Mapanget
Niu Leka dwarf
Thahiti
Thai tall
Bodiri 1
Bodiri 2
Bodiri 3
Pak Chok
Ran thembili 1
Kmandala 1
Kamandala 3
Kamandala 2
Typica (Sri Lanka tall) 1
Typica (Sri Lanka tall) 3
Typica (Sri Lanka tall) 2
Gon thembili 3
Gon thembili 1
Cameroon Kiribi tall
West African tall
Talai Roi
Ran thembili 2
Mosambique tall
Ran thembili 3
Pora poi 1
Pora poi 2
Pora poi 3
Nawasi 1
Nawasi 3
Nawasi 2
Andaman ordinary tall
San ramon (Sri Lanka) 1
San ramon (Sri Lanka) 2
San ramon (Sri Lanka) 3

The second main group consisted of Talls from southern Asia (varieties Bodiri, Ran Thembili, Kamandala, Sri Lanka Ordinary Talls, Gon Thembili, Pora Pol, and Nawasi from Sri Lanka and Andaman Ordinary Tall from the Indian Ocean), eastern Africa (Mozambique Tall from Mozambique), western Africa (Cameroon Kribi Tall and West African Tall from Cameroon and Ivory Coast respectively), and some varieties from Southeast Asia (varieties Thai Tall, Pak Chok, and Thalai Roi from Thailand only). Variety San Ramon sampled from Sri Lanka grouped with the southern Asia/Africa main Tall group, while the same named variety sampled from the Philippines grouped with the rest of the Southeast Asia Tall coconuts. None of the dwarf coconuts grouped with the second main Tall group.

Teulat et al. [41], based on the results of cluster analysis according to UPGMA using the similarity matrix based on the proportion of shared alleles, confirmed the results of Perera et al. [28, 31], stating that there was generally a good grouping of the Tall populations into different geographic regions. West African Tall population grouped with Sri Lanka Tall population and populations from Mozambique and Andaman Islands shared a large number of alleles with them to form a single cluster. Tall populations from Southeast Asia, Papua New Guinea and the Solomon Islands likewise formed a broad cluster including the Panama Tall populations, Panama Tall (Monagre).

Results of genetic relationships obtained by various research groups based on microsatellite markers generally accord with the results of genetic relationships established using other molecular techniques such as RFLP markers [19] and ISTR markers [37a]. According to Harries [15] naturally evolved coconuts, Niu Kafa, predominate in southern Asia, Africa, the Caribbean, and the Atlantic coast of Central America while coconuts selected under cultivation, Niu Vai, were predominant in Southeast Asia, the Pacific, and the west coast of Central America. It is accepted that the coconut palm has existed on the Atlantic coast of Africa, South America, and around the Caribbean region for less than 500 years [6, 33a] and that there is great similarity between these coconuts and those in eastern Africa, India, and Sri Lanka [14]. The grouping of Mozambique Tall, which Harries [14] suggested as the main original source of coconuts to eastern Africa and the Atlantic coast of America, Cameroon Kribi Tall, West African Tall, Sri Lankan Tall and Andaman Ordinary Tall from the Indian Ocean, in a single cluster in the phenetic tree of Perera et al. [28, 31] and others [41] supports Harries theory of natural and human-assisted coconut germplasm dissemination. Grouping of variety San Ramon sampled from Sri Lanka with mainly Sri Lanka Tall cluster suggests that this variety may have been introgressed with Sri Lanka Tall coconuts. Interestingly, variety Comoro Tall, from eastern Africa falls in with the Southeast Asia/Pacific main Tall group and may have originated from Southeast Asia coconuts, the Niu Vai type. Lebrun et al. [19] noted that variety Comoro Tall took an intermediate position between Southeast Asian and Indian Ocean populations based on RFLP markers. Thailand Talls, varieties Thai Tall, Pak Chok, Talai Roi, are grouped with mainly the southern Asia/African Tall group while variety Kalok is grouped with the Southeast Asia/Pacific Tall group. Results of a detailed study on varieties of coconuts in Thailand by Harries et al. [18] based on fruit component analysis of coconut has suggested that both Niu Kafa and Niu Vai types of coconuts are available in Thailand with the former predominant on the Indian Ocean coast. Therefore the variety Kalok is a large fruited coconut type that resembles other large fruited forms such as varieties San Ramon and Tagnanan in the Philippines, variety Bali Tall in Indonesia, variety Rennell Tall in Solomon Islands and variety Panama Tall on the Pacific coast of America. Variety Malayan Tall probably shares common ancestry with variety Kalok and variety Pak Chok could be compared with other Talls that have Niu Kafa type characteristics, for example the Talls from Sri Lanka, India, Mozambique, western Africa, and the Caribbean. Rattanapruk [35] has also stated that variety Pak Chok growing along the Indian Ocean coast resembles coconut from Sri Lanka. Interestingly, variety Kalok is grouped by Perera et al. [28, 31] with varieties San Ramon Tall, Tagnanan Tall, Bali Tall, Rennell Tall, and Malayan Tall in the Southeast Asia/Pacific group of coconuts. Similarly variety Pak Chok is grouped with the varieties from Sri Lanka, Andaman Islands, and Mozambique. However one population of the variety West African Tall obtained from the Malaysian Coconut Gene Bank is grouped with the coconut varieties of the Far East while the same name variety sampled from Ivory Coast is grouped with southern Asia/African coconuts. This is invariance with overall observations and may be due to misidentification of the variety from Malaysia.

The grouping of Panama Talls (varieties Panama Manarge and Panama Aguadulce) with Southeast Asian and Pacific Talls is in agreement with Whitehead's [44] proposition of the eastward movement of coconut from Southeast Asia to the Pacific region and subsequently from there to the

Pacific coast of America. Both types of Panama Talls were from the Pacific coast of Panama. These results are largely in agreement with the results from ISTR (Inverse Sequence-Tagged Repeats) analysis [37a] which grouped Panama Talls with Polynesian varieties/populations.

The grouping of all Dwarfs from different geographic regions in a single cluster within the main Niu Vai type of coconuts and loss of allelic richness observed in Dwarfs suggest that Dwarfs have a common origin and evolved from the Niu Vai type of coconuts in the Southeast Asia/ Pacific region, were domesticated there and later introduced to the other regions. The results of Teulat et al. [41] strongly support a common origin of Dwarfs.

Acknowledgments

The author thanks the Staff of the Genetics and Plant Breeding Division of the Coconut Research Institute of Sri Lanka and the Staff of the Cell and Molecular Genetics Department of the Scottish Crop Research Institute of Dundee, Scotland, UK for their involvement in his coconut molecular work. Special thanks to all the other molecular biologists, who have published their interesting results on coconut biodiversity for the benefit of coconut breeders. The Commonwealth Scholarship Commission of the UK through British Council financed the author's research. Miss Mithila Jayasundara of CRISL kindly proofread the manuscript.

REFERENCES

[1] Allen JB. Geographical variation and population biology in wild *Theobroma cacao*. PhD thesis, University of Edinburgh, UK, 1998.

[2] Ashburner GR, Thompson WK, Halloran GM. RAPD analysis of South Pacific coconut palm populations. Crop Sci 1997b; 37: 992-997.

[3] Ashburner GR, Thompson WK, Halloran GM, Foale MA. Fruit component analyses of South Pacific coconut palm populations. Genet Res Crop Evol 1997a; 44: 327-355.

[4] Burkill IH. A Dictionary of the economic products of the Malay Peninsula. Kuala Lumpur: Ministry of Agriculture and Co-operatives, 1996.

[5] Child R. Coconuts. London: Longmans, UK, 1964.

[6] Child R. Coconuts. London: Longmans, UK, 1974 (2nd ed.).

[7] Cook OF. History of the coconut palm in America. Contribution to US National Herbarium, 1910; 14: 271-342. Cited in Bruman HJ. Some observation on the early history of the coconut culture in the New World. Acta Americana 1994; 2: 220-243.

[8]   Dassanayake N. Molecular characterization of coconut germplasm. PhD thesis, University of Colombo, Sri Lanka, 2002.

[9]   Dennis JV, Gunn CR. Case against trans-Pacific dispersal of the coconut by ocean currents. Econ Bot 1971; 25: 407-413.

[10] Edwards KJE, Barker JHA, Daly A, Jones C, Carp A. Microsatellite libraries enriched for several microsatellite sequences in plants. BioTech 1996; 20: 758-760.

[11] Everard JMDT. Use of molecular markers for breeding of the coconut palm (*Cocos nucifera* L.). MSc thesis, University of New England, Armidale, Australia, 1996.

[12] Fremond Y, Ziller R, de Nuce de Lamothe M. Le cocotier. Paris: Maisonneuve and Larose, 1966.

[12a] Gangolly SR, Satyabalan K, Pandalai KM. Varieties of coconuts. The Indian Coconut Journal 1957; X: 3-28

[13] Guppy HB. Observation of a naturalist in the Pacific between 1896 and 1899. London, UK: Macmillan, 1906.

[14] Harries HC. The Cape Verde region (1499 to 1549); the key to coconut culture in the Western hemisphere? Turrialba 1977; 27: 227-231.

[15] Harries HC. The evolution, dissemination, and classification of *Cocos nucifera* L. Bot Rev 1978; 44: 205-317.

[16] Harries HC. Malesian origin for a domestic *Cocos nucifera* L. In: Baas P, Kalkman K, Geesink R, eds. The plant diversity of Malesia. Kluwer, Dordrecht, Netherlands: Kluwer academic Publications, 1990: 351-357.

[17] Harries HC. Coconut (*Cocos nucifera* L.). In: Smartt J, Simmonds NW, eds. Evaluation of crop plants London, UK: Longman, 1995.

[18] Harries HC, Thirakul A, Rattanapruk V. The coconut genetic resources of Thailand. Proc. Seminar Coconut and Cocoa. Chumphon, Department of Agriculture. Bangkok, Thailand, 19-23 July 1982.

[19] Lebrun P, N'Cho NP, Seguin M, Grivet L. Baudouin L. Genetic diversity in coconut (*Cocos nucifera* L.) revealed by restriction fragment length polymorphism (RFLP) markers. Euphytica 1998; 101: 103-108.

[20] Liyanage DV. Varieties and forms of coconut palms grown in Ceylon. Ceylon Coconut Quart. 1958; 9: 1-10.

[21] Menon KPV, Pandalai KM. The coconut. A monograph. Indian Central Coconut Committee, 1958: 86-102.

[22] Meerow AW, Wisser RJ, Brown JS, Kuhn DN, Schnell RJ, Broschat TK. Analysis of genetic diversity and population structure within Florida coconut (*Cocos nucifera* L.) germplasm using microsatellite DNA, with special emphasis on the Fiji Dwarf cultivar. Theor Appl Genet 2003; 106: 715-726.

[23] Narayana GV, John CM. Varieties and forms of coconut. Madras Agric J 1949; 36: 349-366.

[24] Nei M. Analysis of gene diversity in subdivided populations. Proc Natl Acad Sci USA 1973; 70: 3321-3323.

[25] Nei M. Molecular evolutionary genetics. New York, NY: Columbia University Press, 1987.

[26] Ohler JG. Coconut, tree of life, plant production and protection. Rome: FAO, Paper 57, 1984.

[27] Perera L, Russell JR, Provan J, McNicol JW, Powell W. Evaluating genetic relationships

between indigenous coconut (*Cocos nucifera* L.) accessions from Sri Lanka by means of AFLP profiling. Theor Appl Genet 1998; 96: 545-550.

[28] Perera L, Russell JR, Provan J, Powell W. Identification and characterisation of microsatellites in coconut (*Cocos nucifera* L.) and the analysis of coconut populations in Sri Lanka. Mol Ecol 1999; 8: 344-346.

[29] Perera L, Russell J R, Provan J, Powell W. Use of microsatellite DNA markers to investigate the level of genetic diversity and population genetic structure of coconut (*Cocos nucifera* L.). Genome 2000; 43:15-21.

[30] Perera L, Russell JR, Provan J, Powell W. Levels and distribution of genetic diversity of coconut (*Cocos nucifera* L., var. *Typica* form *typica*) from Sri Lanka assessed by microsatellite markers. Euphytica 2001; 122:381-389.

[31] Perera L. Chloroplast DNA variation of coconut is opposite to its nuclear DNA variation. CORD 2002; XVIII (2): 56-73.

[32] Perera L, Russell J R, Provan J, Powell W. Studying genetic relationships among coconut varieties/populations using microsatellite markers. Euphytica 2003; 132:121-128.

[33] Powell T. On various Samoan plants and their vernacular names. J Bot (London), 1868; (6): 278-285, 242-347, 255-370. Cited in Harries HC. The evolution dissemination and classification of *Cocos nucifera* L. Bot Rev 1978; 44: 205-317.

[33a] Purseglove JW. Tropical Crops: Monocotyledones. London, UK: Longman, 1972.

[34] Purseglove JW. Tropical Crops: Monocotyledones. London, UK: Longman, 1985: 440-450 (5th ed.).

[35] Rattanapruk V. Thailand. In: *Yearly progress report on coconut breeding*. Rome: FAO, 1970: 31-33.

[36] Ridley HN. The dispersal of plants throughout the world. London, UK: Reeves, 1930.

[37] Rivera R, Edwards KJ, Barker JHA, et al. Isolation and characterization of polymorphic microsatellites in *Cocos nucifera* L. Genome 1999; 42: 1-8.

[37a] Rohde W, Kullaya A, Rodrigues J. Genome analysis of Cocos nucifera L. by PCR amplification of spacer sequences separating a subset of Copia-like 16RI repetitive elements. Journal of Genetics and Breeding 1995; 49: 179-186.

[38] Russell JR, Hosein F, Johnson E, Waugh R, Powell W. Genetic differentiation of cocoa (*Theobroma cacao* L.) populations revealed by RAPD analysis. Mol Ecol 1993; 2: 89-97.

[39] Russell JR, Weber JC, Booth A, et al. Genetic variation of *Calycophyllum spruceanum* in the Peruvian Amazon Basin, revealed by amplified fragment length polymorphism (AFLP) analysis. Mol Ecol 1999; 8: 199-204.

[40] Swaminathan MS, Nambiar MC. Cytology and origin of the dwarf palm. Nature 1961; 192: 85-86.

[41] Teulat B, Aldam C, Trehin R, et al. An analysis of genetic diversity in coconut (*Cococs nucifera*) populations from across the geographic range using sequence–tagged microsatellites (SSRs) and AFLPs. Theor Appl Genet 2000; 100: 764-771.

[42] White G, Powell W. Cross-species amplification of SSR loci in the Meliaceae family. Mol Ecol 1997a; 6: 1195-1197.

[43] White G, Powell W. Isolation and characterisation of microsatellite loci in *Swietenia humilis* (Meliaceae): an endangered tropical hardwood species. Mol Ecol 1997b; 6: 851-860.

[44] Whitehead RA. Coconut. In: Simmonds NW, ed. Evolution of crop plants. London, UK: Longman, 1976: 221-225.

Genome Diversity of Triticale (× Triticosecale Witt.) Using the Chromosome Heterochromatin Marker

STANISLAWA MARIA ROGALSKA
University of Szczecin, Chair of Cell Biology, Szczecin, Poland

ABSTRACT

Many forms of triticale are characterized by a varying number of rye chromosomes identified on the basis of the presence of telomeric heterochromatin. In spring forms, high yielding and of good quality, the number of rye chromosome pairs most often was 6 – 7. Substitutions of D chromosome were confirmed in place of defined rye chromosomes; the most frequent was 2R chromosome. Numerous modifications of rye chromosomes were observed with regard to the presence of telomeric heterochromatin as well as structural mutations. Winter triticale was characterized by a more stable karyotype as in most of the varieties and lines studied a complete rye genome with clear bands of telomeric heterochromatin was present.

Thus it may be concluded that depending on the type of triticale, spring or winter, there exists a genotype factor controlling the biological activity of chromatin, which is reflected in the presence or absence of heterochromatin blocks or whole chromosomes of rye. Observation of the course of meiosis in triticale proved that rye chromosomes with clear heterochromatin bands most often participate in disturbances, especially in primary forms. However, in high-yielding forms meiosis is regular and univalent rye chromosomes were observed only in sporadic cells.

Key Words: triticale, heterochromatin, chromosome number, meiosis

Address for correspondence: University of Szczecin; 71-415 Szczecin; Poland. Phone/fax +48(91)4441535, e-mail: strog@univ.szczecin.pl

INTRODUCTION

Evolution of new, wholly man-made species of triticale has taken place in definite stages. Different genotypes of triticale were formed through hybridization between *Triticum* species and *Secale* species and through crossing of the hybrids obtained with each other. Many different forms of triticale had arisen by the mid-1970s. They included primary octoploids and hexaploids, secondary hexaploids, recombinant hexaploids, and wheat lines with added and substituted rye chromosomes. From the end of the 1970s much attention was also paid to tetraploid [2n = 4x = 28] triticale. The primary forms of octoploid and hexaploid triticale as well as the early secondary hexaploids had many disadvantages: low fertility, long, weak and lodging straw, shriveled grains, irregular meiosis and numerous aneuploid plants in the offspring [8,16,20,25,31,32,42, 46,49,50,69,72,74,75,80]. These unfavorable features were attributed to the influence of the rye genome.

Rye chromosomes and wheat chromosomes were primarily identified on the basis of variability in the chromosome morphology or the presence of marker genes located on a particular chromosome. In the 1970s a method of differential staining of heterochromatin in plant chromosomes with Giemsa reagent was elaborated [42,63,79] and since then chromosomes of parental genomes in triticale have been identified by staining them with this reagent. Giemsa dye gives an intense red color to the constitutive heterochromatin present in rye *(Secale cereale* L.) and wheat *(Triticum aestivum* L.) chromosomes. In rye chromosomes it is located mainly in one or both telomeres and forms a characteristic pattern of bands along the chromosome arms [9,73]. Wheat chromosomes are characterized by the presence of heterochromatin in the centromere region; some chromosomes have slight intercalary bands and small telomere bands [11, 66]. According to Jouve and Soler [21], differential staining with Giemsa reagent—so-called C-banding—is a cytogenetic marker useful in physical mapping for observing introgression of alien chromatin and studying the behavior of chromosomes in meiosis; indicates a moderate polymorphism of sites and intensity of heterochromatin bands and they are (the heterochromatin bands) inherited in a co-dominant way.

HETEROCHROMATIN AS A MARKER OF THE PRESENCE OF RYE CHROMOSOMES IN GENOME CONSTITUTION OF WHEAT-RYE HYBRIDS

As there are differences in the distribution of heterochromatin in the chromosomes of triticale parental genomes, it has become a cytogenetic marker which facilitates identification of the presence of particular rye and wheat chromosomes in hybrids, observation of heterochromatin banding pattern, structural changes of chromosomes, and behavior of rye and wheat chromosomes in meiosis. Thanks to cytogenetic analyses of the various forms of triticale and assessment of their value for agriculture, it was established that the most promising and cytogenetically stable are secondary hexaploids—spring and winter triticale. It can be said that evolution of this grain has entered its next stage in which these forms (and in particular their genome composition) are very interesting.

Chromosome Composition of Spring Forms of Triticale

Staining chromosomes with Giemsa has acquired practical value in triticale breeding programs. In numerous populations of wheat-rye hybrids, in the early generations different numbers of rye chromosomes were found. Therefore, it was concluded that some of them were substituted with wheat chromosomes from D genome [17]. A famous variety Armadillo 70HN458 exhibited a pair of D chromosomes substituting rye chromosomes [15, 16]. Other forms of spring hexaploid triticale contained 7 to 5 pairs of rye chromosomes [9]. In 1975, chromosome constitution was studied in 50 forms of triticale bred in CIMMYT [43]. Rye chromosomes were identified in the following lines: Vaca, Cisne-Snoopy, DR-IRA, Beagle, Armadillo 105, Armadillo PM 13, Armadillo Pm 1524, Beaver, Bronco, Bruin, Rosner, MII – Arm "S", Cinnamon, Camel, Camel-Pato, and in other lines marked with respective numbers. A complete rye genome—seven pairs—was found in Vaca, Cisne-Snoopy, DR-IRA, and Beagle. The remaining lines contained a varying number of chromosomes, from six pairs to one pair. Moreover, in DR-IRA line variability occurred in the presence of a big telomeric band on one of the rye chromosomes (C-R) [44]. In the 30 spring lines studied, hexaploid triticale from CIMMYT, obtained by cross-combination of hexaploid triticale with hexaploid wheat, one to seven pairs of rye chromosomes were found [52]. Interestingly, all the studied lines represented a morphological type of wheat and gave no indication that they contained rye chromosomes. Without analysis by means of the Giemsa method, one would have thought these lines were wheat revertants, devoid of rye chromosomes. In another 19 lines of triticale from CIMMYT [53], four pairs of rye chromosomes were present in nine lines, six pairs in one, and seven pairs in one. In two triticale lines with the highest yield, five and four pairs of rye chromosomes were present;

neither of the lines contained 2R and 5R pairs. In the somatic cells of spring forms from Malyszyn (Poland), obtained from crossing Hungarian and Canadian triticales, all pairs of rye chromosomes were present. In 28 substitution lines of triticale [72] no 2R chromosome was detected but it was impossible to find a link between its absence and the appearance of improved features, more useful for farming. Triticale lines analyzed by Gupta and Virk [14], among others Beagle, Mapache, Rahum, and 10 other lines contained 7 pairs of rye chromosomes each. The remaining triticales—Coorong, DTS357, Rosner, Armadillo—lacked the following pairs respectively: 5R, 4R, 2R, and 2R. Heterochromatin was observed in wheat chromosomes and rye chromosomes in 31 triticale lines [22], in 7 lines from the INIA collection (Spain), namely Armadillo, Beagle, Delphine 205 , Manijero, Rosner, Shepherds, Welsh, in six Spanish varieties, and in a series of lines from working collections. In 11 of the lines studied, including Beagle, Cachirulo, and other Spanish varieties, seven pairs of rye chromosomes were found. In Armadillo, Rosner, Shepherds, Welsh and other Spanish lines it was confirmed that 2R chromosome was substituted with 2D chromosome. The absence of 2R chromosome in Rosner and in Armadillo does not seem to be a marker of positive features from the viewpoint of agriculture because in other forms of triticale , e.g. high-yielding Siskiyou, Groquick, Venus, DR-IRA, 6A530 and 6TA876 , complete rye genomes were present [1]. Similarly, in the five analyzed yielding varieties of spring triticale—Sunland, Florida 201, Florico, Topo 121, and Jago—a complete rye genome was present [60].

Analyses of the number of chromosomes in numerous spring forms of triticale proved that specific substitutions of D/R chromosomes occurred very frequently. Most frequently substituted was chromosome 2R, followed by 5R, 4R, and 3R. Nevertheless, most of the forms studied had a complete rye genome that was particularly common in the high-yielding varieties.

Rye Chromosome Modifications in Spring Forms of Triticale

According to Tarkowski and Stefanowska [75], the term "chromosome modification" denotes changes in the morphology of rye chromosomes in the presence of wheat chromosomes in a hybrid wheat-rye cell [75]. Gustafson et al. [19], however, defined the term as changes in quantity and presence of telomeric heterochromatin in rye chromosomes and that is the meaning adopted here.

Considerable polymorphism of heterochromatin in telomeres of rye chromosomes was found in many of the lines and varieties of triticale studied [51,60,66,67]. In some lines all chromosomes had clear, large telomere and intercalary bands, apart from 1R chromosomes in which an intercalary band in the proximity of NOR was only slightly delineated. In other lines, 6R chromosomes had telomere bands of different size and variable intercalary bands on the long arm [53].

Among the other analyzed triticales, the 3R chromosome was characterized by the greatest variability in the presence and size of telomeric heterochromatin [60]. Variability in the pattern of C- bands in rye chromosomes and wheat chromosomes of triticale has been observed by many authors [11, 22, 34, 53, 66]. In some of the studied triticales and in initial forms of wheat the distribution of C-bands in wheat chromosomes was similar and the only chromosome without bands was 1A [66]. Sometimes new heterochromatin bands appeared in wheat chromosomes although they did not exist in the parental wheat forms [22]. Most often they appeared in the 2A chromosome and in all seven chromosomes from the B genome. Measurements of lengths of the heterochromatin sections in wheat chromosomes of triticale indicated that they were longer in hybrids than in the parental wheat forms. Rye chromosomes showed the same range of variability for the presence of heterochromatin bands in wheat-rye hybrids as in parental forms of rye. The most varying number of intercalary bands on the longer arm was observed in 6R chromosome in rye cultivars and in triticale. Loss of intercalary heterochromatin was most often observed in 4R chromosomes of triticale. The absence of heterochromatin in telomeres was most often visible in 3R, 1R, and 6RL chromosomes. It seems that a characteristic feature of the spring forms of triticale is the existence of rye chromosomes modified according to the presence of telomeric heterochromatin and substitution of R chromosomes by the chromosomes from the D genome of hexaploid wheat [9,13,16,17,43,44,45,51].

A visible loss of telomeric heterochromatin in rye chromosomes was also observed in wheat lines with added rye chromosomes [9, 68]. Some observations [41] showed that translocations between wheat and rye chromosomes could contribute to the loss of rye heterochromatin. Such translocations may also bring the new features into triticale, evidence of which is translocation between 1R and 1D wheat chromosomes; the locus of high baking quality of flour—1D5+10—was introduced into 1R [38, 61]. Lukaszewski and Gustafson [36] analyzed numerous generations of the populations of wheat-rye hybrids and among 787 plants found 195 wheat/rye translocations and 64 rye/rye translocations. Moreover, modified rye chromosomes were found in 15 plants. These chromosomes had either deletions or amplifications of telomeric heterochromatin.

Studies were carried out on 24 lines of triticale from Australia and Canada [1] to ascertain the existence of modified rye chromosomes by means of staining C-bands and hybridization *in-situ*. Results showed that the quantity of rye heterochromatin was relatively constant when a heterogeneous probe was used. However, using a specific probe variability was observed which could reflect adjustment of the rye genome to wheat genomes. Some of the studied forms of triticale , e.g. Tayalla, T1245, and T1006 had identical rye chromosomes. On the other hand, in Mapache 5R and 4R/7R chromosomes differed from such chromosomes in the three aforesaid forms. This indicated structural changes of those chromosomes after they had been introduced into the wheat cell.

Many changes of the translocation type between homoeologous chromosomes of wheat and rye and many chromosome substitutions were found in hybrids between primary octoploid triticale and different lines of hexaploid triticale [3,76]. A specific influence of heterochromatin in rye chromosomes on regeneration capabilities of immature embryos cultured *in vitro* was studied in two sister lines of triticale—Rosner and DR-IRA—and in rye lines [7]. The triticale lines differed in the presence of heterochromatin in 6R or 7R chromosomes and had better regeneration capabilities than the compared rye lines.

Studies based on Giemsa C-banding method applied mainly to monitor rye heterochromatin in triticale suggest that its deficiency is a significant condition of genetic stability of spring triticale lines. Loss of telomeric heterochromatin is considered the sign of morphological "adjustment" of rye chromosomes to wheat chromosomes and concomitantly signals favorable genetic collaboration. Neverthless, many stable and high-yielding forms of triticale contained a complete rye genome with distinct heterochromatin bands.

Chromosome Composition of Winter Forms of Hexaploid Triticale and Modifications of Rye Chromosomes

Consequent to numerous studies of cytogenetics and fertility it was concluded that one of the properties which makes the secondary hexaploid forms of triticale superior is substitution of rye chromosomes by genome D chromosomes and the loss of telomeric heterochromatin from particular chromosomes of rye. Subsequent studies of winter forms proved that although the foregoing was a common phenomenon among spring forms, it might not necessarily apply to winter forms (Table 7.1). From cross combinations between winter octoploid lines with hexaploids, with wheat and di- and tetraploid rye, plant material of differentiated chromosome and genetic composition was obtained. Cytogenetic studies of such material made it possible to select the most harmonious chromosome combination of the particular types of plants in defined environmental conditions. Breeding of winter triticale forms sought their adjustment to unfavorable soil and climatic conditions. In 13 lines of triticale under analysis the presence of a complete rye genome without modified chromosomes was confirmed. The studied lines were high yielding and produced plump grains [53]. Of another 90 lines of winter triticale studied, in 76 a complete rye genome without visible rye chromosome modifications was found in generation F5. In lines treated with a mutagen, of 200 plants selected for plump grains, only in four was a deletion of heterochromatin on the long arm of 7R/4R chromosome observed [35]. Another eight lines advanced in breeding, high-yielding with plump grains, were analyzed for the presence of rye chromosomes and their modification. Analyses showed that all the lines contained a complete rye genome whose chromosomes had clear telomeric heterochromatin bands. Moreover, other reports on European winter triticale (high-yielding with plump grains) showed that they contained a complete set of rye chromosomes with characteristic telomere bands [81]. A few winter forms of triticale, in particular octoploids, revealed the presence of modified rye chromosomes. Most frequently modified chromosomes were 3R, 4R, 5R, 6R, and 7R in which a significant increase or decrease in quantity of heterochromatin in telomere sections was found [57]. A pattern of C-bands on rye chromosomes of primary octoploid and hexaploid triticale was compared with the parental varieties of rye. This comparison revealed a change in size of intercalary and centromere heterochromatin bands in 1R, 2R, and 3R chromosomes in the triticale karyotype. It was found that wheat-rye hybridization modified some of the heterochromatin areas of genome R [67]. The analyzed parental form of rye was characterized by high polymorphism in size of heterochromatin bands; the forms of triticale inherited this variant of polymorphism in which the size of telomere bands was average [5].

Table 7.1: Examples of triticale forms with a different number of rye chromosome pairs

Triticale form	Number of chromosome pair of R genome	Missing of certain chromosome of R genome	Reference
Spring types	6	2 R	[15]
Armadillo 70HN458	7	—	
6A530,6A391			
Kangaroo-UM940'S',	6	2R	[9]
Cinnamon	6	2R	
Camel, Armadillo (Inia-rye)	5	2R, 4/7R	
Vaca, Beagle, Cisne-Snoopy, DR-IRA	7		[43]
Beaver, MIIArm"S", Bronco, Bruin, Rosner	6	2R	
Cinnamon, Camel, Camel-Pato	5	2R, 4R	
Beagle, Mapache, Rahum	7		[14]
Coorong	6	5R	
DTS357	6	4R	
Rosner	6	2R	
Armadillo	6	2R	
Beagle, Cachirulo and 9 lines of Spanish collection	7		[22]
Shepherd, Welsh	6	2R	
Siskiyou, Groquick, Venus, DR-IRA, 6A530, 6AT876	7		[1]
Sunland, Florida 201, Florico, Topo121, Jago	7		[60]
Winter types			
Lasko, Grado, Bolero Dagro	7		[59]
Lasko, Presto, Prego, Vero, SZD1740, SZD1834, SZD3801, MAH1590	7		[56]
76 advanced lines (Polish breeding program)	7		[35]
8 advanced lines (Guelph breeding program)	7		[81]

Numerous analyses of the number of chromosomes in winter triticale did not confirm the substitution of chromosomes from genome D/R. In the leading Polish varieties and strains of winter triticale—Lasko, Presto, Prego, Vero, and SZD1740, 1834, 3801, and MAH 1590—a complete rye genome was present with a characteristic pattern of C-bands on the chromosomes. Nevertheless, variability was observed in the size of telomeric bands as well as modifications of 1R and 6R chromosomes [56].

Crossing the winter forms of hexaploid triticale with substitution lines in order to introduce defined substitutions showed that chromosomes of genome A were more frequently substituted than chromosomes of genome B, whereas genome R was complete [3]. Other material was used to check whether there were D/R substitutions. Identification of substitutions proved difficult and in many lines only modified chromosomes 4R and 7R were observed [37]. Results of numerous analyses of chromosome composition of winter forms of triticale have shown its cytogenetic and karyotype stability with regard to C-bands, in contrast to spring forms.

TETRAPLOID TRITICALE (2N = 4X = 28)

Theoretically, tetraploid triticale can be obtained from crossing diploid wheat (*Triticum monococcum* L.) with diploid rye (*Secale cereale* L.). However, in practice no yielding hybrids were obtained in this manner [4,26,27,70]. Crossing hexaploid triticale with diploid rye and wheat-rye hybrids with rye as well as octoploid triticales with tetraploid with a selection of 28-chromosome plants in the offspring proved to be more effective [6,26,27,28] . Studies of the obtained forms of tetraploid triticale revealed that some of them had mixed genomes containing chromosomes of A, B, and D genomes, apart from genome R [10,18,29, 30,39]. In a recent report on obtaining a stable form of secondary, tetraploid triticale of AARR constitution, 14 rye chromosomes showed distinct bands of telomeric heterochromatin [71]. Studies on this form of triticale indicate that it may become a valuable grain for cultivation in the near future.

CYTOGENETIC EFFECT OF HETEROCHROMATIN

One of the problems in the evolution of triticale was its cytogenetic instability, reflected in disturbances in the process of meiosis and aneuploid offspring. Staining heterochromatin in chromosomes with Giemsa reagent facilitated their identification in the process of meiosis and made it possible to observe changes in heterochromatin areas of the chromosomes. Many analyses proved that the main reason for irregular meiosis in triticale is the presence of telomeric heterochromatin in rye chromosomes. Its absence in particular chromosomes made the course of this process more regular. Therefore, it was believed that elimination of telomeric heterochromatin from rye chromosomes was an extremely favorable phenomenon in triticale [8,44,62,77,78]. It was stated that rye chromosomes without telomeric heterochromatin pair better and the loss of heterochromatin is a prerequisite for genetic stability of such lines [23,24,62,64,65]. In the case of 2RS/2BL translocation, heterochromatin on the short arm of 2R had no significant influence on the pairing of chromosomes translocated with the whole chromosome 2B [41].

In three different triticales, advanced in breeding, only rye univalents were found in metaphase I, whereas in cross-combinations of those triticales with different species of rye [Secale cereale and S. montanum] univalent chromosomes of wheat were also observed. It was concluded that in advanced forms with less disturbed meiosis, irregularities were limited only to rye chromosomes [32,33]. In Rosner lines chromosomes 6R with or without telomeric heterochromatin on the short arm were observed. In plants without telomeric heterochromatin 6R chromosome pairing was more efficient that in those containing heterochromatin. Telomeric heterochromatin on the short arm of this chromosome influenced the pairing of all chromosomes, not only the 6R pair [62]. Similar observations were made by Merker [44] in another triticale with rye chromosomes without telomeric heterochromatin. They contained one-third fewer less univalents in metaphase I than those plants in which it was present. It was concluded that this was the main reason for the absence of chromosome pairing and cytogenetic instability of triticale.

By means of the Giemsa technique in some wheat-rye hybrids wheat and rye chromosome pairing was observed, which—as the authors emphasized—could be the route of transfer of genetic variability from wheat to rye [48].

The course of meiosis in PMC was observed in one octoploid and three hexaploid forms of triticale. Rye chromosomes with clear heterochromatin bands were univalents in hexaploids, whereas wheat chromosomes and rye chromosomes were univalents in octoploid triticale. Delayed chromosomes in anaphases I and II were also either only rye chromosomes in hexaploids or rye and wheat chromosomes in octoploids [54].

Studies on the impact of presence or absence of telomeric heterochromatin in 1R chromosomes on their pairing showed that in diploid rye it is insignificant. However, in wheat-rye hybrids the presence of telomeric heterochromatin reduced the number of chiasmata in the bivalent whereas in the heterozygotic condition with regard to C bands, it strongly reduced pairing in the short arms of 1R chromosomes [47]. A significant influence of quantity of telomeric heterochromatin on number of pollen mother cells with disturbed meiosis was detected in other winter and spring forms of triticale [55]. The quantity of heterochromatin could exert this influence through decreasing the number of chiasmata in rye bivalents and reducing the degree of pairing of rye chromosomes. However, as experiments proved, heterochromatin is not the only factor which causes disturbances in meiosis. It turned out that many types of meiotic disturbances depend on a hybrid's genotype, in particular its genomic condition [12, 58].

Consequent to numerous observations, many theories were formulated as to the influence of heterochromatin on the course of meiosis, especially on pairing of rye chromosomes. It was believed that wheat chromosomes could collaborate with telomeric heterochromatin and reduce the degree of pairing of homologue rye chromosomes. Another possibility was the impact of two phenomena, namely late replicating telomeric heterochromatin and a shortened meiotic cycle in triticale which reduced the degree of chromosome pairing. If telomeres are the points of chromosome pairing and if pairing cannot start before the end of replication, these two phenomena can delay initiation of chromosome pairing and cause their asynaptic behavior. According to Merker [44], heterochromatin disturbs the formation of chiasmata and their terminalization as well as chromosome pairing. Thus there is no pairing of the desynaptic rather than asynaptic type. Observations of meiosis in KMP of many winter high-yielding forms of triticale whose rye chromosomes had clear telomere bands revealed a differentiated degree of regularity of the course of this process, depending on the form studied. It may be concluded that specific coexistence of the genes of parental species plays the most important role.

Knowing the biological properties of heterochromatin, one may suspect that its presence may be a reflection of a particular genetic activity influencing many processes, among others a meiotic process. Given the biological characteristics of heterochromatin this seems to be true since heterochromatin occupies peripheral areas in the nucleus and is attached to the nuclear membrane. It replicates late and does not participate in genetic recombination [40]. Moreover, it shows a tendency to aggregate and can maintain bivalents together with an equal quantity of heterochromatin or form the so-called chromocenters in the interphase nucleus. Despite the fact that many biological properties of heterochromatin have been recognized, its precise role in the genome remains a mystery. Heterochromatin can participate in regulating the activity of genes, being a repressive factor. Recent studies make it possible to assume that heterochromatin can be involved in the so-called " genome transcriptability" which means that genes located close to heterochromatin can be "silenced" [40]. Given these properties it may be assumed that in spring and winter forms different specific genes are active. They are associated with the geographic area and environmental conditions in which these forms grow.

[1] Appels R, Gustafson JP, May CE. Structural variation in the heterochromatin of rye chromosomes in triticales. Theor Appl Genet 1982; 63: 235–244.

[2] Apolinarska B. Stabilization of ploidy and fertility level of tetraploid triticale obtained from four cross combinations. Genet Polon 1993; 34: 121-131.

[3] Apolinarska B. Addition, substitutions and translocations of chromosomes in hybrids triticale 8x x triticale 6x substitution lines. Bull Inst Hodowli Aklim Rosl 1995; 195/196: 65-71.

[4] Apolinarska B. Secalotriticum 4x-production, chromosome constitutions and fertility. Cer Res Commun 2001; 1-2: 61-68.

[5] Badaeva ED, Badaev NS, Bolsheva NL, Zelenin AV. Chromosome alterations in the karyotype of triticale in comparison with the parental forms. 1. Heterochromatin regions of R genome chromosomes. Theor Appl Genet 1986; 72: 518-523.

[6] Baum M, Lelley T. A new method to produce 4x triticales and their application in studying the development of a new polyploid plant. Pl Breed 1988; 100: 260-267.

[7] Bebeli P, Karp A, Kaltsikes PJ. Plant regeneration and somaclonal variation from cultured immature embryos of sister lines of rye and triticale differing in their content of heterochromatin. Theor Appl Genet 1988; 75: 929-936.

[8] Bennett M. Meiotic, gametophytic, and early endosperm development in triticale. In: Triticale, eds. MacIntyre and M. Campbell, 1973; pp. 137-148. IDRC, Ottawa.

[9] Darvey NL, Gustafson JP. Identification of rye chromosomes in wheat-rye addition lines and triticale by heterochromatin bands. Crop Sci. 1975; 15: 239-243.

[10] Dubovets NI, Badaev NS, Bolsheva NL, et al. Regularities of karyotype formation in tetraploid triticale. Cer Res Commun 1989; 17: 253-257.

[11] Endo TR, Gill BS. Somatic karyotype, heterochromatin distribution, and nature of chromosome differentiation in common wheat *Triticum aestivum* L. emThell. Chromosoma 1984; 89: 361-369.

[12] Galindo C, Jouve N. C-banding in meiosis. An approach to the study of genome interactions in Triticale. Genome 1989; 32: 1074-1078.

[13] Gupta PK. Role of loss of rye heterochromatin in improvement of hexaploid triticale. Nucleus 1984; 27: 100-107.

[14] Gupta PK, Virk GS. Identification of rye chromosomes in hexaploid triticale (X *Triticosecale* Wittmack). In: Manna GK and Sinha U, eds. Perspectives in cytology and genetics 1984; 4: 535-537.

[15] Gustafson JP, Zillinsky FJ. Identification of D-genome chromosomes from hexaploid wheat in a 42-chromosome triticale. Proc 4th Intl Wheat Genet Symp, Colmbia, Mo. 1973: 225-231.

[16] Gustafson JP, Qualset CO. Genetics and breeding of 42-chromosome Triticale I. Evidence for substitutional polyploidy in secondary Triticale populations. Crop Sci 1974; 14: 248-251.

[17] Gustafson PJ, Bennett MD. Preferential selection for wheat-rye substitution in 42-chromosome triticale. Crop Sci 1976; 16: 688-693.

[18] Gustafson JP, Krolow KD. A tentative identification of chromosomes present in tetraploid triticale based on heterochromatin banding patterns. Can J Genet Cytol 1978; 20: 199-204.

[19] Gustafson JP, Lukaszewski AJ, Bennett MD. Somatic deletion and redistribution of telomeric heterochromatin in the genus *Secale* and in Triticale. Chromosoma (Berl). 1983; 88: 293-298.

[20] Jenkins BC. History of the development of some presently promising hexaploid triticales. Wheat Inf Serv 1969; 28: 18-20.

[21] Jouve N, Soler C. Triticale genomic and chromosomes' history. In: Guedes-Pinto H. Darvey N, Carnide VP, eds. Triticale today and tomorrow. Kluwer Academic Publ, Dordrecht, netherlands: 1996: 91-118.

[22] Jouve N,Galindo C, Mesta M, et al. Changes in triticale chromosome heterochromatin visualized by C-banding. Genome 1989; 32: 735-742.

[23] Kaltsikes PJ, Roupakias DG. Heterochromatin and chromosome pairing in triticale. Crop Sci Soc Amer Ann Meet, Houston, TX 1976; Abstr 54.

[24] Kaltsikes PJ,Gustafson JP, Lukaszewski AJ. Chromosome engineering in triticale. CanJGenet Cytol 1984; 26: 105-110.

[25] Kiss A. Neue richtung in der Triticale-zuechtung. Z Pflanzenzuecht 1966; 55: 309-329.

[26] Krolow KD. 4x-triticale, production and use in Triticale breeding. In: Proc 4th Intl Wheat Genet Symp, Columbia, Mo. 1973; 237-243.

[27] Krolow KD. Research work with 4x-Triticale in Germany (Berlin). Proc Intl Symp 1973, El Batan, Mexico IDRC-024e, 1974; 51-60.

[28] Krolow KD. Selection of 4x-triticale from the cross 6x-triticale x 2x-rye. In: Proc Intl Symp 1973 Leningrad. 1975; 114-122.

[29] Krolow KD, Lukaszewski AJ. Tetraploid triticale—a tool in hexaploid triticale breeding. In: Horn, et al. eds. Genetic manipulation in plant breeding. Berlin: Walter de Gruyter, 1986; 105-118.

[30] Krolow KD, Lukaszewski AJ Gustafson JP. Preliminary results on the incorporation of D and E-genome chromosomes into 4x-triticale. In: EUCARPIA meeting on Triticale, Clermont-Ferrand, France. 1985: 289-295.

[31] Larter EN, Tsuchiya T, Evans L. Breeding and cytology of triticale. In: Proc 3rd Intl Wheat Genet Symp. Aust Acad Sci, Canberra, 1968: 213-221.

[32] Lelley T. Desynapsis as a possible source of univalents in metaphase I of Tritcale. Z Pflanzenzuecht 1974; 73: 249-258.

[33] Lelley T. Identification of univalents and rod bivalents in Triticale with Giemsa. Z Pflanzenzuecht 1975; 75: 252-256.

[34] Lelley T, Josifek K, Kaltsikes PJ. Polymorphism in the Giemsa C-banding pattern of rye chromosomes. CanJGenet Cytol 1978; 20: 307-312.

[35] Lukaszewski AJ, Apolinarska B. The chromosome constitution of hexaploid winter triticale. CanJGenet Cytol 1981; 23: 81-85.

[36] Lukaszewski AJ Gustafson PJ. Translocation and modification of chromosomes in triticale x wheat hybrids. Theor Appl Genet 1983; 64: 239-248.

[37] Lukaszewski AJ, Apolinarska B. Recognition of modified chromosomes and a chromosome substitution in winter triticale. Cer Res Commun 1983; 11: 285-286.

[38] Lukaszewski AJ, Curtis CA. Transfer of the *Glu-D1* gene from chromosome 1D of breadwheat to chromosome 1R in triticale. Pl Breed 1992; 109: 203-210.

[39] Lukaszewski AJ, Apolinarska B, Gustafson JP, Krolow KB. Chromosome constitution of tetraploid triticale. Z Pflanzenzuecht 1984; 93: 222-236.

[40] Mattei MG, Luciani J. Heterochromatin, from chromosome to protein. ECA News Lett 2003; 11: 3-13.

[41] May CE Appels R. The inheritance of rye chromosomes in early generations of triticale x wheat hybrids. CanJGenet Cytol 1982; 24: 285-291.

[42] Merker A. A Giemsa technique for rapid identification of chromosomes in Triticale. Hereditas 1973; 75: 280-282.

[43] Merker A. Chromosome composition of hexaploid triticale. Hereditas 1975; 80: 41-52

[44] Merker A. The cytogenetic effect of heterochromatin in hexaploid triticale. Hereditas 1976; 83: 215-222

[45] Merker A, Rogalska SM. The breeding behavior of a double disomic wheat-rye substitution line. Cer Res Comm 1984; 1-2.

[46] Muntzing A. Experiences from work with octoploid and hexaploid rye-wheat Triticale. Biol Zentralbl 1972; 91: 69-80.

[47] NaranjoT, Lacadena JR. Interaction between wheat chromosomes and rye telomeric heterochromatin on meiotic pairing of chromosome pair 1R of rye in wheat-rye derivatives. Chromosoma (Berl.) 1980; 81: 249-261.

[48] Naranjo T, Lacadena JR, Giraldez &. Interaction between wheat and rye genomes on homologous and homoeologous pairing. Z Pflanzenzuecht 1979; 82: 289-305.

[49] Orlova IN. Causes of mosaicism of the sporogenous tissue with respect to chromosome numbers in microsporocytes of hexaploid Triticale. Genetika 1976; 12: 7-13.

[50] Pieritz WJ. Elimination von Chromosomen in amphidiploiden Weizen-Roggen-Bastarden (*Triticale*). Z Pflanzenzuecht 1970; 64: 90-109.

[51] Pilch J. Analysis of the rye chromosome constitution and amount of telomeric heterochromatin in widely and narrowly adapted hexaploid triticales. Theor Appl Genet 1981; 60: 145-149.

[52] Rogalska SM. Identification of rye chromosomes in hexaploid Triticale. Genet Polon 1977; 18/4: 317-324.

[53] Rogalska SM. The chromosome constitution of plants from selected lines of secondary hexaploid triticale. Hodowla Rosl Aklim Nasienn 1978; 22: 293-301.

[54] Rogalska SM. Identification of univalents in PMC meiosis of different triticale forms by Giemsa technique. Genet Polon 1979; 20: 333-340.

[55] Rogalska SM. The influence of the amount of telomeric heterochromatin on the course of meiosis in PMCs of hexaploid triticale. Genet Polon 1983; 24/3: 227-237.

[56] Rogalska SM. C-banding characteristics of rye chromosomes in winter hexaploid triticale crossed with maize (*Zea mays*). In: Guedes-Pinto H, Darvey N, Carnides VP, eds. Triticale today and tomorrow. Kluwer Acad. Publ. Dordrecht, netherlands: 1996: 203-205.

[57] Rogalska SM, Krupka P. Differentiation in the size of telomeric heterochromatin bands in rye chromosomes of octoploid Triticale. Genet Polon 1982; 23/1-2: 9-15.

[58] Rogalska SM, Prusak L. Genotype-dependent course of meiosis in the pollen mother cells of wheat-rye. In: Tamas Lelley, ed. Current topics in plant cytogenetics related to plant improvement. Wien: WUV-Universitatsverlag, 1998: 162-167.

[59] Rogalska SM, Cybulska-Augustyniak J, Kulawinek A. Aneuploidy in Polish cultivars of winter triticale (X *Triticosecale* Witt.). Genet Polon 1991; 32/1-2: 11-16.

[60] Rogalska SM, Barnett RD, Soffes-Blount A.R. Rye chromosome C-banding and agronomic relationships in hexaploid triticale (X *Triticosecale* Wittmack). Poznan Soc Advancement Arts and Sciences. Sec.Forestry and Agric. Sci 1999; 87: 235-239.

[61] Rogovsky PM, Guidet FLY, Langridge P, Shepherd KW, Koebner RMD. Isolation and characterization of wheat-rye recombinants involving chromosome arm 1DS of wheat Theor Appl Genet 1991; 82: 537-544.

[62] Roupakias DG, Kaltsikes PJ. The effect of telomeric heterochromatin on chromosome pairing of hexaploid triticale. Can J Genet Cytol 1977; 19: 543-548.

[63] Sarma NP, Natarajan AT. Identification of heterochromatic regions in the chromosomes of rye. Hereditas 1973; 74: 233-238.

[64] Schlegel R. Investigations on chromosomal pairing behaviour of hexaploid and octoploid triticale by Giemsa technique. Arch Zuchtungsforsch 1978; 8: 1-11.

[65] Schlegel R, Huelgenhof E. Heterochromatin alterations in chromosomes of hexaploid triticale and their effect on meiotic pairing behaviour. Proc. 3rd EUCARPIA meeting on Triticale, Clermont-Ferrand, France. INRA, 1985: 35-47.

[66] Seal AG. C-banded wheat chromosomes in wheat and triticale. Theor Appl Genet 1982; 63: 39-47.

[67] Seal AG, Bennett MD. The rye genome in winter hexaploid triticales.Can J Genet Cytol 1981; 23: 647-653.

[68] Shigenaga S, Larter EN, McGinnis RC. Identification of chromosomes contributing to aneuploidy in hexaploid Triticale cultivar Rosner. Can J Genet Cytol 1971; 13: 592-596.

[69] Singh RJ, Robbelen G. Giemsa banding technique reveals deletions within rye chromosomes in addition lines. Z Pflanzenzuecht 1976; 76: 11-18.

[70] Sodkiewicz W. Amphiploid *Triticum monococcum* L x *Secale cereale* L(AARR)—a new form of tetraploid triticale. Cer Res Commun 1984; 12: 35-40.

[71] Sodkiewicz W, Apolinarska B. Development of secondary tetraploid triticale with a complete A-genome through crossing primary amphiploids with secondary hexaploid triticale. Cer Res Commun 2000; 28/1-2: 49-56.

[72] Sowa W, Gustafson JP. Relation between chromosomal constitution in triticale and several agronomic characters. Hodowla Rosl Aklim Nasienn 1980; 24/4: 389-396.

[73] Sybenga J. Rye chromosome nomenclature and homoeology relationships. Workshop Report. Z Pflanzenzuecht 1983; 90: 297-304.

[74] Tarkowski Cz. Cytogenetics of hexaploid triticale hybrids with wheat and rye. Genet Polon 1969; 1: 85-86.

[75] Tarkowski Cz, Stefanowska G. Chromosome morphology in the genome rye (*Secale cerale* L.) and in Triticale 6x and 8x. Genet Polon 1972; 13: 83-89.

[76] Tarkowski Cz, Apolinarska B. The use of chromosome substitutions and translocations in the breeding of triticale, wheat and rye. Hereditas. 1992; 116: 281-283.

[77] Thomas JB, Kaltsikes PJ. A possible effect of heterochromatin on chromosome pairing. Proc Natl Acad Sci USA. 1974; 71: 2787-2790.

[78] Thomas JB, Kaltsikes PJ. The genomic origin of the unpaired chromosomes in triticale. Can J Genet Cytol 1976; 18: 687-700.

[79] Verma SC, Rees H. Giemsa staining and the distribution of heterochromatin in rye chromosomes. Hereditas 1974; 32: 118-122.

[80] Weimarck A. Cytogenetic behaviour in octoploid triticale. II. Meiosis with special reference to chiasma frequency and fertility in F1 and parents. Hereditas 1975; 80: 121-130.

[81] Zaiuddin A, Kasha K. Giemsa C-band identification of rye chromosomes in some advanced lines of winter triticales. Can J Genet Cytol 1982; 24: 721-727.

Sequencing *Populus*: An Impetus for Understanding Tree Biodiversity and Evolution

YANRU ZENG[1], JIASHENG WU[1], and *RONGLING WU[1,2]*
[1]*School of Life Sciences, Zhejiang Forestry University, Lin'an, Zhejiang 311300, P. R. China,*

[2]*Department of Statistics, University of Florida, Gainesville, FL 32611 USA*

ABSTRACT

A whole-genome sequencing plan for forest trees was recently initiated on the genome of a tree model system – *Populus*, to understand a number of fundamental evolutionary questions regarding intra- and interspecific differentiations in higher plants. This chapter discusses how the *Populus* genome sequence can be used to gain new insights into the genome organization, structure, and function of woody plants. The sequence data being collected will help in integrating tree genetics and physiology to improve our understanding of tree growth and development from the molecular level to the landscape.

Key Words: Genome sequence, linkage disequilibrium, population structure, *Populus*, single nucleotide polymorphisms

INTRODUCTION

Forest trees are a group of higher plants with tremendous economic and ecological value. Unlike annual herbaceous plants, forest trees have

Address for correspondence: Rongling Wu, Department of Statistics, University of Florida, Gainesville, FL 32611, USA. Tel: (352)392-3806, Fax: (352)392-8555, E-mail: rwu@stat.ufl.edu

several unique biological properties, such as large physical size, complex architecture, dramatic phase change, great life span, long generation interval, and high heterozygosity resulting from outcrossing behavior. These properties have helped forest trees survive in ubiquitous biotic and abiotic competition, but have also thrown significant obstacles to biologists for detailed understanding of tree biology. To unravel the physiological mechanisms underlying developmental and environmental responses in trees, a tree model system is needed. Through its complete genome sequence, such a model system would provide a "toolkit" for other trees [2,3,11,20].

Several attributes of the genus *Populus* (including poplars, cottonwoods, and aspens) make it suitable as the model system for tree molecular biology [2,20,28]. Most importantly, the role of *Populus* as the model tree has been strikingly strengthened by the recent commitment to sequence the *Populus* genome using a female tree of *Populus trichocarpa* (black cottonwood), the largest native angiosperm tree in western North America [3]. As the third publicly available plant genome sequence after *Arabidopsis* and rice, this first forest tree genome sequence will widely serve as raw material for characterizing the genes, regulatory networks, and molecular mechanisms underlying reproduction, development, metabolism, adaptation, and evolution in woody plants.

Sequencing the *Populus* genome would certainly provide more insight into the origin of this genus and the pattern of its intra- and interspecific variation in morphological development and the underlying genetic organization. This chapter discusses the emerging power of sequencing *Populus* to address these fundamental issues and explore basic ideas for analyzing future sequence data in *Populus*. The discussion is largely based on the findings of well-sequenced genomes including humans and annuals. The current status of linkage mapping for quantitative trait loci (QTL) in *Populus* is not reviewed because some of this has been given in Wu et al. [24] and Taylor [20].

TREE DIVERSITY

Genetic diversity in forest tree populations has been studied for about three decades. It was recognized that forest trees preserve higher levels of genetic diversity in terms of number of alleles at a given locus than other organisms. Earlier diversity studies using allozymes [29] provided many insights into population structure and breeding systems. In the last decade, the focus shifted to nucleotide-level surveys of single polymorphic genes. Examples of these DNA-based markers are randomly amplified polymorphic DNA (RAPD), restriction length polymorphism (RFLP), amplified fragment length polymorphism (AFLP), single sequence repeats (SSR), and sequence tag sites (STS). These markers have been employed in genetic fingerprinting studies, linkage analysis, QTL mapping, studies of marker variation and introgression in natural populations and consensus mapping across pedigrees. In the next decade, with advances in genotyping capabilities, nucleotide surveys will surely be more powerful and more precise, allowing thereby more robust analysis of population genetics in forest trees.

POPULUS AS A MODEL SYSTEM TO SEQUENCE

Genomics can be defined as the mapping, sequencing, and functional analysis of genomes, the entire genetic complement of an organism. Before high-throughput DNA sequencing was available, genomes were studied largely by mapping approaches based on molecular markers. Availability of the genome sequence will accelerate development of high-density genetic markers, e.g. single nucleotide polymorphisms (SNPs), thereby aiding assessment of genetic variation in natural populations, comparing different genomes between several species, and clarifying the evolutionary history of a species and its closest relatives.

The choice of *Populus* as the first tree genome to sequence was based on its considerable variation within and among ~30 species [5], rapid growth rate [23], small genome size [1], easy genetic manipulation using transgenic plants [7] and widespread use in environmental protection by the forestry industry [18]. Phylogenetic studies using expressed sequence tags (ESTs) and regulatory proteins put *Populus* in an important position, as indicated in Figure 8.1a [17] and Figure 8.1b [10]. As stated by Brunner et al. [3], the phylogenetic position of *Populus* gives it an advantage over other forest trees in comparative analysis with other intensively studied and sequenced genomes, such as *Arabidopsis* and rice. First, it facilitates identification of putative orthologues to genes characterized in those well-studied species, thus providing clues to gene function. In general, there is extensive genome colinearity between species in the same family and some mircosynthetic relationships between distantly related species. For example, based on comparative sequence analysis of orthologous regions, Stirling et al. [19] revealed substantial synteny and microcolinearity between *Populus* and *Arabidopsis*.

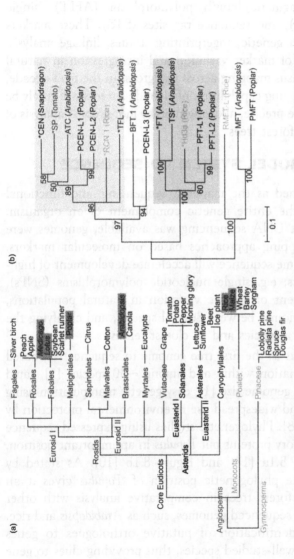

Fig. 8.1: Phylogenetic relationships of *Populus* (poplar) as a model system for tree biological studies (**a**) Angiosperm phylogeny based on Soltis et al. [17] showing only orders that include genera with >10,000 expressed sequence tags (ESTs) in dbEST. Also included are trees for which >10,000 ESTs are in the process of being generated or for which there are large, currently proprietary EST data bases (i.e., eucalypts and radiata pine). Woody plants in blue type and species for which complete genome sequences are available, or soon will be, in orange boxes. (**b**) Phylogenetic tree of CENTRORADIALIS (CEN)/TERMINAL FLOWER1(TFL1)-like regulatory proteins [10] produced by the neighbor-joining method. Bootstrap values (percentages based on 1,000 replications) are shown at nodes. All six *Arabidopsis* family members, *Populus trichocarpa* members present in GenBank or identified in the genome sequence, and selected genes from other species are included. The subclade highlighted in yellow includes genes (indicated by asterisks) that have been shown to maintain the indeterminate inflorescence or vegetative phases, whereas the subclade highlighted in pale blue includes genes that promote the transition to flowering. Figure and caption adapted from Brunner et al. [3].

Second, with an increasing number of trees being studied with ESTs (Fig 8.1a), the complete genomic toolkit of *Populus* provides a valuable reference species for the study of other tree species. Based on phlyogenetic conservation and similarity in expression, *Populus* genomic sequences have already been used to identify regulatory motifs in the floral development gene AGAMOUS from various eudicots [6]. In other studies, comparative analyses have been done for the regulatory networks controlling flowering between two distantly related angiosperms, rice and *Arabidopsis*, using their complete genome sequences, together with genetic studies [8].

Population History and Population Structure

DNA sequencing of *Populus* or its resultant high-throughput markers, such as SNPs, enables understanding the extent of linkage disequilibrium (LD), i.e., nonrandom association of nonalleles at different loci, across the genome. Figure 8.2 illustrates the occurrence/absence of LD between two SNPs. A variety of statistics have been used to measure LD. The

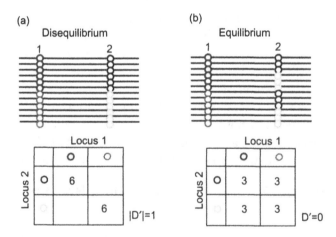

Fig. 8.2: Schematic representation of **(a)** linkage disequilibrium and **(b)** linkage equilibrium between two loci. **(a)** When LD is present, all individuals that have a red allele at locus 1 have a blue allele at locus 2. **(b)** When there is linkage equilibrium, individuals with a red allele at locus 1 could have any allele at locus 2. The corresponding contingency tables and values of the LD index D' are shown below. The hypothetical example shows only a few individuals and statistical significance may be calculated by the Fisher Exact Test: P value for (a) 0.001, for (b) 0.716. Figure and caption adapted from Rafalski [14].

relative advantages and disadvantages of each statistical approach was reviewed by Jorde [9].

LD observed in a population is a result of the interplay of many factors, including selection, mutation, recombination rate, and mating systems. Thus, the distribution of LD in the genome expressed as the relationship between LD and genetic distance is highly dependent on the population in which it is measured [15]. In a pioneering study, Reich et al. [15a] found that LD in humans is highly population dependent. Whereas in some populations (e.g., Europeans), LD extends to 60 kb, in others (e.g., Yoruba Africans), it declines within a few kilobases. This difference can be explained by different population histories. Population bottlenecks, associated with geographic expansion and population isolation, resulting from glaciation in European populations, may have led to increased LD in these populations. Similar analyses of LD reflecting its distribution and pattern across the genomes could be pursued in *Populus* using different species and different natural populations within a species.

Theory predicts that LD decays with increasing recombination and increasing effective population size that makes random mating more probable [15]. Thus, interpretation of the extent of LD should be based on population structure and the possible effect of admixture. For some *Populus* species, such as *P. nirga* naturally distributed in the Eurasian continent, in which anthropological influences result in the consistent loss of rare allelic combinations, increased LD levels would be expected. Comparative analysis among several different populations should be based on identical genetic regions in order to avoid the effect of genomic location. With such comparative analyses, any difference found would be due solely to the differences between the populations under study.

Haplotype Blocks and Recombination Hot Spots in *Populus*?

Haplotype is defined as a linear arrangement of nonalleles on a single chromosome. In humans, meiotic crossover events, which contribute to the breakdown of LD, are distinctly not randomly distributed [22]. As a result of this, the human genome can be broken into a series of discrete haplotype blocks. In each haplotype block, consecutive sites are in complete (or nearly complete) LD with each other and there is limited haplotype diversity due to little (coldspot) intersite recombination. Adjacent blocks are separated by sites that show evidence of historical recombination (hotspot). It has generally been assumed that the presence of haplotype blocks provides evidence for fine-scale variation in recombination rates, with blocks corresponding to regions of reduced recombination, separated by recombination hotspots. Based on a study of human chromosome 21,35,989 observed SNPs can be classified into different blocks with very low haplotype diversity and 80% of the variation in this chromosome can be described by only three SNPs per block [13].

In plants, although haplotype structure across longer genome segments remains to be explored, several lines of evidence suggest that such blocks can be common. However, as pointed out by Rafalski and Morgants [15], plants display different features from humans in terms of physical distances over which recombination affects LD. There is no doubt that availability of the *Populus* genome-sequencing data will enable systematic exploration of haplotype structure throughout the entire genome and highlight possible mechanisms of maintenance of the genome.

ASSOCIATION STUDIES

Knowledge of the pattern of LD distribution will help in designing a molecular experiment for associating DNA sequence variants with biological characteristics that best determine population diversity in *Populus*. To do so, tree biologists create experimental populations by crossing contrasting genotypes in which linkage analysis between a number of markers and phenotypes are performed. However, the requirement for a considerable large sample size for high-resolution mapping is hardly met in forest trees given their large physical size and possible outcrossing or inbreeding depression. Association studies relying on the amount of LD across the genome can take advantage of the high degree of recombination in natural populations that has occurred over hundreds of generations.

In association mapping, the minimum number of markers required to scan the genome depends on the extent of LD, but it will differ both between species and between different populations within species. If LD declines slowly with increasing distance from the gene responsible for the phenotype on a chromosome, even a low density of markers suffices to identify associated markers (Fig. 8.3a). On the other hand, if LD declines very rapidly around the causative gene, a much greater density of markers is required to identify an associated marker (Fig. 8.3b). In *Populus*,

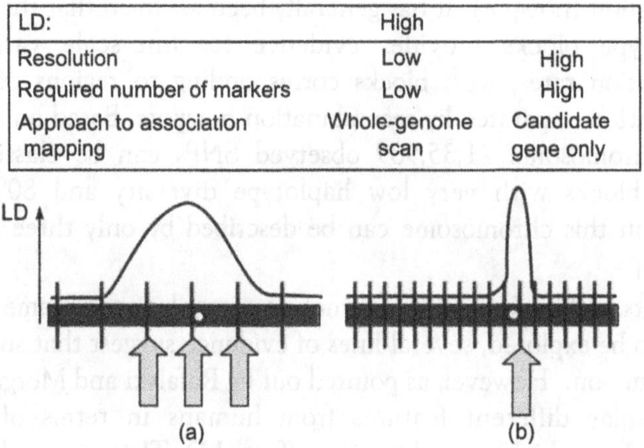

LD:	High	
Resolution	Low	High
Required number of markers	Low	High
Approach to association mapping	Whole-genome scan	Candidate gene only

(a)   (b)

Fig. 8.3: Relationship between extent of LD and resolution of association studies. In **(a)**, LD declines slowly with increasing distance from the gene responsible for the phenotype (red oval) on a chromosome. In this case, even a low density of markers (shown as red vertical bars) suffices for identifying associated markers (yellow arrows). In **(b)**, LD declines very rapidly around the causative gene and a much greater density of markers is required to identify an associated marker (yellow arrow). Figure and caption adapted from Rafalski [14].

construction of a low- or high-density map of markers depends on detailed knowledge about the distribution of LD over the genome.

The use of LD mapping in plants can be affected by the presence of population stratification and an unequal distribution of alleles within these groups. Due to these factors, two polymorphic sites on different chromosomes may produce highly significant LD. This spurious association can be eliminated by a joint analysis of linkage and LD mapping. Just recently, a general framework for integrating these two mapping approaches was proposed [27]. This framework can also be used to enforce the efficiency and effectiveness of high-resolution mapping of complex phenotypes. It makes use of genetic information from the family pedigree and population data by adopting a two-stage hierarchical sampling scheme as proposed in Wu and Zeng [24] and Wu et al. [25]. The upper hierarchy of this scheme describes the gene distribution in a natural population from which a sample is randomly drawn (LD analysis), and the lower hierarchy describes gene segregation in the progeny population derived from matings (linkage analysis). These two hierarchies are connected through population and Mendelian genetic theories.

Forest trees, including *Populus*, have a long history of breeding. In the past decades, a number of open-pollinated progeny have been selected from natural populations [26]. These materials contain information about the selected mother trees and the progeny derived from these mother trees and hence can be used directly for joint linkage and LD mapping.

IMPACT OF MATING SYSTEMS

Different mating systems have a dramatic impact on LD [5a]. In predominantly selfing species, such as *Arabidopsis thaliana* [12,16] or soybean [30], high individual homozygosity can make recombination ineffective, which does not lead to reduction of LD. The LD in both these species was found to persist over tens to hundreds of kilobases. Although selfing species are composed of homozygous individuals as a result of their mode of reproduction, genetic diversity can be high at the population level. In predominantly outcrossing species, such as human, maize or many conifers and poplars [4], a much more rapid decline in LD has been observed or can be expected. In maize, at least in some populations, LD declines over a few hundred base pairs [21].

PERSPECTIVES

Without doubt, an understanding of forest tree biology can contribute significantly to overall sciences. In the past, it was impeded by the many recalcitrant properties of forest trees. Today, the choice of *Populus* as a widely accepted model system will remarkably improve our comprehensive insights into growth and development in forest trees as well as their roles in natural and managed ecosystems. The recent development of a complete genome sequence in *Populus* will accelerate studies of tree biology.

Until the *Populus* genome is completely sequenced, nucleotide diversity must be extensively surveyed for a wide range of *Populus* species and loci. Then, fundamental evolutionary questions, such as evolution and speciation, can be addressed using this wealth of data. New discoveries for the genome organization, structure, and function in humans and *Arabidopsis* can provide scientific guidance for designing molecular experiments for *Populus* and other forest tree species.

Acknowledgments

This work was partially supported by an Outstanding Young Investigators Award (No. 30128017) of the National Natural Science Foundation of China and the University of Florida research Opportunity Fund (No. 02050259) to R. W.

[1] Bradshaw HD, Jr., Stettler RF. Molecular genetics of growth and development in *Populus*. I. Triploidy in hybrid poplars. Theor App Gene1993; 86: 301-307.

[2] Bradshaw HD, Jr, Ceulemans R. Davis J, Stettler R. Emerging model systems in plant biology: Poplar (*Populus*) as a model forest tree. J Pl Growth Regul 2000; 19: 306-313.

[3] Brunner AM, Busov BV, Strauss SH. Poplar genome sequence: functional genomics in an ecologically domain plant species. Trends Pl Sci 2004; 9: 49-56.

[4] Dvornyk V, Sirvio A, Mikkonen M, Savolainen O. Low nucleotide diversity at the pall locus in the widely distributed *Pinus sylvestris*. Molec Biol Evol 2002; 19: 179-188.

[5] Eckenwalder JE. Systematics and evolution of *Populus*. In: Stettler RF, et al., eds. Biology of Populus and its implications for management and conservation. Ottawa, CAN: NRC Research Press, 1996: 7-32.

[5a] Flint-Garcia SA, Thornsberry JM, Buckler ES 4th. Structure of linkage disequilibrium in plants. Annu Rev Plant Biol. 2003; 54: 357-374.

[6] Hong RL, Hamaguchi L, Busch MA, Weigel D. Regulatory elements of the floral homeotic gene AGAMOUS identified by phylogenetic footprinting and shadowing. Pl Cell 2003; 15: 1296-1309.

[7] Hu WJ, Harding SA, Lung J, et al. Regression of lignin biosynthesis promotes cellulose accumulation and growth in transgenic trees. Nature Biotech 1999; 17: 808-812.

[8] Izawa T, Takahashi Y, Yano M. Comparative biology comes into bloom: genomic and genetic comparison of flowering pathways in rice and Arabidopsis. Curr Opin Pl Biol 2003; 6: 113-120.

[9] Jorde LB (2000) Linkage disequilibrium and the search for complex disease genes. Genome Res 2000; 10: 1435-1444.

[10] Kobayashi Y, Kaya H, Goto K, Iwabuchi M, Araki T. A pair of related genes with antagonistic roles in mediating flowering signals. Science 1999; 286: 1960-1962.

[11] Lev-Yadun S, Sederoff R. Pines as model gymnosperms to study evolution, wood formation, and perennial growth. J Pl Growth Regul 2003; 19: 290-305.

[12] Nordborg M, Borevitz JO, Bergelson J, et al. Extent of linkage disequilibrium in *Arabidopsis thaliana*. Nature Genet 2002; 30: 190-193.

[13] Patil N, Berno AJ, Hinds DA, et al. Blocks of limited haplotype diversity revealed by high-resolution scanning of human chromosome 21. Science 2001; 294: 1719-1723.

[14] Rafalski A. Application of single nucleotide polymorphisms in crop genetics. Curr Opin Pl Biol 2002; 5: 94-100.

[15] Rafalski A, Morgante M. Corn and humans: recombination and linkage disequilibrium in tow genomes of similar size. Trends Genet. 2004, 20: 103-111.

[15a] Reich DE, Cargill M, Bolk S et al. Linkage disequilibrium in the human genome. Nature, 2001, 411: 199-2004.

[16] Shepard KA, Purugganan MD. Molecular population genetics of the Arabidopsis CLAVATA2 region. The genomic scale of variation and selection in a selfing species. Genetics 2003; 163: 1083-1095.

[17] Soltis DE, Soltis PS, Chase MW, et al. Angiosperm phylogeny inferred from 18S rDNA, rbcL, and atpB sequences. Bot J Linn Soc 2000;. 133: 381-461.

[18] Stettler RF, Bradshaw HD, Jr., Heilman PE, Hinckley TM, eds. Biology of *Populus* and its implications for management and conservation. Ottawa, CAN: NRC Research Press, 1996, 539 pp.

[19] Stirling B, Yang ZK, Gunter LE, Tuskan GA, Bradshaw HD, Jr. Comparative sequence analysis between orthologous regions of the Arabidopsis and Populus genomes reveals substantial synteny and colinearity. Can J For Res 2003; 33: 2245-2251.

[20] Taylor G. *Populus*: Arabidopsis for forestry. Do we need a model tree? Ann of Bot 2002; 90: 681-689.

[21] Tenaillon MI, Sawkins MC, Long AD, et at. Patterns of DNA sequence polymorphism along chromosome 1 of maize (*Zea mays* ssp. Mays L.). Proc Natl Acad Sci USA 2001; 98: 9161-9166.

[22] Wall JD, Pritchard JK. Haplotype blocks and linkage disequilibrium in the human genome. Nature Rev Genet 2003; 4: 587-597.

[23] Wu R, Stettler RF. The genetic dissection of juvenile canopy structure and function in a three-generation pedigree of *Populus*. Trees—Structure Function 1996; 11: 99-108.

[24] Wu RL, Zeng ZB. Joint linkage and linkage disequilibrium mapping in natural populations. Genetics 2001; 157: 899-909.

[25] Wu RL, Ma CX, Casella G. Joint linkage and linkage disequilibrium mapping of quantitative trait loci in natural populations. Genetics 2002; 160: 779-792.

[26] Wu RL, Zeng ZB, McKend SE, O'Malley DM. The case for molecular mapping in forest tree breeding. Pl Breed Rev 2000; 19: 41-68.

[27] Wu RL, Lou XY, Todhunter RJ, et al. A general statistical framework for unifying interval and linkage disequilibrium mapping: Towards high-resolution mapping of quantitative traits. J Amer Stat Assoc 2004 (accepted).

[28] Wullschleger SD, Tuskan GA, Difazio SP. Genomics and the tree physiologist. Tree Physiol 2000; 22: 1273-1276.

[29] Yeh FC, ElKassaby YA. Enzyme variation in natural populations of Sitka spruce (*Picea sitchensis*). 1. Genetic variation patterns among trees form 10 IUFRO provenances. Can J For Res 1980; 10: 415-422.

[30] Zhu YL, Song QJ, Hyten DL, et al. Single-nucleotide polymorphisms in soybean. Genetics 2003; 163: 1123-1134.

Origin, Evolution, and Diversity of the Coffee (*Coffea arabica* L.) Genome

FRANÇOIS ANTHONY and *PHILIPPE LASHERMES*
Institut de Recherche pour le Développement, "Plant resistance to pests" Unit, Montpellier, France

ABSTRACT

The recent development of DNA-based techniques has broadened the horizon for investigating the composition and organization of the coffee genome. It was found to be small according to both nuclear DNA content data and estimates of physical and genetic size. The origin of the allotetraploid (2n = 4x = 44) species *Coffea arabica* was confirmed and the diploid (2n = 2x = 22) progenitors of the species were identified. High affinity was noted between the two subgenomes of *C. arabica* and between *C. arabica* and the diploid species, suggesting that speciation of *C. arabica* was a relatively recent event. Coffee cultivation, begun about 1,500 years ago in Ethiopia, has been characterized by successive reductions in genetic diversity within the wild gene pool. The transfer of genes from diploid coffee, in particular resistance genes, was also studied. There seemed to be a substantial extent of alien DNA in introgressed lines after several selfed generations. Disomic inheritance of molecular markers was found in the *C. arabica* genome while tetraploid interspecific hybrids of crosses between *C. arabica* and a duplicated diploid coffee presented tetrasomic inheritance. The results could be used to enhance breeding strategies based on gene introgression without lowering cup quality.

Key Words: Allotetraploid, *C. arabica*, coffee, diversity, genome, polymorphism

Address for correspondence: IRD, BP 64501, 34394 Montpellier cedex 5, France. Tel: +33 (0) 467416289, Fax. +33 (0) 467416320, E-mail: anthony@mpl.ird.fr

Abbreviations:

AFLP: amplified fragment length polymorphism; cpDNA: chloroplastic DNA; rDNA: ribosomal DNA; GISH: genomic *in-situ* hybridization; ITS: internal transcribed spacer; Mb: million base pairs; NIR: near-infrared spectroscopy; RAPD: random amplified polymorphic DNA; RFLP: restriction fragment length polymorphism; SSR: simple sequence repeat

INTRODUCTION

Coffee species belong to family Rubiaceae, one of the largest tropical angiosperm families. Variations in cpDNA classified the Coffeeae tribe into the Ixoroideae monophyletic subfamily, close to Gardenieae, Pavetteae, and Vanguerieae [17]. Two genera, *Coffea* L. and *Psilanthus* Hook. f., were distinguished on the basis of flowering and flower criteria [19,57]. Each genus was divided into two subgenera based on growth habit (monopodial vs. sympodial development) and type of inflorescence (axillary vs. terminal flowers). Approximately 100 coffee species have been identified to date [20,29,76] and new taxa are still being discovered [28,77]. All species are perennial woody bushes or trees in intertropical forests of Africa and Madagascar for the *Coffea* genus, and Africa, Southeast Asia and Oceania for the *Psilanthus* genus. They differ greatly in morphology, size and ecological adaptations. Some species, e.g. *C. canephora* Pierre and *C. liberica* Hiern, are widely distributed from Guinea to Uganda. Other species display specific adaptations, e.g. *C. congensis* Froehner to seasonally flooded areas in the Zaire basin and *C. racemosa* Loureiro to very dry areas in the coastal region of Mozambique [27]. However, many characters considered in coffee taxonomy are weak and variable, and many taxa have not yet been fully characterized, so it is hard to draw valid conclusions about their relationships [18]. All species are diploid (2n = 2x = 22) and generally self-incompatible, except for *C. arabica* Linne, which is tetraploid (2n = 4x = 44) and self-fertile [27].

Coffee is one of the world's most valuable export commodities, ranking second on the world market after petroleum products. The total retail sales value exceeded US$70 billion in 2003 and about 125 million people depend on coffee for their livelihood in Latin America, Africa, and Asia [69]. Commercial production relies on two species, *C. arabica* and *C. canephora*. The better quality (low caffeine content and fine aroma) of *C. arabica* makes it by far the most important species, representing 70% of world production. Cultivation of *C. arabica* started in southwestern Ethiopia about, 1,500 years ago [56,82]. Modern coffee cultivars are derived from two base populations—known as Typica and Bourbon—that were spread worldwide in the eighteenth century [25,45]. Historical data indicate that these populations were composed of progenies of very few plants, i.e., only one for the Typica population [21,30]. Breeders exploited these narrow genetic bases, resulting in Typica- and Bourbon-derived cultivars with homogeneous agronomic behavior characterized by high susceptibility to many pests and low adaptability [12].

The transfer of desired characters, especially pest resistance, from diploid species has become a priority in *C. arabica* breeding programs [24]. Crossing *C. arabica* with diploid coffee plants gives rise to triploid hybrids, which are vigorous but very low in fertility [78]. Other interspecific hybrids have been produced after duplication of diploid progenitors by colchicine treatment of germinating seeds [61] or shoot apices [9]. The hybrids obtained were tetraploid and presented relatively good fertility [55], even when the diploid coffee parent belonged to genus *Psilanthus* [32]. Natural interspecific hybrids between *C. arabica* and *C. canephora* or *C. liberica* provided the first sources of resistance to leaf rust [15,75]. Progenies from controlled interspecific hybridizations have also been selected for leaf rust resistance, e.g. Icatu hybrids (*C. arabica* x *C. canephora*) in Brazil [23]. Introgression from wild species into the *C. arabica* genome can nevertheless lead to transfer of undesirable genes that may lower beverage quality. Conventional coffee-breeding methods require a minimum of 25-30 years after interspecific hybridization to eliminate undesirable genes and restore the genetic makeup of the recipient cultivars. Therefore it is a very difficult task to combine various desired traits without reducing coffee quality and do so within an acceptable timeframe. Development of a selection method assisted by DNA-based markers could overcome present conventional coffee-breeding limitations [52].

The results presented here are related to the origin, evolution, and diversity of the *C. arabica* genome. Several DNA analysis-based studies have been carried out to estimate the size of coffee genomes, determine genetic relationships between species, identify the diploid progenitors of *C. arabica* and investigate the organization of genetic diversity in the allotetraploid genome. The knowledge generated has been used to design breeding strategies based on gene introgression from diploid species and assisted by molecular markers.

SIZE OF THE COFFEE GENOME

Genome size was estimated by DNA content analyses and mapping activities. Considering the range of plant genome sizes (100-100,000 Mb), the coffee genome appears to be rather small.

Physical Size

Evaluation of the nuclear DNA content of 13 *Coffea* species was carried out using flow cytometry [34]. Three groups of accessions with increasing DNA content were defined by aggregative clustering. The lowest values (0.9-1.3 pg) were found in East African species (C. *pseudozanguebariae* Bridson, C. *racemosa*, C. *sessiliflora* Bridson) and the highest values (1.61-1.8 pg) detected in species native to West and Central Africa (C. *congensis*, C. *liberica*, C. *humilis* Chevalier, *Coffea* sp. "Moloundou"). The intermediate group comprised species from all African regions. In line with its tetraploid constitution, the nuclear DNA content of C. *arabica* (2.61 pg) was estimated to be about twice as high as the content of diploid species. Vis-a-vis the nuclear DNA content of other plants, *Coffea* species could be considered to have a low DNA content, similar to species *Acacia heterophylla* (1.60 pg), *Dioscorea alata* (1.47 pg), and *Vigna unguiculata* (1.20 pg) [34].

The mean coffee haploid genome was estimated at 700 Mb on the basis of flow cytometry data [34]. For C. *canephora* and C. *arabica*, the size was estimated at 800 and 1,200 Mb respectively [47]. However, intraspecific variations in nuclear DNA content induce variations in such estimations. This explains why the dihaploid genome of C. *arabica* could also be estimated at 1,300 Mb [66]. As compared with plants whose genome has recently been sequenced, the size of the haploid coffee genome appears to be around twofold larger than the rice genome and sevenfold larger than the *Arabidopsis thaliana* genome, but much smaller than the wheat genome (16,000 Mb).

Genetic Length

Correspondence with genetic length could only be attempted for C. *canephora*, using the first genetic map developed for this species [54]. The 11 linkage groups identified in the map were estimated to measure about 1,400 cM and so 1 cM could correspond to approximately 570 kb.

PHYLOGENETIC RELATIONSHIPS OF COFFEE SPECIES

The molecular phylogeny of coffee species was established through an analysis of cpDNA [35, 46] and rDNA [48] polymorphism. No major difference between coffee and other plants was observed in the arrangement of the chloroplast genome and in the structure of the ITS 2 region in nuclear rDNA.

Relationships of Diploid Species

Variations in cpDNA were assessed by a RFLP analysis on total cpDNA and the atpB-rbcL intergenic region [46], and by sequencing the trnL-trnF intergenic spacer [35]. Other data were obtained by an analysis of the ITS 2 sequence of rDNA [48]. All of these studies revealed few differences between coffee species. The *Psilanthus* species did not differ from *Coffea* species, suggesting that the present division into two genera should be revised. The dendrograms generated by either cpDNA sequence data or rDNA sequence data showed a strong geographical correspondence (i.e., Madagascar, East Africa, Central Africa, Central and West Africa) (Fig. 9.1).

Phylogeny of *C. arabica*

Regarding cpDNA polymorphism, *C. arabica* appeared to be similar to two species from Central Africa [35]: *C. eugenioides* Moore, widely distributed in Uganda, Kenya, and Tanzania between 1,000 and 2,100 m elevation [20], and *Coffea* sp. "Moloundou", a new taxon growing in lowland areas (< 500 m) near the Cameroon-Congo Brazzaville border [3,63]. rDNA polymorphism data confirmed the genetic relationships between *C. eugenioides* and *Coffea* sp. "Moloundou", but *C. arabica* was not grouped with these species [48]. The ITS 2 sequence of *C. arabica* appeared similar to those of *C. canephora*, *C. brevipes* Hiern, and *C. congensis*. These species are native to lowlands in Central Africa [29]. Based on cytological observations and fertility of interspecific hybrids, *C. canephora* and *C. congensis* were already suggested to have a common ancestor with *C. arabica* [22,33]. M oreover, *C. congensis* can be considered an ecotype of *C. canephora* in light of the success of interspecific hybridizations [58], fertility of interspecific hybrids [59], and genetic diversity detected by molecular markers [72].

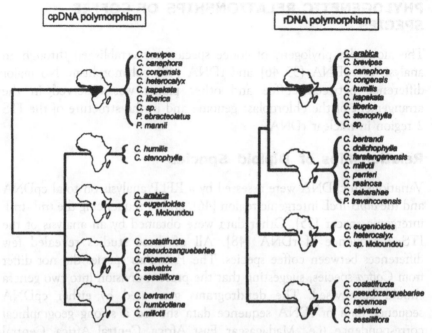

Fig. 9.1: Phylogeny of *Coffea* species based on cpDNA (left) [35] and rDNA (right) polymorphism [48].

ORIGIN OF *C. ARABICA* GENOME

The phylogenetic study identified two groups of diploid species which presented strong genomic similarities with the tetraploid species *C. arabica*: i) *C. eugenioides* and *Coffea* sp. "Moloundou" and ii) *C. canephora*, *C. congensis*, and *C. brevipes*. Further studies were undertaken to specify the origin of the *C. arabica* genome.

Allotetraploid Origin

The RFLP patterns of potential diploid progenitor species were compared with those of *C. arabica* at 23 loci [50]. In agreement with the results of the phylogenetic studies, the diploid species were divided into two groups (Fig. 9.2): Canephoroid type (i.e., *C. canephora*, *C. congensis*) and Eugenioides type (i.e., *C. eugenioides*, *Coffea* sp. "Moloundou"). The *C. arabica* accessions formed a third group, separately from the diploid accessions. Eighteen RFLP loci (75%) were present in duplicate in the *C. arabica* accessions as expected for an allotetraploid genome. Furthermore,

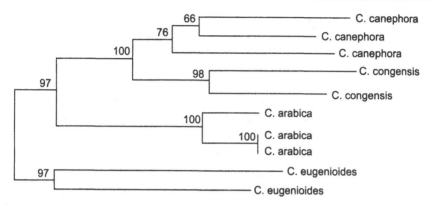

Fig. 9.2: Neighbour-joining tree of coffee accessions from *C. arabica* and potential diploid progenitor species using RFLP-based genetic distances [49]. Numbers on the branches are bootstrap values obtained from 100 replicate analyses.

80% of the alleles found in *C. arabica* were detected in at least one of the diploid accessions included in the study. For 11 of the 23 loci analyzed, genotypes of *C. arabica* accessions could be obtained by combination of alleles from the Canephoroid and Eugenioides groups.

Maternal and paternal progenitors of *C. arabica* could be identified since cpDNA has strict maternal inheritance in coffee [46]. The *C. arabica* genome arose from hybridization between *C. eugenioides* as maternal donor and *C. canephora* as paternal donor, or ecotypes related to these species. To further specify the *C. arabica* genome donors, new collections are required in order to increase the genetic diversity of germplasm collections. No extensive prospecting missions have been undertaken in Uganda where *C. canephora* and *C. eugenioides* were reported to be sympatric [81]. In addition, *C. canephora* is a very polymorphic species, which has led to the identification of several genetic groups [36]. *C. canephora* samples were previously identified under seven *Coffea* species names and 18 variety names [8]. Nevertheless, none of these varieties can be readily distinguished on the basis of the characters of herbarium specimens [20].

The GISH technique was successfully applied to discriminate between *C. arabica* chromosomes simultaneously using two probes from total DNA of *C. canephora* and *C. eugenioides* labeled with digoxigenin and biotin respectively [50]. Two sets of 22 chromosomes were clearly distinguished and the allotetraploid origin of the *C. arabica* genome thus confirmed.

Genome Dating

The low polymorphism found by phylogenetic analyses indicated a recent origin of the *Coffea* genus and supported a radial mode of speciation for coffee species [30,46,48]. The lack of cpDNA divergence between *C. arabica* and the cytoplasmic donors (i.e., *C. eugenioides*, *Coffea* sp. "Moloundou") suggests that speciation of the allotetraploid species took place relatively recently [30]. On the contrary, both RFLP analysis and GISH revealed substantial differences between the genomes of *C. canephora* and *C. eugenioides* and between the two genomes of *C. arabica* [50]. Assuming that the *Coffea* species started to diverge from a common ancestor about 5-25 mya [46], speciation of *C. arabica* is a relatively recent event that occurred from historical times to 1 mya [50]. Other data (unpublished) indicated that the genome of *C. arabica* could have diverged from the diploid species only a few thousand years ago.

Consequences of Polymorphism

As a consequence of the evolutive process and reproductive biology, the allotetraploid genome of *C. arabica* is characterized by low polymorphism compared to diploid species. Screening of 41 loci in *C. arabica* revealed that the proportion of polymorphic RFLP-defined loci was only 7% in *C. arabica* vs. 74% and 68% in *C. canephora* and *C. congensis* respectively [50]. Furthermore, no heterozygous loci were detected in *C. arabica* while they represented 23% in *C. eugenioides*, 27% in *C. congensis*, and 36% in *C. canephora*. Such low polymorphism is a major drawback for developing genetic maps of the *C. arabica* genome. Nevertheless, the genetic diversity of *C. arabica* cannot be considered insignificant since the genome has a considerable extent of fixed heterozygosity in relation to its allotetraploid origin. Such fixed heterozygosity provides *C. arabica* with a twofold higher level of internal genetic variability than in the diploid species, which show the highest heterozygosity at individual loci [49].

CHROMOSOME BEHAVIOR

Studies on coffee chromosomes were first undertaken in the 1930s using cytological methods. The basic chromosome number was established at 11, as in most genera of the Rubiaceae family [78]. Observations of chromosome morphology showed that coffee chromosomes have a similar and relatively small size (1-3 μm) [16]. Structural similarities were observed in the pachytene patterns of *C. arabica* chromosomes [70].

Recent development of molecular markers has provided new tools for investigating inheritance, recombination, and segregation mechanisms in the genome of C. *arabica* and interspecific hybrids. Results have highlighted marked affinity between the two constitutive genomes of C. *arabica* and between C. *arabica* and the diploid species.

Tetraploid Genomes

Results of classical genetic analyses suggest that the allotetraploid genome of C. *arabica* behaves like a diploid genome for most morphological characters [26]. Such inheritance was recently confirmed by a segregation analysis of RFLP markers [53]. Segregation in an F_2 progeny of C. *arabica* indicated disomic inheritance with regular bivalent pairing of homologous chromosomes in the F_1 hybrid. On the other hand, RFLP-defined loci followed a tetrasomic inheritance pattern in tetraploid interspecific hybrids derived from crosses between C. *arabica* and duplicated diploid coffee [53]. Finally, C. *arabica* chromosomes appear to behave in two different ways when present in duplicate in C. *arabica* or when combined in a single copy with C. *canephora* chromosomes in tetraploid interspecific hybrids. Nonpairing of homeologous chromosomes of C. *arabica* was interpreted as the result of functioning of pair-regulating genes rather than a consequence of structural differentiation.

The recombination fractions of C. *canephora* chromosome segments were estimated in two BC1 populations derived from a tetraploid interspecific hybrid (C. *arabica* x C. *canephora* 4x) [43]. The recombination frequencies were similar in the BC_1 populations and in C. *canephora*, suggesting that recombination in the tetraploid interspecific hybrids was not significantly restricted by genetic differentiation between chromosomes belonging to different genomes. These tetraploid hybrids therefore appear to be particularly favorable for intergenomic recombination events and gene introgressions.

Triploid Genomes

Crossing C. *arabica* as maternal parent with a diploid coffee as paternal parent produced triploid interspecific hybrids which showed vigorous growth but low fertility [27]. Attempts to produce hybrids using C. *arabica* as pollen donor were far less successful [25]. Gene introgression by way of triploid hybrids (C. *arabica* x C. *canephora*) was evaluated in six BC_1 populations using interspecific hybrids as male parents [44]. All triploid hybrids exhibited a nuclear DNA content midway between the two parental species, while the DNA content of the BC_1 plants was estimated to be close to that of C. *arabica*. This showed that most BC_1 plants were nearly tetraploid with around 44 chromosomes. AFLP and SSR markers were used to estimate the amount of introgression in these BC_1 plants [44]. Only a few plants (13.8%) presented C. *canephora*-specific markers, showing a marked deficiency in C. *canephora* alleles. This deficiency was attributed to mechanisms specific to triploid hybrids, associated with the formation of diploid gametes, rather than the consequence of counterselection related to the expression of lethal or sublethal C. *canephora* genes. The low level of introgression in triploid hybrids could be an advantage by facilitating recovery of the recurrent parent and possibly reducing the number of required backcrosses. On the other hand, this could substantially limit the transfer of complex traits or several monogenic traits.

Factors regulating gene introgression into C. *arabica* from related diploid species were studied in BC_1 populations [42]. Triploid hybrids involving either C. *canephora* or C. *eugenioides* were backcrossed with C. *arabica* as male parent. The mean proportion of introgressed markers per plant was significantly lower in populations derived from C. *eugenioides* (0.39) than from C. *canephora* (0.58) (Fig. 9.3). This might have been due to a selective mechanism resulting from the accumulation of unfavorable alleles from C. *eugenioides*. Another effect on gene introgression was deduced from an analysis of reciprocal backcross progenies involving C. *canephora*: a severe reduction (1.7% *vs.* 55.0%) in the frequency of C. *canephora* introgressed markers was observed when the triploid hybrids were used as male parent. This contrasting situation suggested a dissimilar behavior of male and female meiosis in the triploid hybrids.

GENETIC DIVERSITY IN C. ARABICA

Genetic diversity was evaluated in two gene pools of C. *arabica* corresponding to wild and cultivated coffee. Forests containing wild coffee trees have been reported in Ethiopia [2,79], on the Boma Plateau in Sudan [80] and on Mount Marsabit in Kenya [5]. Coffee was first cultivated in southwestern Ethiopia about 1,500 years ago [56,82]. Seeds or plants were exported to Yemen as early as 575 A.D. [82] or possibly just three or four centuries ago [37]. Extension of coffee cultivation

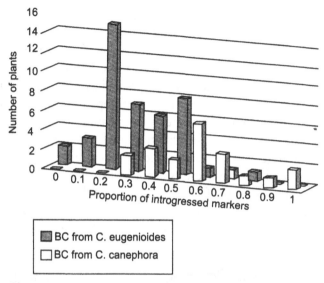

Fig. 9.3: Histograms of the number of BC$_1$ plants in which the particular proportion of either *C. canephora* or *C. eugenioides* introgressed markers was detected [42].

throughout the tropical world originated in the dissemination of two base populations from Yemen between 1690 and 1718. These populations were described as two distinct botanical varieties, i.e., *C. arabica* var. Typica and *C. arabica* var. Bourbon [26,45]. Historical evidence suggests that the Typica population originated from a single plant from Indonesia that was subsequently cultivated in Amsterdam in 1706 [21,30]. The Bourbon population was composed of the few plants that survived the introductions on Bourbon Island (now Réunion) in 1715 and 1718 [41]. Since isozymes failed to reveal polymorphism among *C. arabica* accessions [10], genetic diversity studies remained weak until the development of DNA markers. Results of molecular studies confirmed that coffee domestication has involved successive reductions of diversity in the wild gene pool.

Diversity of Wild Coffee

RAPD markers were used to assess the genetic diversity of genotypes collected in Ethiopia and conserved in a field gene bank [6]. Preliminary studies [49,68] revealed a low level of polymorphism: only 11% of tested 10-mer oligonucleotides gave reproducible bands and detected polymorphism. The 88 accessions (119 genotypes) were classified into four groups on the basis of genetic similarity estimates (Fig. 9.4): that with the most genotypes from southwestern Ethiopia (i.e., Ethiopian 1) and three with genotypes from southern and southeastern Ethiopia (i.e., Ethiopian 2, 3, and 4). The genotypes classified as Ethiopian 1 presented 28 of a total of 29 markers, while those of Ethiopian 2, 3, and 4 presented only 13, 12, and 8 markers respectively. Based on genetic distance estimates, Ethiopian 1 is not ably differentiated with respect to the other Ethiopian groups (0.09-0.18) and could be considered equidistant from Ethiopian 2, 3, and 4. These results accord with the hypothesis of recent introductions of coffee trees from southwestern Ethiopia to southern and southeastern Ethiopia. A recent report on the origin of the species *C. arabica* [50] suggested that coffee colonization of Ethiopia occurred after

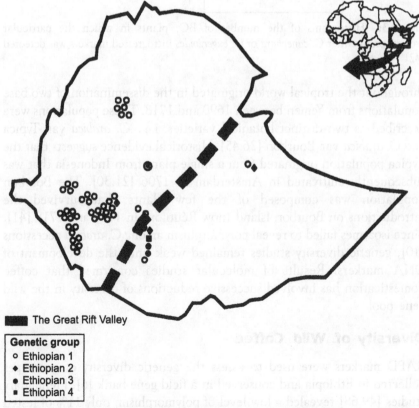

Fig. 9.4: Distribution of the genetic groups of wild coffee collected in Ethiopia based on RAPD polymorphism [6].

formation of the tectonic fault, the Great Rift Valley, which crosses Ethiopia from northeast to southwest. The distinction between coffee trees from the western and eastern sides of the Great Rift Valley could not thus be interpreted as a consequence of their genetic isolation.

Diversity of Cultivated Coffee

Genetic diversity was evaluated among coffee cultivars using AFLP and SSR approaches [7]. Apart from one AFLP marker, all molecular markers present in the cultivars were also detected in the wild accessions, thus confirming the Ethiopian origin of the Typica and Bourbon populations. Accessions derived from the base populations were grouped separately according to their genetic origin and distinguishable from the wild genotypes. Genetic diversity, as expressed by the number of markers detected, and polymorphism appeared to be much weaker in the cultivars than in wild coffee. Only about half of the 107 AFLP markers in the study were observed in the cultivars, whereas 90% of the markers were found in the wild accessions. Polymorphism was much higher within the wild accessions (94 AFLP markers) than within the Typica- (5) and Bourbon-derived (14) cultivars. No cultivar-specific DNA markers were identified in the study. Dissemination of coffee and the selection that followed have therefore strongly reduced the genetic diversity present in the diversity center of the species. In addition, polymorphism was reduced during the selection cycles due to the homogenization of genetic structure, favored by the reproductive mode (autogamy) of the species.

Genetic distances calculated from the AFLP data were low between the Typica- and Bourbon-derived cultivars, viz. 0.17 on average [7]. This weak differentiation does not explain the vigor observed in hybrids (Typica x Bourbon) bred in Brazil under the name cv. Mundo Novo [26]. Heterosis of the hybrids could be a result of the complementary action of certain genes that govern the whole plant rather than genetic differentiation of their parents. Higher genetic distance values were estimated between cultivars and wild accessions, ranging from 0.40 to 0.71 for Typica-derived cultivars and 0.37 to 0.70 for Bourbon-derived [7]. The highest values were generated by a wild accession collected in southern Ethiopia, thus confirming the division of wild coffee into two groups separated by the Great Rift Valley [6].

In conclusion, genetic diversity of C. *arabica* species is structured in two main groups corresponding to wild and cultivated coffee, with each group divided into two subgroups (Fig. 9.5). Base populations of cultivated coffee clearly originated from wild coffee collected in southwestern Ethiopia. Genetic differentiation was evaluated as low between the Typica and Bourbon groups and much higher between wild and cultivated coffee. These results should increase interest in wild coffee for the purpose of broadening the genetic base of cultivars. Few resistance genes have been identified in wild coffee from Ethiopia, however, which has limited their use in breeding programs [4].

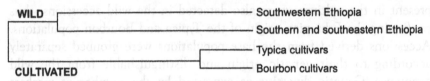

Fig. 9.5: Summary representation of the genetic diversity in *C. arabica*, based on DNA polymorphism analyses.

BREEDING STRATEGIES BASED ON GENE INTROGRESSION

Many characters of interest for *C. arabica* breeding are present in diploid coffee species [4]. The most relevant traits are resistance to leaf rust (*Hemileia vastatrix*) in *C. canephora* and *C. liberica*, coffee berry disease (*Colletotrichum kahawae*) in *C. canephora*, root-knot nematodes (*Meloidogyne* spp.) in *C. canephora* and *C. racemosa*, and the leaf miner (*Perileucoptera coffeella*) in *C. racemosa*. Up to now, only a few *C. canephora* and *C. liberica* genotypes have been used as sources of resistance in *C. arabica* breeding programs. Progenies of interspecific hybrids have been intensively selected for leaf rust resistance inherited from the diploid progenitor. Studies of introgressed lines reported below sought to specify the extent of introgression after several selfed generations and the impact on coffee quality. The results are useful for designing breeding strategies based on gene introgression.

Introgression from *C. canephora*

The presence of *C. canephora* DNA fragments was sought in accessions derived from a tetraploid interspecific hybrid known as Timor Hybrid. Timor Hybrid originated from a spontaneous cross between *C. arabica* and *C. canephora* on the island of Timor [15]. Following a backcross with a C. *arabica* cultivar (i.e., cv. Caturra or Villasarchi), progenies were selfed and selected over three to five generations in several important coffee-producing countries such as Brazil, Colombia and Kenya. The introgressed accessions were distinguished from the C. *arabica* accessions by 178 AFLP markers consisting of 109 additional bands and 69 missing bands [51]. The additional bands were detected only in C. *canephora* accessions and none of the C. *arabica* accessions, so they could be considered introgressed markers. Similarly, the missing bands were detected in C. *arabica* but not in C. *canephora*. These markers were considered related to the introgression process, but not as introgressed markers per se. The number of additional and missing bands ranged from 18 to 59 and 0 to 32 respectively, among Timor Hybrid-derived accessions (Fig. 9.6). The introgressed fragments were estimated to represent from 8% to 27% of the C. *canephora* genome. Assuming that a single genotype of C. *canephora* was involved in formation of the Timor Hybrid, the overall 109 introgressed fragments identified in the Timor Hybrid derived accessions were estimated to represent 51% of the C. *canephora* genome. Many introgressed DNA were neither eliminated nor counterselected during the selfing and selection process.

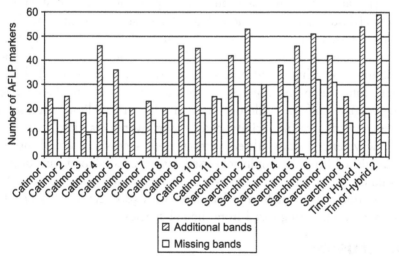

Fig. 9.6: Number of AFLP markers attributable to introgression in accessions derived from a natural interspecific hybrid (C. *arabica* x C. *canephora*), called Timor Hybrid [51].

Introgression from *C. liberica*

A similar study was undertaken to analyze the S.288 progeny of a putative natural hybrid (C. *arabica* x C. *liberica*) and accessions (F_2 and F_4) derived from the cross (S.288 x Kent) [71]. The introgression markers were composed of 65 additional bands and 37 missing bands. Differences in the introgression level between the introgressed parent S.288, F_2 and F_4 progenies were not marked. The limited number of introgressed markers generally indicated that introgression was restricted to a few chromosome segments. Considering the 36 AFLP combinations common to this study and to the analysis of Timor Hybrid-derived accessions [51], the number of introgression markers was found to be similar in the C. *liberica* and C. *canephora* introgressed accessions.

Impact of Introgression on Coffee Quality

C. *canephora* coffee has poorer beverage quality than C. *arabica*, C. *canephora* beans are characterized by higher caffeine and chlorogenic acid contents than C. *arabica* beans, and lower fat, sucrose and trigonelline contents [31]. The incidence of C. *canephora* introgression on biochemical components of beans and cup quality was investigated in Timor Hybrid-derived accessions [13]. These accessions presented 1 to 37 introgressed AFLP markers. No correlation was found between the amount of introgression and the biochemical contents or the beverage attributes. Some highly introgressed accessions (> 25 markers) did not differ from the nonintrogressed controls for any of the biochemical components (caffeine, chlorogenic acids, fat, trigonelline, sucrose) or beverage attributes (i.e. acidity, body, flavor, overall standard). Based on NIR wavelength data, three introgressed accessions could not be distinguished from the nonintrogressed controls. Combined with data on resistance to leaf rust and a root-knot nematode (*Meloidogyne exigua*), it appears that it is possible to select lines with good cup quality and the resistance traits introgressed from diploid species.

CONCLUSION

Recent development of molecular markers has opened new avenues for investigating the composition and organization of the coffee genome. The allotetraploid origin of C. *arabica* was clearly confirmed and the diploid progenitors of the species were found to be genetically close to the actual parent species, i.e., *C. eugenioides* as maternal donor and *C. canephora* as paternal donor. The high level of chromosome affinity between the two genomes of *C. arabica* and between *C. arabica* and the diploid species suggests that speciation of the allotetraploid species took place recently. This recent origin and the biological characteristics of the species, such as autogamy and perenniality, have helped to homogenize the genetic structure of the *C. arabica* genome, thus substantially reducing polymorphism. On the contrary, the genome of diploid species was characterized by high polymorphism and heterozygosity. They represent a valuable source of genes, especially resistance genes, for potential *C. arabica* cultivar improvement. Genes from diploid species can be transferred into the *C. arabica* genome by controlled hybridization. Analysis of segregation and recombination in interspecific hybrids has led to the development of adapted strategies for gene introgression. When gene transfer is attempted through triploid hybrids, the introgressed fragments are severely counterselected, which could facilitate recovery of the *C. arabica* phenotype. It would be more efficient to produce a large number of crosses rather than a large number of plants per progeny to ensure the transfer of desired genes using triploid hybrids. On the other hand, tetraploid interspecific hybrids appear to be particularly suitable for intergenomic recombination, which could enhance polymorphism. There is still a substantial extent of alien DNA after four or five selfed generations. However, introgressed genotypes can still be obtained that combine both desired genes and cup quality similar to the nonintrogressed standard.

Conventional coffee-breeding programs have been handicapped by considerable difficulties in gene introgression, in terms of time, labor, and cost. Around 30 years were spent in the selection of introgressed lines resistant to leaf rust derived from the Timor Hybrid (e.g. Variedad Colombia, Iapar 59, Obata). Development of molecular marker-assisted selection promises to increase the efficiency of breeding programs by: i) enabling selection at early stages and on a large number of breeding lines, ii) reducing the number of backcross cycles required to restore the quality of traditional cultivars, and iii) combining selection for various traits in one step, in particular for pest resistance [52]. Several research programs have been implemented to identify genes of interest for *C. arabica* breeding. Thus the first resistance gene was recently identified in coffee [65] and fine mapping of the region containing the gene is under way. Genes of interest for quality have also been identified in the biosynthetic pathways of caffeine [62,67] and ethylene [64], and in the synthesis of seed storage proteins [1,60,73].

Other recent advances in coffee biotechnological techniques could benefit coffee breeders for the development of new cultivars [40]. The high performance already achieved in the *in-vitro* propagation process by somatic embryogenesis has facilitated mass propagation of superior heterozygous genotypes (e.g. F_1 hybrids) [38, 39]. The technique could be applied to C. *canephora* to produce rootstocks resistant to soil pests [11, 14]. Following the advances achieved in *in-vitro* regeneration of plantlets, genetic transformation has been successfully developed by several research teams for C. *arabica* and C. *canephora* but it is still a tedious process. Only a few genes have been transferred so far into coffee genotypes through *Agrobacterium*-mediated transformation [74]. Studies are currently under way to define promoters driving gene expression in specific tissues or inducible promoters.

REFERENCES

[1] Acuña R, Bassüner R, Beilinson V, et al. Coffee seeds contain 11S storage proteins. Physiol Pl 1999; 105: 122-131.

[2] Aga E, Bryngelsson T, Bekele E, Salomon B. Genetic diversity of forest arabica coffee (*Coffea arabica* L.) in Ethiopia revealed by random amplified polymorphic DNA (RAPD) analysis. Hereditas 2003; 138: 36-46.

[3] Anthony F, Couturon E, de Namur C. Résultats d'une mission de prospection effectuée par l'ORSTOM en 1983. Proc 11th Intl Sci Coll Coffee. Vevey: ASIC, 1985: 495-505.

[4] Anthony F, Astorga C, Berthaud J. Los recursos genéticos: las bases de una solución genética a los problemas de la caficultura latinoamericana. In: Bertrand B, Rapidel B, eds. Desafíos de la caficultura centroamericana. San José (Costa Rica): IICA, 1999: 369-406.

[5] Anthony F, Berthaud J, Guillaumet JL, Lourd M. Collecting wild *Coffea* species in Kenya and Tanzania. Pl Genet Res Newsl 1987; 69: 23-29.

[6] Anthony F, Bertrand B, Quiros O, et al. Genetic diversity of wild coffee (*Coffea arabica* L.) using molecular markers. Euphytica 2001; 118: 53-65.

[7] Anthony F, Combes MC, Astorga C, et al. The origin of cultivated *Coffea arabica* L. varieties revealed by AFLP and SSR markers. Theor Appl Genet 2002; 104: 894-900.

[8] Berthaud J, Charrier A, Guillaumet JL, Lourd M. Les caféiers. In: Pernès J, ed. Gestion des ressources génétiques des plantes. Tome 1: Monographies. Paris: Lavoisier, 1984: 45-104.

[9] Berthou F. Méthode d'obtention de polyploïdes dans le genre *Coffea* par traitements localisés de bourgeons à la colchicine. Café Cacao Thé 1975; 19: 197-202.

[10] Berthou F, Trouslot P. L'analyse du polymorphisme enzymatique dans le genre *Coffea*: adaptation d'une méthode d'électrophorèse en série, premiers résultats. Proc. 8th Intl Sci Coll Coffee. Vevey: ASIC, 1977: 373-383.

[11] Berthouly M, Etienne H. Somatic embryogenesis of coffee. In: Jain SM, Gupta PK, Newton RJ, eds. Somatic embryogenesis in woody plants, vol. 5. Dordrecht, Netherlands: Kluwer Academic Publishers, 1999: 259-288.

[12] Bertrand B, Aguilar G, Santacreo R, Anzueto F. El mejoramiento genético en América Central. In: Bertrand B, Rapidel B, eds. Desafíos de la caficultura centroamericana. San José (Costa Rica): IICA, 1999: 407-456.

[13] Bertrand B, Guyot B, Anthony F, Lashermes P. Impact of *Coffea canephora* gene introgression on beverage quality of C. *arabica*. Theor Appl Genet 2003; 107: 387-394.

[14] Bertrand B, Anzueto F, Moran MX, Eskes AB, Etienne H. Creation and diffusion by somatic embryogenesis of the 'Nemaya' rootstock variety (*Coffea canephora*) in Central America. Plantations Rech Dev 2002; special issue: 105-107.

[15] Bettencourt A. Consideraçoes gerais sobre o 'Hibrido de Timor'. Circular n°31 of Instituto Agronômico de Campinas 1973: 256.

[16] Bouharmont J. Recherche sur les affinités chromosomiques dans le genre *Coffea*. INEAC; 1959: Série scientifique 77.

[17] Bremer B, Jansen RK. Comparative restriction site mapping of chloroplast DNA implies new phylogenetic relationships within *Rubiaceae*. Amer J Bot 1991; 78: 198-213.

[18] Bridson D. Studies in *Coffea* and *Psilanthus* (*Rubiaceae* subfam. *Cinchonoideae*) for Part 2 of 'Flora of Tropical East Africa': *Rubiaceae*. Kew Bull 1982; 36: 817-859.

[19] Bridson D. Nomenclatural notes on *Psilanthus*, including *Coffea* sect. *Paracoffea* (*Rubiaceae* tribe *Coffeae*). Kew Bull 1987; 42: 453-460.

[20] Bridson D, Verdcourt B. *Coffea*. In: Polhill RM, ed. Flora of Tropical East Africa. Rubiaceae (Part 2). Rotterdam: A.A. Balkema 1988: 703-727.

[21] Carvalho A. Distribuição geografica e classificação botânica do gênero *Coffea* com referência especial à espécie Arabica. Bolétim da Superintendência dos Serviços do café 1946; 21: 174-180.

[22] Carvalho A. Taxonomia de *Coffea arabica* L. Caracteres morfologicos dos haploides. Bragantia 1952; 12: 201-212.

[23] Carvalho A. Melhoramento do cafeeiro. Cruzamentos entre C. *arabica* e c. *canephora*. Proc. 10th Intl Sci Coll Coffee. Vevey: ASIC, 1982: 363-368.

[24] Carvalho A. Coffee agronomy principles and practice of coffee plant breeding for productivity and quality factors: *Coffea arabica*. In: Clarke RJ, Macrae R, eds. Coffee, volume 4: Agronomy. London: Elsevier Applied Science, 1988: 129-165.

[25] Carvalho A, Monaco LC. Relaciones genéticas de especies seleccionadas de *Coffea*. Café IICA 1968; 9: 1-19.

[26] Carvalho A, Ferwerda FP, Frahm-Leliveld JA, et al. Coffee. In: Ferwerda FP, Wit F, eds. Outlines of perennial crop breeding in the Tropics. Wageningen: Veenman & Zonen NV, 1969: 189-241.

[27] Charrier A, Berthaud J. Botanical classification of coffee. In: Clifford MN, Willson KC, eds. Coffee botany, biochemistry and production of beans and beverage. London: Croom Helm, 1985: 13-47.

[28] Cheek M, Csiba L, Bridson D. A new species of *Coffea* (*Rubiaceae*) from western Cameroon. Kew Bull 2002; 57: 675-680.

[29] Chevalier A. Les vrais et les faux caféiers. Nomenclature et systématique. In: Les caféiers du globe. Paris: P. Lechevalier, 1947 (Encyclopédie biologique n° 28).

[30] Chevalier A, Dagron M. Recherches historiques sur les débuts de la culture du caféier en Amérique. Comm Actes Acad Sci Col (Paris) 1928; 5: 1-38.

[31] Clifford MN. Chemical and physical aspects of green coffee and coffee products. In: Clifford MN, Wilson KC, eds. Coffee: botany, biochemistry and productions of beans and beverage. London: Croom Helm, 1985: 305-374.

[32] Couturon E, Lashermes P, Charrier A. First intergeneric hybrids (*Psilanthus ebracteolatus* Hiern x *Coffea arabica* L.) in coffee trees. Can J Bot 1998; 76: 542-546.

[33] Cramer PJS. Review of literature of coffee research in Indonesia. Turrialba: Interamer Inst Agric Sci, 1957 (misc publ. no. 15).

[34] Cros J, Combes MC, Chabrillange N, et al. Nuclear content in the subgenus *Coffea* (Rubiaceae): inter- and intra-specific variation in African species. Can J Bot 1995; 73: 14-20.

[35] Cros J, Combes MC, Trouslot P, et al. Phylogenetic relationships of *Coffea* species: new evidence based on the chloroplast DNA variation analysis. Mol Phyl Evol 1998; 9: 109-117.

[36] Dussert S, Lashermes P, Anthony F, et al. Hamon S. Coffee (*Coffea canephora*). In: Hamon P, Seguin M, Perrier X, Glaszmann C, eds. Genetic diversity of cultivated tropical plants. Plymouth: Science Pub, Inc. 2003: 239-258.

[37] Eskes AB. Identification, description and collection of coffee types in P.D.R. Yemen. Rome: IPGRI, 1989.

[38] Etienne H. Somatic embryogenesis protocol: coffee (*Coffea arabica* L. and C. *canephora* P.). In: Jain SM, Gupta P, eds. Protocols of somatic embryogenesis—woody plants. Dordrecht, Netherlands. Kluwer Academic Publishers, 2004 (in press).

[39] Etienne H, Etienne-Barry D, Vásquez N, Berthouly M. Aportes de la Biotecnología al mejoramiento genético del café: el ejemplo de la multiplicación por embriogénesis somática de híbridos F1 en America Central. In: Bertrand B, Rapidel B, eds. Desafíos de la Caficultura Centroamericana. San José (Costa Rica): IICA, 1999: 457-496.

[40] Etienne H, Anthony F, Dussert S, et al. Biotechnological applications for the improvement of coffee (*Coffea arabica* L.). In Vitro Cell Dev Biol Pl 2002; 38: 129-138.

[41] Haarer AE. Modern coffee production. London: Leonard Hill Limited, 1956.

[42] Herrera JC. Amélioration génétique de l'espèce polyploïde *Coffea arabica* : maîtrise de son introgression par les espèces diploïdes de caféiers. PhD thèse, Ecole Nationale Supérieure Agronomique de Montpellier, Montpellier 2003.

[43] Herrera JC, Combes MC, Anthony F, Charrier A, Lashermes P. Introgression into the allotetraploid coffee (*Coffea arabica* L.): segregation and recombination of the C. *canephora* genome in the tetraploid interspecific hybrid (C. *arabica* x C. *canephora*). Theor Appl Genet 2002; 104: 661-668.

[44] Herrera JC, Combes MC, Cortina H, Alvarado G, Lashermes P. Gene introgression into *Coffea arabica* by way of triploid hybrids (C. *arabica* x C. *canephora*). Heredity 2002; 89: 488-494.

[45] Krug CA, Mendes JET, Carvalho A. Taxonomia de *Coffea arabica* L. Campinas: Instituto Agronômico do Estado, 1939 (Bol Técn no. 62).

[46] Lashermes P, Cros J, Combes MC, et al. Inheritance and restriction fragment length polymorphism of chloroplast DNA in the genus *Coffea* L. Theor Appl Genet 1996; 93: 626-632.

[47] Lashermes P, Bertrand B, Anthony F. Current status of molecular research on coffee with special reference to marker assisted breeding: scope and limitations. Proc Intl scientific symposium on coffee. Bangalore: Central Coffee Research Institute, 2000: 31-39.

[48] Lashermes P, Combes MC, Trouslot P, Charrier A. Phylogenetic relationships of coffee tree species (*Coffea* L.) as inferred from ITS sequences of nuclear ribosomal DNA. Theor Appl Genet 1997; 94: 947-955.

[49] Lashermes P, Trouslot P, Anthony F, Combes MC, Charrier A. Genetic diversity for RAPD markers between cultivated and wild accessions of *Coffea arabica*. Euphytica 1996; 87: 59-64.

[50] Lashermes P, Combes MC, Robert J, et al. A. Molecular characterisation and origin of the *Coffea arabica* L. genome. Mol Gen Genet 1999; 261: 259-266.

[51] Lashermes P, Andrzejewski S, Bertrand B, et al. Molecular analysis of introgressive breeding in coffee (*Coffea arabica* L.). Theor Appl Genet 2000; 100: 139-146.

[52] Lashermes P, Combes MC, Topart P, et al. Molecular breeding in coffee (*Coffea arabica* L.). In: Sera T, Soccol CR, Pandey A, Roussos S, eds. Coffee biotechnology and quality. Dordrecht, Netherlands: Kluwer Academic Publishers, 2000: 101-112.

[53] Lashermes P, Paczek V, Trouslot P, et al. Single-locus inheritance in the allotetraploid *Coffea arabica* L. and interspecific hybrid C. *arabica* x C. *canephora*. J Heredity 2000; 91: 81-85.

[54] Lashermes P, Combes MC, Prakash NS, et al. Genetic linkage map of *Coffea canephora*: effect of segregation distortion and analysis of recombination rate in male and female meioses. Genome 2001; 44: 589-595.

[55] Le Pierrès D. Etude des hybrides interspécifiques tétraploïdes de première génération entre *Coffea arabica* et les caféiers diploïdes. PhD thèse, Université Paris XI, Orsay 1995.

[56] Lejeune JBH. Rapport au Gouvernement Impérial d'Ethiopie sur la production caféière. Rome: FAO, 1958.

[57] Leroy JF. Evolution et taxogenèse chez les caféiers: hypothèse sur l'origine. C R Acad Sci Paris 1980; 291: 593-596.

[58] Louarn J. Bilan des hybridations interspécifiques entre caféiers africains diploïdes en collection en Côte d'Ivoire. Proc. 12th Intl Sci Coll Coffee. Vevey: ASIC, 1982: 375-384.

[59] Louarn J. Structure génétique des caféiers africains diploïdes basée sur la fertilité des hybrides interspécifiques. Proc 15th Intl Sci Coll Coffee. Vevey: ASIC, 1993: 243-252.

[60] Marraccini P, Deshayes A, Pétiard V, Rogers WJ. Molecular cloning of the complete 11S seed storage protein gene of *Coffea arabica* and promoter analysis in transgenic tobacco plants. Pl Physiol Biochem 1999; 37: 273-282.

[61] Mendes AJT. Duplicaçao do numero de cromosomicos em café, algodao e fumo, pela açao da colchicina. Campinas: Instituto Agronômico do Estado, 1939 (Bol Técn no. 75).

[62] Moisyadi S, Neupane KR, Stiles JI. Cloning and characterization of a cDNA encoding xanthosine-N^7-methyltransferase from coffee (*Coffea arabica*). Acta Hort 1998; 461: 367-377.

[63] Namur C de, Couturon E, Sita P, Anthony F. Résultats d'une mission de prospection des caféiers sauvages du Congo. Proc. 12th Intl Sci Coll Coffee. Vevey: ASIC, 1987: 397-404.

[64] Neupane KR, Moisyadi S, Stiles JI. Cloning and characterization of fruit-expressed ACC synthase and ACC oxidase from coffee. Proc 18th Intl Sci Coll Coffee. Vevey: ASIC, 1999: 322-326.

[65] Noir S, Anthony F, Bertrand B, Combes MC, Lashermes P. Identification of a major gene (Mex-1) from Coffea canephora conferring resistance to Meloidogyne exigua in Coffea arabica. Pl Path 2003; 52: 97-103.

[66] Noir S, Patheyron S, Combes MC, Lashermes P, Chalhoub B. Construction and characterisation of a BAC library for genome analysis of the allotetraploid coffee species (Coffea arabica L.). Theor Appl Genet 2004; 109: 225-230 .

[67] Ogawa M, Herai Y, Koizumi N, Kusano T, Sano H. 7-Methylxanthine methyltransferase of coffee plants: gene isolation and enzymatic properties. J Biol Chem 2001; 276: 8213-8218.

[68] Orozco-Castillo C, Chalmers KJ, Waugh R, Powell W. Detection of genetic diversity and selective gene introgression in coffee using RAPD markers. Theor Appl Genet 1994; 87: 934-940.

[69] Osorio N. The global coffee crisis: a threat to sustainable development. London: ICO, 2002.

[70] Pinto-Maglio CAF, da Cruz ND. Pachytene chromosome morphology in Coffea L. II. C. arabica L. complement. Caryologia 1998; 51: 19-35.

[71] Prakash NS, Combes MC, Somanna N, Lashermes P. AFLP analysis of introgression in coffee cultivars (Coffea arabica L.) derived from a natural interspecific hybrid. Euphytica 2002; 124: 265-271.

[72] Prakash NS, Combes MC, Dussert S, Naveen S, Lashermes P. Analysis of genetic diversity in Indian robusta coffee genepool (Coffea canephora) in comparison with a representative core collection using SSRs and AFLPs. Genet Res Crop Evol 2004 (in press).

[73] Rogers WJ, Bézard G, Deshayes A, et al. Biochemical and molecular characterization and expression of the 11S-type storage protein from Coffea arabica endosperm. Pl Physiol Biochem 1999; 37: 261-272.

[74] Spiral J, Leroy T, Paillard M, Pétiard V. Transgenic coffee (Coffea species). In: Bajaj YPS, ed. Biotechnology in agriculture and forestry, vol. 44: Transgenic trees. Berlin/ Heidelberg: Springer-Verlag, 1999: 55-76.

[75] Sreenivasan MS, Ram AS, Prakash NS. Tetraploid interspecific hybrids in coffee breeding in India. Proc 15th Intl Sci Coll Coffee. Vevey: ASIC, 1993: 226-233.

[76] Stoffelen P. Coffea and Psilanthus (Rubiaceae) in tropical Africa: a systematic and palynological study, including a revision of the West and Central African species. PhD thesis, Katholieke Universiteit Leuven, Faculteit Wetenschappen, Louvain, Belgium, 1998.

[77] Stoffelen P, Robbrecht E, Smet E. Coffea (Rubiaceae) in Cameroon: a new species and a nomen recognized as species. Belg J Bot 1996; 129: 71-76.

[78] Sybenga J. Genetics and cytology of coffee (a literature review). Bibliografia Genetica 1960; 19: 217-316.

[79] Sylvain PG. Some observations on Coffea arabica L. in Ethiopia. Turrialba 1955; 5: 37-53.

[80] Thomas AS. The wild arabica coffee on the Boma Plateau, Anglo-Egyptian Sudan. Empire J Exper Agric 1942; 10: 207-212.

[81] Thomas AS. The wild coffee of Uganda. Empire J Exper Agric 1944; 12: 1-12.

[82] Wellman FL. Coffee: botany, cultivation and utilization. London: Leonard Hill Books, 1961.

Molecular Diversity of *Carica papaya* and Related Species

RAY MING[1], BART VAN DROOGENBROECK[2], PAUL H. MOORE[3], FRANCIS T. ZEE[4], TINA KYNDT[2], XAVIER SCHELDEMAN[5], TERRY SEKIOKA[6], and GODELIEVE GHEYSEN[2]

[1]Hawaii Agriculture Research Center, 99-193 Aiea Heights Drive, Aiea, HI 96701, USA

[2]Department of Molecular Biotechnology, Faculty of Agricultural and Applied Biological Sciences, Ghent University, Coupure Links 653, 9000 Ghent, Belgium

[3]USDA-ARS, Pacific Basin Agricultural Research Center, 99-193 Aiea Heights Drive, Aiea, HI 96701, USA

[4]USDA-ARS, Pacific Basin Agricultural Research Center, Tropical Plant Genetic Resource Management Unit, Hilo, HI 96720, USA

[5]International Plant Genetic Resources Institute (IPGRI), Office for the Americas, A.A. 6713, Cali, Colombia

[6]Department of Tropical Plant and Soil Sciences, University of Hawaii, Honolulu, HI 96822, USA

ABSTRACT

Molecular diversity of Caricaceae species has been a frequent subject of research over the past decade. Genetic relationships among Caricaceae species were established using DNA markers derived from nuclear, chloroplast, and mitochondria genomes. The accumulated molecular data contributed to

Address for Correspondence: Ray Ming, Hawaii Agriculture Research Center, 99-193 Aiea Heights Drive, Aiea, HI 96701, USA Tel: 808-486-5374 Fax: 808-486-5020 E-mail: rming@harc-hspa.com

reinstating *Vasconcellea* as a genus independent of the genus *Carica* in the family Caricaceae. These data provided evidence for the interspecific origin of some species, and support for a reticulate evolution for this genus. Within the cultivated species *Carica papaya*, limited genetic variations were detected among commercial cultivars, improved but not released breeding lines, and unimproved germplasm, reflecting the consequence of inbreeding from a limited gene pool. Genetic diversity among dioecious cultivars was similar to that of the hermaphrodite cultivars, possibly due to the narrow genetic base from which both dioecious and hermaphrodite cultivars were derived. The recent identification of a primitive Y chromosome in papaya offered a new and unique aspect for studying the divergence and evolution among the 35 species of Caricaceae and promises new knowledge in the origin of dioecy in this family. Future large-scale sequencing of papaya genomic DNA will revolutionize methods for assessing genetic variation within the species *Carica papaya*.

Key words: Caricaceae, germplasm, genetic relationship, molecular phylogeny, polymorphism

INTRODUCTION

Papaya (*Carica papaya* L.) is a principal fruit crop of tropical and subtropical regions worldwide. Papaya trees are grown for both fruit and papain, a commercially valuable proteolytic enzyme [15]. The fruit is consumed and fruit, stems, leaves, and roots of papaya are used in a wide range of medical applications [35, 36, 38]. Papaya is one of the few plant species that bear fruit throughout the year and can produce ripe fruit in as little as 9 months from planting. A papaya tree may live for 25 years or longer, bearing continuously one or more fruit in each leaf axil and each fruit containing about 1,000 seeds. Hand pollination is easily done by transferring pollen from male or hermaphrodite flowers to the stigmata of female or hermaphrodite flowers. These characteristics – easy in making crosses, producing large numbers of offspring, short crop cycle, and continuous flowering throughout the year – make papaya an attractive tree model for genetic research.

The economic importance of Caricaceae is largely due to the fruit production of *C. papaya* L. However, some species of the genus *Vasconcellea*, the so-called 'highland papayas', are regarded as unexploited crops despite their taste, high-quality fruits and being a source of proteolytic enzymes [34]. The babaco (*V. × heilbornii* 'Babaco') is one such *Vasconcellea* species. Thanks to its high yields, the babaco is of great commercial interest. It is at this moment the only *Vasconcellea* species cultivated intensively, albeit only in Ecuador and New Zealand [53]. The National Research Council [34] classifies the potential of *Vasconcellea* species at three levels: (i) direct use of the fruits; (ii) use of genetic variability as "raw" material for the creation of new *Vasconcellea* fruits; and (iii) use in breeding programs for papaya improvement in order to extend cultivation range, by using genetic endowment for cold adaptability and papaya production, by using disease resistance genes from highland papayas. In addition, a preliminary study on *Vasconcellea* species showed that they, especially *V. stipulata* and a number of as yet unclassified varieties of *V. × heilbornii*, contain latex showing a proteolytic activity up to 17 times higher than that of the reference *Carica papaya*. Taking into account the clearly higher proteolytic activity, *Vasconcellea* species can thus constitute a good alternative for papain extraction [41].

Papaya is a diploid with nine pairs of chromosomes and a small genome size of 372 Mbp [3,46]. Over the past 15 years, several major advancements have enhanced papaya research and improvement. (i) Application of DNA markers in germplasm evaluation clarified the phylogenetic relationships of papaya and its wild relatives [2, 24, 44, 51, 52]. (ii) Transgenic papaya with the papaya ringspot virus (PRSV) coat protein gene was successfully developed and released to save the Hawaiian papaya industry from collapse [16, 19]. (iii) Sex-linked DNA markers were developed and used in selection of preferred sex types at the seedling stage [13,37,50]. (iv) Genetic linkage maps of papaya were constructed for mapping major genes and quantitative trait loci (QTL) [28,42]. (v) A deep coverage bacterial artificial chromosome (BAC) library was constructed for gene cloning and physical mapping [33]. (vi) The youngest Y chromosome in eukaryotes known to date was discovered in papaya [27]. Papaya is gradually being recognized as a good model system for genomic and biological studies.

TAXONOMY

Papaya is a member of a small dicotyledonous family, Caricaceae, comprising six genera of herbaceous or semiwoody plants [4, 5, 6]. The taxonomy of genera *Carica* and *Vasconcella* has changed over time. *Vasconcella*, once a genus, later classified as a section of *Carica*, was recently reinstated as a genus based on molecular evidence of the chloroplast DNA and morphological observations [2, 6]. Shortly after his 2000 publication [6], Badillo [7] published a corrective note for the generic name of genus *Vasconcella*, suggesting that it should be *Vasconcellea*. Following the reinstatement of *Vasconcellea*, *Carica papaya* is the only species in genus *Carica*, while *Vasconcellea* comprises the other 21 species.

There are major differences between genera *Carica* and *Vasconcellea*. (i) *C. papaya* is reproductively isolated. *C. papaya* could not be hybridized with any *Vasconcellea* species, except when embryo rescue techniques were used [14], while the *Vasconcellea* species hybridize freely, and there is a complex of natural hybrids in South America with unknown parentage. (ii) *C. papaya* originated and adapted to lowland tropical area, while *Vasconcellea* species are highland papayas and relatively cold tolerant [5, 6]. (iii) Chloroplast and nuclear DNA between these two genera have distinctive fingerprints [2, 24, 51]. (iv) The stems of *C. papaya* are hollow, while those of *Vasconcellea* species are solid. (v) *C. papaya* is characterized by a unilocular ovary, whereas *Vasconcellea* species possess pentalocular ovaries. Other morphological differences between the two economically most important Caricaceae genera are summarized in Table 10.1.

Table 10.1: Morphological differences between *Vasconcellea* and *Carica* [6]

	Vasconcellea	*Carica*
Ovary	(almost) completely 5-locular	1-locular
Stem	Medullary stem	Hollow
Number of veins	≤ 7 (-11)	≥ 7 (-11)
Trichomes on filaments	If present, never moniliform	Moniliform
Stigmata	Repeatedly and irregularly divided	Integer or forked 1-2 times
Sclerotesta	Without lamelliform protuberances	With lamelliform protuberances

The small family Caricaceae consists of six genera and 35 species. *Vasconcellea* is the largest genus with 21 species, followed by *Jacaratia* seven species, *Jarilla* three species, *Cylicomorpha* two species, and *Horovitzia* and *Carica* with one species each (Table 10.2). Within the family, only two species, *C. papaya* and *Vasconcellea cundinamarcensis* (formerly *Carica pubescens*), are described by Badillo [5] as polygamous with all three sex types; *V. monoica* is monoecious and the other 32 species are all described as dioecious. However, during field trips in Ecuador monoecious specimens have also been observed in other *Vasconcellea* spp., such as *V. parviflora* and *V. × heilbornii* (pers. obs., X. Scheldeman, B. Van Droogenbroeck, and T. Kyndt).

Table 10.2: Overview of the 35 species of Caricaceae, their status (wild or cultivated) common name, origin and use (based on Badillo, [5, 7] and pers. obs.)

Genus Carica

Species	Status	Common Name	Origin	Uses
C. papaya L.	Cultivated, wild	papaya	Tropical Central America	Edible fruit

Genus Cylicomorpha

Species	Status	Common Name	Origin	Uses
Cylicomorpha parviflora Urban	Wild	mtonto, milanyemba	Eastern Africa: Kenya, Malawi, Tanzania	
Cylicomorpha solmsii (Urban) Urban	Wild		Western Africa : Nigeria, Congo, Cameroun, Central African Republic	

Genus Horovitzia

Species	Status	Common Name	Origin	Uses
H. cnidoscoloides (Lorrence et Torres) V. Badillo	Wild	Sp: mala mujer	Southern Mexico (Oaxaca)	

Genus Jacaratia

Species	Status	Common Name	Origin	Uses
J. chocoensis A. Gentry et Forero	Wild		Colombia	
J. corumbensis O. Kuntze	Wild	Sp: sipoy, yacónPor: mamaozinho do mato	Bolivia, Paraguay, Argentina, Brazil	

Contd.

Table 10.2   (Contd.)

Species	Status	Common Name	Origin	Uses
J. digitata (Poeppig et Endl.) Solms-Laub.	Wild	Sp: papaya de monte, chamburo, toronche, papaya de monte espinuda, jacaratia, papaya caspi Por: mamao brabo	Bolivia, Brazil, Colombia, Ecuador, Peru	Edible fruit
J. dolichaula (J.D. Smith) Woodson	Wild	Sp: papaya montana, sapota, palo de barril, papaillo	Mexico to Panama	
J. heptaphylla (Vellozo) A. DC.	Wild, cultivated	Por: mamaozinho	Brazil	
J. mexicana A. DC.	Wild	Sp: bonete, orejona, yucatec, kunché, coahuayote, coalsuayote, guayote, julepe	Mexico, Nicaragua, El Salvador, Guatemala, Belize	
J. spinosa (Aublet) A. DC.	Wild	Sp: higo, jacaratia, nacaratia, papayón, papayillo de venado Por: mamao de mato, bravo	Nicaragua to Argentina	Edible fruit

Genus Jarilla

Species	Status	Common Name	Origin	Uses
J. caudata (Brandegee) Standley	Wild	Sp: jarilla, guanadilla, chiguitera	Mexico	
J. chocola Standley	Wild	Sp: chocola, baretillo, kapiah	Mexico, Guatemala	
J. heterophylla (Cerv.) Rusby	Wild	Sp: jarilla, guanadilla, chiguetera, machicua	Mexico	

Contd.

Table 10.2 (Contd.)

Genus *Vasconcellea*

Species	Status	Common Name	Origin	Uses
V. candicans (A. Gray) A.DC.	Wild	Sp: chungay, toronche chicope; mito,	Southern Ecuador to Peru	Edible fruit
V. cauliflora (Jacq.) A.DC.	Wild	Sp: tapaculo, papayo de montaña, zonzapote	Southern Mexico to northern part of South-America	Edible fruit
V. chilensis (Planch. ex A.DC.) A.DC.	Wild	Sp: palo gordo, monte gordo	Central Chile	Fodder
V. crassipetala (V. Badillo) V. Badillo	Wild		Colombia, Ecuador	Edible fruit
V. cundinamarcensis (Solms-Laub.) V. Badillo	Wild, cultivated	Sp: toronche, chamburo, papaya de tierra fria, siglalon	Panama, Colombia, Venezuela, Ecuador, Peru, Bolivia	Edible fruit
V. glandulosa A.DC.	Wild		Peru, Brazil, Bolivia, Argentina	
V. goudotiana Triana et Planch.	Wild, cultivated		Colombia	Edible fruit
V. horovitziana (V. Badillo) V. Badillo	Wild	Sp: papayuela de bejuco, badea del monte	Ecuador	
V. longiflora (V. Badillo) V. Badillo	Wild		Colombia,	Ecuador
V. microcarpa (Jacq.) A.DC. 4 subspecies: baccata microcarpa	Wild	Sp: col de monte, sapiro, lechoso de monte, higuillo Po: mamao rana	Panama, Venezuela, Colombia, Ecuador, Brazil, Peru, French Guyana, Bolivia	Edible fruit

Contd.

Table 10.2 (Contd.)

pilifera heterophylla				
V. monoica (Desf.) A.DC.	Wild	Sp: col de monte, peladera, col de montaña, yumbo papaya, brenjena, toronche, chamburo	Ecuador, Peru, Bolivia	Edible fruit, edible leaves
V. omnilingua (V. Badillo) V. Badillo	Wild	Sp: col de monte	Ecuador	
V. palandensis (V. Badillo et al.) V. Badillo	Wild	Sp: papaillo	Southern Ecuador	Edible fruit
V. parviflora A.DC.	Wild	Sp: papaya de monte, coral, papayillo, yuca del campo	Ecuador, Peru	Edible fruit
V. pulchra (V. Badillo) V. Badillo	Wild	Sp: col de monte	Ecuador	
V. quercifolia (St.-Hil.) A.DC.	Wild	Sp: higuera del monte, sacha higuera, higuerón, calasacha, gargatea, orto karalau, mamón del monte Po:mamaosinho	Peru, Brazil, Argentina, Bolivia, Paraguay, Uruguay	Edible fruit
V. sphaerocarpa (García-Barr. et Hern.) V. Badillo	Wild	Sp: papaya de monte, higuillo negro, higuillo, papayo silvestre, papayuela	Colombia	Edible fruit
V. sprucei (V. Badillo) V. Badillo	Wild		Ecuador	
V. stipulata (V. Badillo) V. Badillo	Wild	Sp: toronche, siglalón silvestre	Ecuador, Peru	Edible fruit

Contd.

Table 10.2 (Contd.)

V. *weberbaueri* (Harms) V. Badillo	Wild	Sp: mausha	Ecuador, Peru	
V. × *heilbornii* (V. Badillo) V. Badillo Dif. Varieties: fructifragrans chrysopetala 'Babacó'	Wild, Cultivated	Sp: toronche, chihualcán, babaco, chamburo	Ecuador	Edible fruit

CENTER OF ORIGIN

The center of origin for five of the six genera in Caricaceae is South and Central America. The single exception, *Cylicomorpha*, originated in Africa. Papaya believed to have originated in Central America, probably in the region of southern Mexico to Costa Rica [10, 39, 49], is now distributed throughout the tropical and subtropical regions worldwide. Highland papayas, a group of underexploited *Vasconcellea* species, occur in South America with the greatest concentration of species in Ecuador, Colombia, and Peru [5, 52, pers. obs., X. Scheldeman].

MOLECULAR DIVERSITY

Genetic Diversity and Relationships among Caricaceae Species

Genetic relationships between papaya and related wild species have been investigated using random amplified polymorphic DNA (RAPD) and isozyme markers [23], restriction fragment length polymorphism (RFLP) in a cpDNA intergenic spacer region [2], and amplified fragment length polymorphism (AFLP) markers [51]. The most extensive molecular analysis of Caricaceae was performed by Van Droogenbroeck et al. [51]. In this study 95 accessions belonging to three genera including *C. papaya*, at least eight *Vasconcellea* species and two *Jacaratia* species, all collected in Ecuador were investigated. The cluster analysis (Fig. 10.1), based on 496 polymorphic AFLP markers generated with five primer combinations, clearly separated the species of the three genera and illustrated the large genetic distance between *C. papaya* accessions and the *Vasconcellea* group. The mean genetic similarity value (0.24), based on the pairwise comparison among all species from *Jacaratia* and *Vasconcellea*, was slightly higher than the average similarity value derived from the comparison of

---

Fig. 10.1: Dendrogram showing the genetic relationships among 95 Caricaceae accessions based on AFLP data, using Dice coefficient of similarity and UPGMA clustering. Accessions are labeled with a species name abbreviation (*V. stipulata*: stip; *V.* × *heilbornii* var. *chrysopetala*: chrys; *V.* × *heilbornii* 'Babacó': bab; *V.* × *heilbornii*: nm??; *V. cundinamarcensis*: pub; *V. weberbaueri*: web; *V. parviflora*: parv; *V. palandensis*: pal; *V. monoica*: mon; *V. sp.??*: carv or carp; *J. digitata*: dig; *J. spinosa*: spin; *C. papaya*: pap) and with a collection number. Numbers shown at the different nodes indicate bootstrap confidence values (1,000 bootstrap replicates). Nodes without numbers had bootstrap values of less than 50.

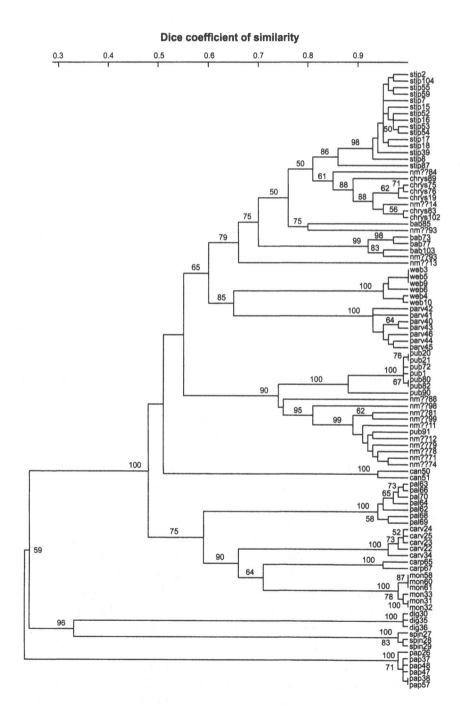

Dice coefficient of similarity

all accessions of *Carica* and *Vasconcellea* (0.23). Within species, all groups displayed very high genetic similarity values (more than 0.90), except for the group of unidentified varieties of *V.* × *heilbornii* (0.78) and the 'Babaco' group (0.78). The average genetic similarity among all *Vasconcellea* taxa was 0.54 [51].

In this study some undescribed *Vasconcellea* specimens collected in Ecuador were investigated with molecular markers, the so-called 'carv' genotypes in Figure 10.1. After morphological and molecular comparison with other described *Vasconcellea* spp., it was discovered that these specimens belong to the species *V. goudotiana*, which had not been previously described in Ecuador (Van Droogenbroeck and Kyndt, unpubl. data).

PCR-RFLP and AFLP results suggested that C. *papaya* diverged from the highland papayas (*Vasconcellea*) early in its evolution. Moreover, Aradhya et al. [2] and Van Droogenbroeck et al. [51] noted that the wild South American species of *Vasconcellea* are more closely allied with a member of the related genus *Jacaratia* than with C. *papaya*. Considering the extent of divergence between papaya and the rest of the genus, and the lack of any closely related South American relatives, the arrival of a papaya progenitor on the North American continent must have occurred considerably before the relatively recent formation of the Isthmus of Panama only 3 million years ago [2]. A likely possibility is that an ancestral *Carica*, as well as progenitors of the Central American genera *Jarilla* and *Horovitzia*, made the leap from South to North America at an earlier time, presumably via dispersal across an island chain between continents, and continued to evolve in isolation from the rest of the members of Caricaceae after the migration route was severed [2]. The timing of these events is speculative, but biogeographers have extensively documented movements of other groups of plants and animals between North and South America, in particular during the late Cretaceous and early Tertiary periods (about 70 to 60 million years ago) and again from late Miocene (about 10 mya) to the present [9, 43]. The pattern displayed by the Caricaceae, in which a predominantly tropical South American family has divergent Central American members, is common among the Gondwanan floral groups [40], as demonstrated in the plant family Bignoniaceae [17].

More recently, the chloroplast and mitochondrial DNA diversity of 79 genotypes belonging to genera *Vasconcellea* (17 species), *Carica* (1 species), and *Jacaratia* (2 species), were studied by PCR-RFLP analysis of noncoding cpDNA and mtDNA regions [52]. Figure 10.2 shows a typical PCR-RFLP pattern obtained after restriction of the PCR-amplified noncoding chloroplast DNA region K1K2 with the enzyme *AfaI*. Intraspecific variability within *V. microcarpa* chloroplast DNA was detected and the morphologically similar *V. longiflora* and *V. pulchra* genotypes displayed an identical restriction pattern with this fragment/ enzyme combination. It was evident that after a detailed cluster analysis of the polymorphic restriction fragments that an insufficient number of markers discriminated among the different Caricaceae genera. Hence inference of intergeneric relationships within Caricaceae was not possible. However, some conclusions could be noted at a lower taxonomic level (Fig. 10.3). Several species shared the same haplotype, others had a unique haplotype. The putative progenitors of the economically important, natural sterile hybrid *V. × heilbornii*, (i.e., *V. stipulata* and *V. cundinamarcensis*), were only distantly related. This indicates that probably none of these species was involved as the maternal progenitor in the origin of *V. × heilbornii*. Surprisingly, *V. × heilbornii* had organellar genome patterns identical with *V. weberbaueri*, suggesting a possible involvement of this species in the origin of *V. × heilbornii*. On the basis of discrepancy between morphological traits and the cpDNA profiles of some pairs of *Vasconcellea* species, we believe that in addition to *V. × heilbornii*, some other species also originated through interspecific

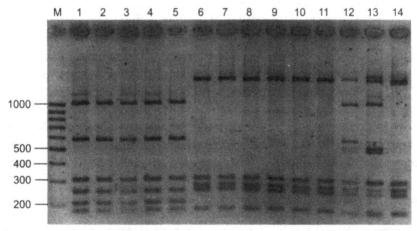

Fig. 10.2: Restriction fragment patterns of chloroplast DNA from *Vasconcellea* spp. detected by the fragment|enzyme combination K1K2|*AfaI*. M = Massruler™ DNA ladder (size in base pairs). Lanes 1-14 represent the following taxa: 1-5 *V. goudotiana*, 6-8 *V. longiflora*, 9-11 *V. pulchra*, 12-14 *V. microcarpa*

hybridization. Intraspecific cpDNA variation was detected in *V. microcarpa*, *V. stipulata*, and *V.* × *heilbornii*, providing molecular evidence for the high diversity previously indicated by morphological observations in these gene pools and also suggesting the occurrence of interspecific hybridization events with subsequent chloroplast capture. A reticulate evolution for *Vasconcellea*, the largest genus within Caricaceae, has therefore been suggested. To further elucidate phylogeny of Caricaceae and to study the extent of hybridization in the genus *Vasconcellea*, nuclear and chloroplast sequences are being currently being investigated [26].

The mode of inheritance of chloroplast and mitochondrial DNA has been determined in intergeneric hybrids between *C. papaya* and four different *Vasconcellea* species using PCR-RFLP. These hybrids were generated to obtain introgression of desired *Vasconcellea* traits into papaya. These traits include resistance to papaya ringspot virus (PRSV), found in several species, the pleasant fragrance of *V. stipulata* or babaco, the monoecious habit of *V. monoica*, the cold tolerance of *V. cundinamarcensis* and *V. stipulata*, the high sugar content of *V. quercifolia*, and the ornamental qualities of pink-flowered *V. parviflora* [31]. In the initial attempts, recovery of intergeneric hybrids was limited due to ovule abortion and/or faulty endosperm development [30, 31]. Introduction of embryo-rescue techniques helped avoid these incompatibility barriers and significantly enhanced the pace and scope of wide hybridization [14, 29]. Nowadays, these biotechnological approaches are routinely used to overcome constraints to sexual hybridization and currently several of these hybrids are under evaluation to verify introgression of desired traits. The restriction fragment patterns of all hybrids correspond, in all cases, to that of *C. papaya*, and differ from that of the *Vasconcellea* species which have acted as paternal parent, indicating the maternal inheritance of cpDNA and mtDNA in intergeneric hybrids between *C. papaya* and wild relatives of the *Vasconcellea* genus. These results accord with the commonly observed

---

Fig. 10.3: UPGMA cluster analysis based on polymorphic fragments obtained with PCR-RFLP derived from two chloroplast regions and one mitochondrial region. Genotypes are labeled as in Fig. 1, with a species name abbreviation and a collection number. Species not included in Fig. 1: *V.* × *heilbornii* var.*fructifragrans*: fruct; *V. quercifolia*: quer; *V. crassipetala*: crass; *V. chilensis*: chil; *V. candicans*: can; *V. cauliflora*: cau; *V. microcarpa*: micro; *V. omnilingua*: omni; *V. longiflora*: lon; *V. pulchra*: pul; *V. horovitziana*: hor

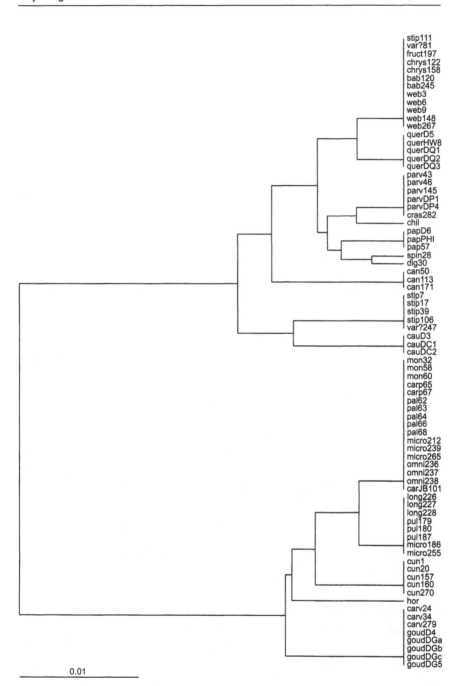

0.01

maternal inheritance of the chloroplast and mitochondrial genomes in angiosperms (Van Droogenbroeck et al., unpubl. data).

Genetic Variation within *Carica papaya*

Papaya germplasm shows considerable phenotypic variation from pigmentation of stem, leaves, and flowers to important horticultural traits, such as fruit size, fruit shape, flesh color, flavor, and sweetness, length of juvenile period, plant stature, stamen carpellody, and carpel abortion. Moreover, commercial papaya cultivars may be inbred gynodioecious lines, typified by Hawaiian Solo lines; outcrossing dioecious populations, such as the Australian papaws from southern Queensland; F_1 hybrids, including the Tainung series (Taiwan), Eksotica II (Malaysia), and Rainbow (Hawaii); or occasionally even clones, such as Hortus Gold in South Africa.

The first study of genetic variation within the species *Carica papaya* included a small group of seven Hawaiian Solo cultivars and three unrelated lines, using RAPD markers [44]. The results from 102 RAPD markers validated the relationships among a group of known pedigree. The minimum genetic similarity detected in this study was 0.7 suggesting the narrow genetic base of domesticated papaya germplasm.

The second study of *C. papaya* genetic variation included 63 accessions of papaya, six accessions of related species of genus *Vasconcellea*, and two accessions of interspecific *Vasconcellea* hybrids, to provide a broad overview of papaya variation [24]. Among the 63 papaya accessions, 18 were commercial cultivars, 15 improved but not released breeding lines, and 30 unimproved germplasm. This collection represents cultivars and breeding lines adapted to different climates in tropical, subtropical, and temperate regions, and unimproved germplasm from the center of origin. Nine pairs of *Eco*R I/*Mse* I primers produced 186 polymorphic markers and 259 monomorphic markers. Genetic variation among this diverse group of *C. papaya* germplasm was tested by pairwise comparison of genetic similarities based on simple matching coefficients. The overall average genetic similarity among 63 papaya accessions was 0.88 and ranged from 0.74 to 0.98. About 82% of the pairwise comparisons showed genetic similarity greater than 0.85; less than 4% were lower than 0.80. Considering that only polymorphic markers were used for statistical analysis, the actual genetic similarity is even higher than the numbers represent. This limited genetic variation did not correspond to the wide range of morphological characteristics observed in the field. However, it is well documented that one or a few genes can sometimes significantly change plant morphology. The genetic variation of Hawaiian Solo papaya cultivars and breeding lines was even narrower with an average genetic similarity of 0.92 among the 15 Solo accessions, ranging from 0.86 to 0.98, reflecting the consequence of many generations of inbreeding from a limited gene pool. Four Hawaiian Solo cultivars, Kamiya, Line 8, Waimanalo, and Maunawili Sweet, appeared to be the most closely related, sharing genetic similarity of 0.98. It is known that Line 8 was one of the parental lines of Waimanalo and that Kamiya was selected from Waimanalo. Maunawili Sweet, selected by a farmer in Maunawili, is possibly derived from one of these three cultivars. The lack of DNA polymorphism among Hawaiian Solo cultivars was confirmed by a large-scale high-density genetic mapping project using an F_2 mapping population derived from Solo cultivars Kapoho and SunUp. Among the 987 pairs of AFLP primers used, 106 generated no polymorphic markers and the remaining 881 primer pairs generated an average 2.0 markers per primer pair [28]. This is very low for the multiplex AFLP markers system compared to the average 20 markers per primer pair detected in a sorghum mapping population [32].

An unexpected conclusion from this study is that the level of genetic variation among dioecious cultivars was similar to that of the hermaphrodite cultivars. Among eight dioecious cultivars collected from tropical, subtropical, and temperate regions in Australia, the average genetic similarity was 0.91, ranging from 0.84 to 0.96. One might think that genetic variation among dioecious cultivars should be higher than the self-pollinated hermaphrodite cultivars, due to their enforced outcrossing. However, data suggest that the narrow genetic base was already established in the papaya progenitor from which both dioecious and hermaphrodite cultivars are derived.

Cluster analysis of 71 accessions of papaya and related species illustrated the genetic relationship among individual cultivars developed in different regions and between papaya and its wild relatives (Fig. 10.4). Discrete clusters separate cultivars derived from the same or similar gene pool. One cluster included all 15 Solo cultivars and breeding lines, plus Eksotica I derived from a cross between Solo cultivar Sunrise and a Malaysian variety, Subang-6, then subsequently backcrossed to Sunrise [11]. Another cluster included dioecious Australian cultivars and Indian cultivars that grow in subtropical or temperate regions. The materials

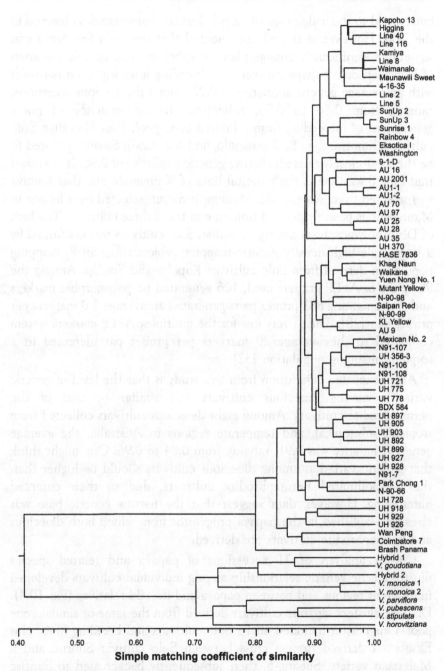

Fig. 10.4: Dendrogram showing subtle difference among papaya clusters, but substantial difference between *Carica* and *Vasconcellea*. The cluster analysis was based

from Central America, mostly unimproved germplasm, grouped into four different clusters, reflecting the greater genetic variation in the center of origin and their potential for broadening the gene pool in the papaya breeding program.

SEX DETERMINATION

Papaya (*Carica papaya* L.) is a polygamous angiosperm exhibiting distinct male, female, and hermaphrodite forms. Sex determination in this tropical fruit species has been the subject of extensive genetic analyses because it is directly related to optimizing cultivation for efficient fruit production. Hofmeyr [20] and Storey [45] made crosses among the three sex forms and concluded that sex in papaya is determined by a single gene with three alleles: M, male; M^h, hermaphrodite; m, female. Females (mm) are homozygous recessive. Males (Mm) and hermaphrodites (M^hm) are enforced sex heterozygotes, whose fruits will contain 25% dominant-allelic nonviable seeds because all combinations of the dominant alleles, MM, MM^h, M^hM^h, are embryo-lethal. With additional field data and many years of observation, Hofmeyr [21] suggested that M and M^h represent genetically inactive regions of the sex chromosomes from which vital genes are missing. Over the years, Storey [47, 49] developed a sophisticated hypothesis that a complex of genes controlling or suppressing stamen and carpel development is located in the sex determination region along with a lethal factor and another factor that prevents recombination. Based on the fact that males and hermaphrodites are heterogametic while females are homogametic and the observation of high frequency of precocious segregation for a pair of chromosomes [25,47], sex determination in papaya has been regarded by some scientists to be the XY chromosome type [22,25,54], even though heteromorphic chromosomes have not been reported.

Unraveling the sex determination process could have profound application in production of dioecious plants such as papaya. For example, it is common practice in Hawaii to plant at least five papaya seedlings per hill and later on remove four of these plants to keep the number of female plants in the field lower than 3% (each hybrid cultivar

---

on 186 polymorphic AFLP markers from 63 accessions of *Carica papaya*, 6 related species, and two interspecific hybrids, excluding 259 monomorphic AFLP markers. The parentages of the two hybrids are: Hybrid 1: (*V. parviflora* × *V. goudatiana*) × *V. goudatiana*; Hybrid 2: *V. parviflora* × *V. goudatiana*.

Rainbow seed has 50% chance to be hermaphrodite or female). The five plants in a hill must be grown for 4 - 6 months until sexes can be determined and one hermaphrodite per hill is kept for fruit production. This process is inefficient utilization of time, labor, water, and nutrients, and furthermore results in delayed production due to competition among the plants in early growth. Better understanding of the sex determination process in papaya may lead to the development of methods to eliminate the undesired seeds or plants of a particular sex automatically before seeds are sown or seedlings transplanted.

A genetic map with 63 RAPD markers placed the sex determination locus on linkage group (LG) 1 [42]. To further characterize the sex determination locus, a high-density genetic map of papaya (*Carica papaya* L.) was constructed using 54 F_2 plants derived from cultivars Kapoho and SunUp with 1,501 markers, including 1498 AFLP markers, the papaya ringspot virus coat protein marker, morphological sex type, and fruit flesh color. These markers were mapped into 12 LGs covering a total length of 3,294.2 cM, with an average distance of 2.2 cM between adjacent markers. This map revealed severe suppression of recombination around the sex determination locus with a total of 225 markers cosegregating with sex types. The cytosine bases in this region appear to be highly methylated based on the distribution of methylation sensitive and insensitive markers. This high-density genetic map is essential for cloning of specific genes of interest such as the sex determination gene, and for integration of genetic and physical maps [28].

The sex determination gene in papaya was fine mapped using 4,380 informative chromosomes (two each from 2,190 female and hermaphrodite plants from three F_2 and one F_3 population) and six DNA markers. No recombinants were detected. The finding that a region so recombinationally small contains such a large percentage of polymorphic markers (see above) indicates that the two homologues are highly differentiated in this region. The nonrecombining (NR) region was then physically mapped using a 13.7 X BAC library [33]. The sex-linked SCAR marker W11 identified four positive BACs. A BAC contig assembled by chromosome walking spanned 990 kb. Forty-two *Carica papaya* sex markers (cpsm) and three previously available sex cosegregating markers, T12, PSDM, and Nafp, hybridized to additional BACs. Among these 42 cpsm markers, 24 (57%) could be placed on BAC contig maps, three identified individual contigs, and 15 contained repetitive sequences that could not be mapped. Ninety-two BAC ends were (iteratively) cloned from contig-terminal BACs and used to close the gaps. The resultant 2.5

Mb physical map includes two major and three smaller contigs containing 4 SCAR, 82 cpbe (*Carica papaya* BAC ends), and 24 cpsm loci. Random subclone sequences (628 reads totaling 513 kb) from 25 nonredundant BACs of the NR region were analyzed to determine the nature of the sequences and estimate the gene density. Sequencing results revealed that the male-specific region (MSY) consists of a mosaic of conserved, X-degenerated, and ampliconic sequences. Compared to a genome-wide sample of papaya DNA based on 684 reads totaling 517 kb (GenBank Accessions CG026197 to CG026996), the MSY region showed 37.7% lower gene density, 27.6% higher retroelement density, and 188.9% higher inverted repeat (IR) density. High frequencies of sequence duplications and transposable element insertions contributed to the degeneration of the MSY resulting in low gene density. Based on the size of the present contig map (2.5 Mb) and the 57% cpsm markers accounted for, the physical size of the MSY is estimated at 4 – 5 Mb or 10% of papaya's primitive Y chromosome. This finding provides direct evidence for the origin of sex chromosomes from autosomes [27].

Hermaphrodite and male plants of papaya appeared to share identical DNA sequences in most parts of the MSY region, since 12 of the 13 hermaphrodite specific markers showed near identical sequences in male and hermaphrodite plants [27]. These two sex types appear to share a haplotype for most of this region which differs from that of females and is recently derived from a common ancestral chromosome. This suggests that divergence between male and hermaphrodite is a second step of sex chromosome evolution after the recessive mutation resulting in female (see [12]). Thus, gynodioecy might be an intermediate step to dioecy within Caricaceae.

Papaya provides an extraordinary opportunity to elucidate the evolutive process of dioecy. *Vasconcellea* is particularly valuable in this aspect since polygamous (*V. cundinamarcensis*, *V. parviflora*, and *V. x heilbornii*), monoecious (*V. monoica*), and dioecious (the remaining 17 species) all exist within this genus. Comparative analyses of the MSY region in three closely related species, *V. cundinamarcensis*, *V. monoica*, and *V. goudotian*, will shed light on the evolution of sex chromosomes and dioecy.

PROSPECTS

The application of molecular markers to assess the genetic diversity of papaya and related species has resulted in a better understanding of phylogenetic relationships among them and contributed directly to the reinstatement of genus *Vasconcellea*. Finding limited genetic variation within the species *Carica papaya* prompted researchers to adapt multiplex high-throughput DNA marker systems for genetic mapping and marker-assisted selection. The discovery of a primitive Y chromosome in papaya resolved a longstanding question regarding the mechanism of sex determination and opened a new vista for studying the biodiversity and evolution in family Caricaceae. Complete sequencing of the MSY region in papaya will provide additional tools for assessing the divergence of Caricaceae species.

In addition to resolving issues of plant sex chromosome evolution, papaya is an especially promising system in which to nucleate exploration of tropical tree genomes. As a genomic model, papaya is appealing for a long list of reasons, including:

- A small genome of 372 mbp [3], about 10% smaller than the rice genome, has been fairly well sequenced [18, 55];
- Diploid inheritance, with nine chromosomes in its gametes;
- A detailed genetic map of 1,501 loci (based on 54 F_2 plants) [28];
- An existing 13.7 X papaya BAC library comprising about 40,000 clones with average insert size of 132 kb [33];
- A family (Caricaceae, which includes papaya) that is closely related to family Brassicaceae (which includes *Arabidopsis* whose genome is completely sequenced) [1, 8];
- A well-established transformation system [16].

Additional features of papaya biology that add to its functionality as a research system include:

- A short juvenile phase (3 to 8 months) and generation time (9 to 15 months);
- Continuous flowering throughout the year;
- An efficient breeding system, each fruit producing about 1,000 seeds, and a single tree producing hundreds of fruits in its lifetime. Hand pollination using pollen from male or hermaphrodite flowers is easily done on the stigmata of female or hermaphrodite flowers. The perennial nature of papaya is advantageous in this regard;
- A well-established clonal propagation system that allows testing of individual plants in multiple environments;

- Numerous types of flowers ranging from complete female, to hermaphrodite, to complete male with gradual variations in between;

Papaya genetic research and improvement have benefited from the rapid advances in biotechnology. It is reasonable to expect that the papaya whole genome will be sequenced in the foreseeable future because of the advantages listed above and its economic importance in tropical and sub-tropical regions. A sequenced papaya genome will be instrumental in papaya improvement and studying biodiversity and evolution of this tropical crop. It will open the door to comparative genomic analyses among species within the family and outside the family, such as completely sequenced *Arabidopsis* and rice. The molecular diversity of papaya and its wild relatives can be evaluated in a systematic and genomic fashion.

Acknowledgments

The authors thank Henrik Albert, Blake Vance, Melinda Moore, and Stephanie Whalen for reviewing the manuscript. This work was supported by a USDA-ARS Agreement for Minor Crops Research via the University of Hawaii, and USDA-ARS Cooperative Agreements (no. CA 58-5320-3-460 and CA 58-5320-9-105) with the Hawaii Agriculture Research Center. Part of this work was funded by the Flemish Fund for Scientic Research (FWO-Vlaanderen Project no. 3G005100) and by the IWT-Vlaanderen grant of a PhD fellowship to Tina Kyndt.

REFERENCES

[1] *Arabidopsis* Genome Initiative. Analysis of the genome sequence of the flowering plant *Arabidopsis thaliana*. Nature 2000; 408: 796-815.

[2] Aradhya MK, Manshardt RM, Zee F, Morden CW. A phylogenetic analysis of the genus *Carica* L. (Caricaceae) based on restriction fragment length variation in a cpDNA intergenic spacer region. Genet Res Crop Evol 1999; 46: 579-586.

[3] Arumuganathan K, and Earle ED. Nuclear DNA content of some important plant species. Pl Mol Biol Rept 1991; 9: 208-218.

[4] Badillo VM. Monografia de la familia Caricaceae. Publicada por la Associacion de Profesores, Venezuela: Univ. Centr. Venez. 1971, pp. 220.

[5] Badillo VM. Caricaecae, Segundo Esquema. Publicada por la Associacion de Profesores, Alcance 43, Maracay: Univ. Centr. Venez., 1993, pp. 111.

[6] Badillo VM. *Carica* L. vs. *Vasconcella* St. Hil. (Caricaceae): con la rehabilitación de este último. Ernstia 2000; 10: 74-79.

[7] Badillo VM. Nota correctiva *Vasconcellea* St. Hil y no *Vasconcella* (Caricaceae). Ernstia 2001; 11: 75-76.

[8] Bremer K, Chase MW, Stevens PF. An ordinal classification for the families of flowering plants. Ann Mol Bot Gard 1998; 85: 531-553.

[9] Briggs J. Biogeography and plate tectonics. New York, NY: Elsevier, 1987.

[10] Candolle A de. Origin of cultivated plants. New York, NY; D. Appleton & Co, 1908.

[11] Chan YK. Backcross method in improvement of papaya (*Carica papaya* L.) Malays Appl Biol Rept 1987; 1: 19-21.

[12] Charlesworth B, Charlesworth D. A model for the evolution of dioecy and gynodioecy. Amer Nat 1978; 112: 975-997.

[13] Deputy JC, Ming R, Ma H, et al. Molecular marker for sex determination in papaya (*Carica papaya* L.). Theor Appl Genet 2002; 106: 107-111.

[14] Drew RA, O'Brien CM, Magdalita PM. Development of interspecific *Carica* hybrids. Acta Hort 1998; 461: 285-292.

[15] Dunne J, Horgan L. Meat tenderizers, In: Hui YH, ed. Encyclopedia of food science and Technology, vol. 3. New York; NY: Wiley, 1992: 1745-1751.

[16] Fitch MMM, Manshardt RM, Gonsalves D, Slightom JL. Virus resistance papaya derived from tissue bombarded with the coat protein gene of papaya ringspot virus. Bio/Tech. 1992; 10: 1466-1472.

[17] Gentry, A. Distributional patterns of Central American and West Indian Bignoniaceae. In: Darwin S, Welden A, eds., Biogeography of Mesoamerica, New Orleans, LA: Tulane University, 1992; 111–125.

[18] Goff SA, Ricke D, Lan T-H, et al. A draft sequence of the rice genome (*Oryza sativa* L. ssp. *japonica*). Science 2002; 296: 92-100.

[19] Gonsalves D. Control of papaya ringspot virus in papaya: a case study. Annu Rev Phytopath 1998; 36: 415-437.

[20] Hofmeyr JDJ. Genetical studies of *Carica papaya* L. I. The inheritance and relation of sex and certain plant characteristics. II. Sex reversal and sex forms. S, Afric Dept Agric Sci Bull 1938; 187; 64.

[21] Hofmeyr JDJ. Some genetic and breeding aspects of *Carica papaya*. Agron Trop 1967; 17: 345-351.

[22] Horovitz S, Jiminez H. Cruzamientos interespecificos e intergenericos en Caricaceaes y sus implicaciones fitotecnias. Agron Trop 1967; 17: 353-359.

[23] Jobin-Décor MP, Graham GC, Henry RJ, Drew RA. RAPD and isozyme analysis of genetic relationship between *Carica papaya* and wild relatives Genet Res Crop Evol 1997; 44: 471-477.

[24] Kim MS, Moore PH, Zee F, et al. Genetic diversity of *Carica papaya* L as revealed by AFLP markers. Genome 2002; 45: 503-512.

[25] Kumar LSS, Abraham A, Srinivasan VK. The cytology of *Carica papaya* Linn. Indian J Agric Sci 1945; 15: 242-253.

[26] Kyndt T, Van Droogenbroeck B, Gheysen G. Phylogenetic analysis of the Caricaceae based on nuclear and chloroplast DNA sequences. Abstracts PAGXII Conference San Diego. 2004; p220.

[27] Liu Z, Moore PH, Ma H, et al. A primitive Y chromosome in papaya marks incipient sex chromosome evolution. Nature 2004; 427: 348-352.

[28] Ma H, Moore PH, Liu Z, et al. High-density genetic mapping revealed suppression of recombination at the sex determination locus in papaya. Genetics 2004 (in press).

[29] Magdalita PM, Adkins SW, Godwin ID, Drew RA. An improved embryo-rescue protocol for a *Carica* interspecific hybrid. Austr J Bot 1996; 44: 343-353.

[30] Magdalita PM, Drew RA, Adkins SW, Godwin ID. Morphological, molecular and cytological analyses of *Carica papaya* × C. *cauliflora* interspecific hybrids. Theor Appl Genet, 1997; 95: 224-229.

[31] Manshardt RM, Wenslaff TF. Inter-specific hybridization of papaya with other species. J Amer Soc Hort Sci 1989; 114: 689-694.

[32] Menz MA, Klein RR, Mullet JE, et al. A high-density genetic map of *Sorghum bicolor* (L.) Moench based on 2926 AFLP, RFLP and SSR markers. Pl Mol Biol 2002; 48: 483-499.

[33] Ming R, Moore PH, Zee FT, et al. A papaya BAC library as a foundation for molecular dissection of a tree-fruit genome. Theor Appl Genet 2001; 102: 892-899.

[34] National Research Council. Highland papayas. In: Ruskin FR, ed. Lost crops of the Incas: little-known plants of the Andes with promise for worldwide cultivation. Washington DC; National Academy Press, 1989; 252-261.

[35] Ockerman HW, Harnsawas S, Yetim H. Inhibition of papain in meat by potato protein or ascorbic acid. J Fod Sci 1993; 58: 1265-1268.

[36] Osato JA, Santiago L, Remo G, Cuadra M, Mori A. Antimicrobial and antioxidant activities of unripe papaya. Life Sci 1993; 53: 1383-1389.

[37] Parasnis AS, Ramakrishna W, Chowdari KV, Gupta VS, Ranjekar PK. Microsatellite (GATA)$_n$ reveals sex-specific differences in papaya. Theor Appl Genet 2000; 99: 1047-1052.

[38] Purina A, Sandhya B. Genotypic differences of in vitro lateral bud establishment and shoot proliferation in papaya. Curr Sci 1988; 7: 440-442.

[39] Purseglove JW. Tropical Crops. London, UK: Longman Inc., 1968: 45-51.

[40] Raven PH, Axelrod DI. Angiosperm biogeography and past continental movements. Ann Mo Bot Gard 1974; 61: 539–673.

[41] Scheldeman X. Distribution and potential of cherimoya (*Annona cherimola* Mill.) and highland papayas (*Vasconcellea* spp.) in Ecuador. PhD thesis, Ghent University, Belgium, 2002; pp. 176.

[42] Sondur SN, Manshardt RM, Stiles JI. A genetic linkage map of papaya based on randomly amplified polymorphic DNA markers. Theor Appl Genet 1996; 93, 547-553.

[43] Stehli F, Webb S. A kaleidoscope of plates, faunal and floral dispersals, and sea level changes. In: Stehli F, Webb S, eds., The great American biotic interchange. New York, NY: Plenum Press, 1985: 3-16.

[44] Stiles JI, Lemme C, Sondur S, Morshidi MB, Manshardt RM. Using randomly amplified polymorphic DNA for evaluating genetic relationships among papaya cultivars. Theor Appl Genet 1993; 85: 697-701.

[45] Storey WB. Segregations of sex types in Solo papaya and their application to the selection of seed. Proc Amer Soc Hort Sci 1938; 35: 83-85.

[46] Storey WB. The botany and sex relations of the papaya. Hawaii Agric Exper Sta Bull 1941; 87: 5-22.

[47] Storey WB. Genetics of the papaya. J Hered 1953; 44: 70-78.

[48] Storey WB. Pistillate papaya flower: A morphological anomaly. Science, 1969; 163: 401-405.

[49] Storey WB. Papaya. In: Simmonds NW ed. Evolution of Crop Plants. London, UK: Longman Inc, 1976: 21-24.

[50] Urasaki N, Tokumoto M, Tarora K, et al. A male and hermaphrodite specific RAPD marker for papaya (Carica papaya L.). Theor Appl Genet 2002; 104: 281-285.

[51] Van Droogenbroeck B, Breyne P, Goetghebeur P, et al. AFLP analysis of genetic relationships among papaya and its wild relatives (Caricaceae) from Ecuador. Theor Appl Genet 2002; 105: 289-297.

[52] Van Droogenbroeck B, Kyndt T, Maertens I, et al. Phylogenetic analysis of the highland papayas (Vasconcellea) and allied genera (Caricaceae) using PCR-RFLP. Theor Appl Genet 2004 (in press).

[53] Villareal L, Dhuique-Mayer C, Dornier M, Ruales J, Reynes M. Evaluation de L' intérêt du babaco (Carica pentagona Heilb.). Fruits 2003; 58: 39-52.

[54] Westergaard M. The mechanism of sex determination in flowering plants. Adv Genet 1958; 9: 217-281.

[55] Yu J, Hu S, Wang J, et al. A draft sequence of the rice genome (Oryza sativa L. ssp. indica). Science 2002; 296: 79-92.

Artemisia and Its Allies: Genome Organization and Evolution and their Biosystematic, Taxonomic, and Phylogenetic Implications in the Artemisiinae and Related Subtribes (Asteraceae, Anthemideae)

JOAN VALLÈS[1] and *TERESA GARNATJE*[2]

[1]Laboratori de Botànica, Facultat de Farmàcia, Universitat de Barcelona, Av. Joan XXIII s/n, 08028 Barcelona, Catalonia, Spain

[2]Institut Botànic de Barcelona (CSIC-Ajuntament de Barcelona), Passeig del Migdia s/n, 08038 Barcelona, Catalonia, Spain

ABSTRACT

With around 500 species, *Artemisia* is the largest genus of tribe Anthemideae and one of the largest of family Asteraceae. This genus and its more or less closely related genera—some of them segregated from *Artemisia*—constitute subtribe Artemisiinae. Other genera with conflictive taxonomic relationships with *Artemisia* are included in the current classification, in other subtribes such as Chrysantheminae, Handeliinae, Leucantheminae, and Tanacetinae. Many of the taxa belonging to this complex are of economic importance in various fields, such as medicinal, food, and ornamental. This pool of about 30 genera and 650 species has been profusely, but not completely studied using, among others, cytogenetic and molecular approaches. This paper presents a synthetic

Address for correspondence: Joan Vallès, Laboratori de Botànica, Facultat de Farmàcia, Universitat de Barcelona, Av. Joan XXIII s/n, 08028 Barcelona, Catalonia, Spain. Tel: +34-934024490, Fax +34-934035879, E-mail: joanvalles@ub.edu

approach, with new, original data, to our contributions and those of other authors to the knowledge of genome structure and evolution in these taxa and its systematic implications. Data available suggest that classical and current classifications, mostly based on morphological characters, do not reflect natural groups, either at subtribal, or at generic or infrageneric taxonomic levels. Completion of genome studies in the Artemisiinae and related subtribes would lead to a structuring representative of the phylogenetic relationships within this group.

Key Words: *Absinthium, Ajania,* Anthemideae, *Artemisia, Artemisiastrum,* Artemisiinae, Asteraceae, *Brachanthemum,* Chrysantheminae, chromosome banding, *Chrysanthemum,* Compositae, *Dendranthema,* DNA sequencing, *Dracunculus, Elachanthemum, Filifolium,* FISH, Handeliinae, *Hippolytia,* karyology, *Kaschgaria, Lepidolopsis,* Leucantheminae, *Mausolea,* molecular cytogenetics, molecular phylogeny, *Neopallasia, Nipponanthemum,* nuclear DNA amount, *Oligosporus, Picrothamnus,* rDNA ITS, *Seriphidium, Sphaeromeria,* Tanacetinae, *Tanacetum, Tridentatae, Turaniphytum.*

INTRODUCTION

Biosystematic studies in plants, especially those dealing with genome evolution, are the object of renewed interest because the combination of cytogenetic and molecular analyses and their comparison with other data engender an understanding of the nature and extension of evolutionary mechanisms as well as the species phylogeny within a plant group. In addition, this kind of investigation contributes to the evaluation of plant biodiversity necessary for a rational exploitation of natural resources.

Family Asteraceae or Compositae, with more than 20,000 species, is of special interest because, despite its undeniable unity and easy characterization, it has undergone very different speciation and differentiation processes and mechanisms. This and the fact that it contains many useful plants focused the attention of the international botanical community on Asteraceae. International congresses devoted only to this family (Reading, 1975; Kew, 1994; Pretoria, 2003; Barcelona in 2006) and the group of synantherologists, The International Compositae Alliance (TICA), attest its relevance.

Our research group has been studying genus *Artemisia* for the last 20 years, and in the last five began an investigation of its satellite genera. Our first focus was general biosystematics but soon we specialized in genome organization and evolution. We present here in a synthetic approach some new, original data pertaining to subtribe Artemisiinae and other genera which should probably be included in this group. Our ultimate objective is to solve questions linked to systematics, phylogeny, and evolution in order to achieve a natural and stable classification of *Artemisia* and its allies.

Genus *Artemisia* and subtribe Artemisiinae

First postulated by Lessing [60] and redefined by Bremer and Humphries [16], subtribe Artemisiinae (Anthemideae, Asteraceae) encompasses more than 650 species, grouped in about 30 genera, most of them oscillating around the type genus, *Artemisia*, to which they are very closely related.

Artemisia L. is the largest genus in tribe Anthemideae and one of the largest genera in Asteraceae, with taxa numbering from 300 to 500 depending on the author [16,63,64,66,67,69,74,128]. It is largely distributed in the Northern Hemisphere and very scarce in the Southern Hemisphere. Several of its species dominate arid zones in various regions, with Middle and Central Asia (Uzbekistan, Tadzhikistan, Turkmenistan, Kazakhstan, Kirguizistan, parts of Russia and China, Mongolia) its principal center of diversity and speciation. Other centers are in or lie in Irano-Turanian, Mediterranean, and North American regions. Perennial plants dominate the genus, only 10-15 being annual or biennial. Many species are useful in various fields (medicinal, food, forage, ornamental, soil stabilization), while some taxa are toxic or allergenic and some others weeds that can negatively affect agronomic harvests [89,111,140].

Taxonomic treatments of genus *Artemisia* and other genera of the subtribe Artemisiinae and related taxa in large geographic areas are presented in [3,36,56,62,63,64,65,66,67,91,92,93,101]. Since Tournefort [119], many attempts at infrageneric classification in *Artemisia* have been made. Five large groups—*Absinthium, Artemisia, Dracunculus, Seriphidium,* and *Tridentatae*—treated as sections or subgenera are classicaly recognized [11,12,13,14,19,78,101] (Table 11.1). Some of these groups are considered unnatural by authors working on the genus [90, 128, and references therein] and a complete, detailed classification has yet to be reached.

Many proposals for splitting the genus have been made since Cassini [20], who segregated the taxa of subgenus *Dracunculus* into an independent genus, *Oligosporus*. More recently, some authors [15, 16, 62, 63, 65, 66] have supported the separation of *Seriphidium* as a genus. Apart from these large taxa, a number of small, often monotypic genera have been separated based on or containing former *Artemisia* species: *Ajania*

Table 11.1: Comparison of various infrageneric classifications in *Artemisia* [adapted from 116, 125]

Category	Infrageneric taxa					Reference
Genera	*Absinthium*		*Artemisia*		*Artemisia*	[119]
Genera	*Abrotanum*		*Artemisia*			[68]
Genera		*Artemisia*			*Oligosporus*	[20, 60]
Sections	*Absinthium*	*Abrotanum*	*Artemisia*	*Seriphidium*	*Dracunculus*	[11, 12, 13, 14, 19]
Subgenera		*Euartemisia*		*Seriphidium*	*Euartemisia*	[100]
Subgenera / Sections	*Absinthium*	*Abrotanum*	*Artemisia*	*Seriphidium* / *Tridentatae*	*Dracunculus*	[101]
Subgenera			*Artemisia*	*Seriphidium*	*Dracunculus*	[93]
Subgenera	*Absinthium*		*Artemisia*	*Seriphidium*	*Dracunculus*	[90]
Sections			*Artemisia*		*Dracunculus*	[120]
Subgenera			*Artemisia*	*Seriphidium* / *Tridentatae*	*Dracunculus*	[78]
Subgenera			*Artemisia*	*Seriphidium*	*Dracunculus*	[91]
Genera / Subgenera			*Artemisia* / *Artemisia*	*Seriphidium* / *Seriphidium*	*Artemisia* / *Dracunculus*	[63, 64]

Poljakov, *Artemisiastrum* Rydb., *Artemisiella* A. Ghafoor, *Crossostephium* Less., *Elachanthemum* Y. Ling et Y. R. Ling, *Filifolium* Kitam., *Hippolytia* Poljakov, *Kaschgaria* Poljakov, *Lepidolopsis* Poljakov, *Mausolea* Poljakov, *Neopallasia* Poljakov, *Picrothamnus* Nutt., *Stilpnolepis* H. Kraschen, and *Turaniphytum* Poljakov. Finally, some genera are very close to *Artemisia*, but have never been included in it, such as *Sphaeromeria* Nutt. The taxonomy of many taxa of these genera has been very controversial: they have either been described in *Artemisia* and combined in other genera [such as *Hippolytia megacephala* (Rupr.) Poljakov, described as *Artemisia megacephala* Rupr.], or described in other genera and combined in *Artemisia* [such as *Artemisia incana* (L.) Druce, described as *Tanacetum incanum* L.], or even combined in more than one genus [such as *Lepidolopsis turkestanica* (Regel et Schmalh.) Poljakov, described as *Crossostephium turkestanicum* Regel et Schmalh. and also combined as *Artemisia turkestanica* (Regel et Schmalh.) Franch. and *Tanacetum turkestanicum* (Regel et Schmalh.) Poljakov].

Genus *Artemisia* and most of its satellite genera form the core of subtribe Artemisiinae Less. emend. Bremer et Humphries (Table 11.2). Delimitation of this taxon is not yet clear. On the one hand, some genera closely related to *Artemisia* are included in different subtribes of tribe Anthemideae, such as Handeliinae and Tanacetinae. On the other hand, in the Bremer and Humphries' [16] proposal, Artemisiinae include genera which are close to *Chrysanthemum*, and hence with affinities with subtribes Chrysantheminae and Leucantheminae.

MATERIALS AND METHODS

Plant Material

We studied from various points of view and to different depths more than 130 populations belonging to about 100 taxa, most of them (around 85) of genus *Artemisia* and the remaining from other related genera. Details on many populations investigated can be found in our publications cited in the References [37,38,58,72,73,84,85,86,113,114,115,116,117,118, 122,123,124,125,126,127,128,129,130,131,132,133], and vouchers of all of them are deposited in the herbarium of the Centre de Documentació de Biodiversitat Vegetal de la Universitat de Barcelona (BCN). In addition, we did an extensive survey of the literature related to cytogenetic, molecular and other biosystematic aspects of genus *Artemisia* and its relatives (see References).

Table 11.2: *Artemisia* and its related genera [lists elaborated from 15,16,43,72,73,133]

• *Genera of subtribe Artemisiinae. Those asterisked (*) are rarely accepted and usually considered only in synonymy*

Ajania Poljakov	Kaschgaria Poljakov
Ajaniopsis Shih	Mausolea Poljakov
Arctanthemum (Tzvelev) Tzvelev	Neopallasia Poljakov
Artemisia L.	*Oligosporus Cass.
*Artemisiastrum Rydb.	Phaeostigma Muld.
Artemisiella A. Ghafoor	Picrothamnus Nutt.
Brachanthemum DC.	*Seriphidium (Besser ex Hook.) Fourr.
*Chamartemisia Rydb.	Sphaeromeria Nutt.
Crossostephium Less.	Stilpnolepis H. Kraschen.
Dendranthema (DC.) Des Moul.	Tridactylina (DC.) Schultz-Bip. in
Elachanthemum Y. Ling et Y.R. Ling	Webb et Berthelot
Filifolium Kitam.	Turaniphytum Poljakov
*Hulteniella Tzvelev	*Vesicarpa Rydb.

• *Genera of other subtribes of tribe Anthemideae containing species described or combined in* Artemisia *or in other genera of Artemisiinae*

Subtribe Chrysantheminae	Subtribe Leucantheminae
Chrysanthemum L.	Nipponanthemum L.
Subtribe Handeliinae	Subtribe Tanacetinae
Lepidolopsis Poljakov	Hippolytia Poljakov
	Tanacetum L.

Methodology

Cytogenetics

— Chromosome number counting and karyotype elaboration after conventional staining (orcein, Feulgen) [86,127].

— Giemsa and fluorochrome banding to detect constitutive heterochromatin and AT- and GC-rich regions [86,127].

— Fluorescence *in-situ* hybridization (FISH) to reveal in chromosomes the 18S-5.8S-26S and 5S regions of ribosomal DNA [21,117] linked to the nucleolar organizers. This technique is a contribution to the genome physical mapping.

— Genome size assessment by flow cytometry [38,113].

Molecular phylogeny

— DNA sequencing. Some DNA regions particularly used in systematic and evolutionary studies were sequenced, and the sequences obtained analyzed with the program PAUP [110] to obtain the phylogeny of the groups studied. We sequenced the nuclear ribosomal DNA internal transcribed spacers (ITS) [115, 132] and also studied the chloroplast DNA *trn*L-F intron, which was not useful for our purposes. We are currently studying the nuclear ribosomal DNA external transcribed spacer (ETS).

Population biology

— We assessed the populational genetic variability of a narrow endemic species by means of isozyme electrophoresis [114].

Complementary techniques

— We complemented the genome organization investigations with morphometrical studies. Among them, palynological ones are particularly relevant, including pollen grain morphometry [4,5,7,27,72,73,98,129].

— Chorological, ecological, phenological, taxonomical and nomenclatural aspects were also treated.

RESULTS AND DISCUSSION

Chromosome Number

We published [130,131,132, and references therein] the chromosome number for about 130 populations of 90 *Artemisia* taxa and nine more genera of the pool considered and hold unpublished counts on about 90 populations of 15 *Artemisia* taxa and three more genera. These results and those found in the literature [chromosome number indices cited in 130, 131, and 134] mean that ca. 350 *Artemisia* species or subspecies have been studied from this viewpoint, i.e., over 50%. Analysis of these chromosome numbers led to the conclusion that two of the most powerful evolutionary cytogenetic mechanisms, dysploidy and polyploidy, occurred in the plant group studied, in addition to other cytogenetic causes for differentiation or microevolution, such as aneusomaty and the presence of accessory chromosomes.

Dysploidy

The two basic chromosome numbers in Artemisiinae are $x = 9$ and $x = 8$ [128, and references therein]. The most frequent, $x = 9$, is also the most common in tribe Anthemideae and in the entire Asteraceae family [86, 102, 107, 127, 128]. This is the only basic number found in all the satellite genera. Conversely, *Artemisia* is a dysploid genus, with a minority but large group of taxa with $x = 8$, belonging to subgenera or sections *Absinthium*, *Artemisia*, and *Dracunculus*, which have both basic numbers. The only two infrageneric groups without dysploidy (only with $x = 9$) are the large *Seriphidium* and the small, North American endemic *Tridentatae*.

An outstanding question is whether dysploidy is ascending or descending in *Artemisia*. Kawatani and Ohno [48] postulated a descending sense, based only on the great proportion of $x = 9$ under taxa. In addition, as stated by Siljak-Yakovlev [106], this is the most common dysploidy in plants. We found evidence of a centric or Robertsonian chromosomal fusion in *A. vulgaris* leading from 18 to 16 chromosomes, thus from $x = 9$ to $x = 8$ [127]. In this taxon, the first chromosome pair is perfectly metacentric and much longer than the remaining, suggesting an origin by fusion. Banding results revealed pericentromeric heterochromatin in this pair as an exception, because the banding pattern of *Artemisia* is largely dominated by telomeric heterochromatin. This chromosome pair often presents centromeric fragility, which indicates that the fusion is relatively recent. We also confirmed that the fusion event has probably occurred several times during *Artemisia* evolution, because we detected this phenomenon in diploid and polyploid taxa belonging to different sections of the genus, such as *A. granatensis*, *A. splendens* (*Absinthium*), *A. judaica*, *A. lucentica*, *A. reptans*, and *A. vulgaris* (*Artemisia*) (Fig. 11.1A, B, F).

In addition to these most common basic numbers, $x = 9$ and $x = 8$, we can mention two more not yet confirmed with certainty. Only a count of $2n = 14$ has been reported, which could indicate a basic number $x = 7$, in a high mountain North American plant, *A. pattersonii* [138]. Further studies on this and other related taxa are needed to confirm or deny this hypothesis. Some species have $2n = 34$ chromosomes. There are several ways to reach this number, which implies, according to some authors, a dysploid process leading to a basic number of polyploid origin, $x = 17$, but can also be interpreted as the product of polyploidy followed by hypoaneuploidy or hyperaneuploidy [26,42,86,128 and unpublished data] (Fig. 11.2).

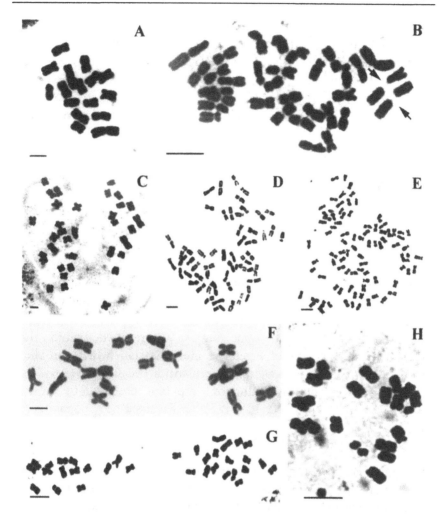

Fig. 11.1: Metaphase plates of various *Artemisia* species stained with orcein. A-A. *judaica*; B-A. *splendens*; C-A. *umbelliformis* subsp. *umbelliformis*; D-A. *crithmifolia*; E-A. *dracunculus*; F-A. *reptans*; G-A × *wurzellii*; H-A; *chamaemelifolia* subsp. *chamaemelifolia*. Note the large metacentric chromosome pairs originated by centric fusion in A, B and F; arrows in B indicate chromosomes showing centromeric fragility. C, D, E, and G are examples of polyploidy, including aneusomaty in E. Note the B chromosomes in H. Scale bars = 5 μm.

Polyploidy

This evolutionary mechanism, very common in plants but not in animals [17,29,30,31,32,87,95], has proven particularly active in different groups of genus *Artemisia* and in other genera of subtribe Artemisiinae

[25,26,28,37,57,58,76,77,78,86,88,90,115,118,122,123,125,128,130,131,132]. Polyploidy, contrary to dysploidy, appears in all the large subgroups of *Artemisia* and in many related genera. Both basic numbers present a polyploid series, up to x = 12 (ploidy level reported only once, in a Russian population of A. *macrantha* [70]) for x = 9 and up to 6x for x = 8 (Fig. 11.1C, D, E, G). In some taxa a positive correlation between ploidy level and pollen grain and stomata size was found [123]. Genome size also correlated well with ploidy level [38, 113].

Aneusomaty

Aneusomaty or intrapopulational or even intraindividual chromosome number variation [24] is not very common but relatively frequent in the plant kingdom. This phenomenon was seen in two species with high ploidy level: A. *verlotiorum* (6x) and A. *dracunculus* (10x) [58, 122] (Fig. 11.1E). These intrademic and intraindividual imbalances must create meiotic instability. Weinedel-Liebau [136] in his pioneering work on *Artemisia* cytogenetics, reported such meiotic troubles in A. *dracunculus*, and so did Rousi [99] in the same species. We are of the opinion that plants with a high degree of vegetative multiplication (which can allow themselves to suffer these meiotic irregularities because meiosis is not very relevant for their reproduction) are best prepared to present aneusomaty ([22, 24, 61, 90, 121, 122]; C. Favarger, pers. comm.).

Accessory chromosomes

Apart from A chromosomes, normal in a chromosome complement, quite often B, supernumerary or accessory chromosomes are found. In a large revision of subgenus *Tridentatae* of *Artemisia*, McArthur and Sanderson [77] reported their presence in 3% of the populations. We found one to five B chromosomes in three diploid taxa (A. *chamaemelifolia*, Fig. 11.1H), A. *herba-alba* subsp. *valentina* (see Fig. 11.4A) and A. *tridentata* subsp. *spiciformis* (Fig. 11.4F) and one tetraploid (A. *barrelieri*, Fig 11.4C) taxa. Although they do not appear in a constant number in the plants or cells of one species or population, B chromosomes can interact with A and may be a relevant evolutionary factor [18, 47]. In a comparison of A. *chamaemelifolia* and A. *molinieri*, B chromosomes are an essential element of karyotypic differentiation [114]. The two species have a quite similar karyotype but A. *chamaemelifolia* possesses, while A. *molinieri* lacks, accessory chromosomes.

Chromosome number evolution

A combination of polyploidy and dysploidy provided the evolutive pathway for Artemisiinae chromosome number that we synthesize in Fig. 11.2. It is remarkable that a large x = 9-based polyploid series, from 2x to 10x, can be illustrated with only one species, *A. dracunculus*, including its wild and cultivated populations [58,82,99,130]. The x = 8-based polyploid series is exemplified by the *A. vulgaris* group, which, in addition, also shows dysploidy (Fig. 11.3). According to our experience, the somatic number 2n = 34, thus the putative basic number x = 17, can be reached from 2n = 18 (Fig. 11.2) either via polyploidy (from 2n = 18 to 2n = 36) followed by aneuploidy or dysploidy (from 2n = 36 to 2n = 34; the case for the *A. umbelliformis* complex; [26,42,86,126]) or via descending dysploidy (from 2n = 18 to 2n = 16) followed by polyploidy (from 2n = 16 to 2n = 32) and, after that, by ascending dysploidy or aneuploidy (from 2n = 32 to 2n = 34; the case of *A.* x *wurzellii* [132 and unpublished data] (Fig. 11.1G)—a taxon described as a hybrid [46],

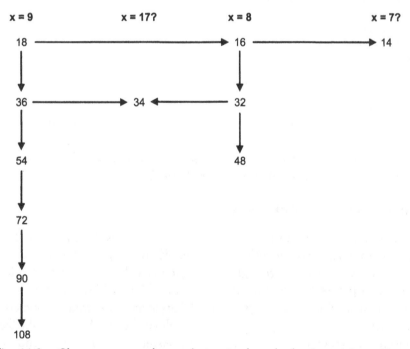

Fig. 11.2: Chromosome number evolution in the subtribe Artemisiinae. Figures related with arrows represent somatic chromosome numbers (2n) [adapted from 125].

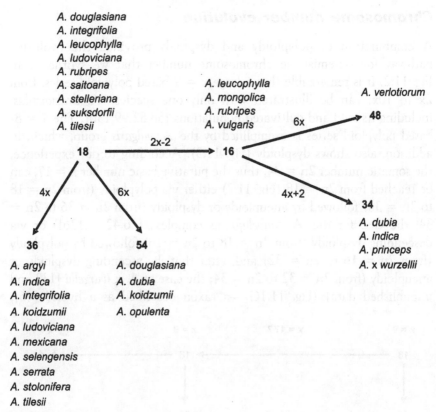

Fig. 11.3: Chromosome number evolution in the *Artemisia vulgaris* complex. Figures related with arrows represent somatic chromosome numbers (2n).

which should be verified by means of genomic *in situ* hybridization—and Japanese populations of A. *vulgaris* or A. *princeps*).

Karyotype Morphometry

Chromosomes of Artemisiinae are of mean dimensions, ranging from 2 to 10 µm. Total karyotype length ranges from 40 to 500 µm. The karyotypes are rather symmetrical, although some asymmetric have been reported for a few taxa [33,34,35]. Taking into account Stebbins' [108] statements on the ancestral character of symmetrical karyotypes, we would conclude that Artemisiinae are a primitive group. Nevertheless, being a subtribe considered rather derived within family Asteraceae [15], we tend to believe that *Artemisia* and its satellite genera underwent a process of secondary karyotype symmetrization, a mechanism Stebbins [108] considered frequent.

We could confirm the karyotype evolutionary trends postulated by Stebbins [108] in some groups within genus *Artemisia*. *Artemisia umbelliformis* subsp. *umbelliformis* is tetraploid and has a more asymmetrical karyotype than its diploid ancestor, A. *umbelliformis* subsp. *eriantha* [86,126 and unpublished data]. In the A. *campestris* complex, a similar correlation between increase in ploidy level and karyotype asymmetry has been found [116].

Banding and *in situ* Hybridization Patterns

Following animal cytogenetics, plant cytogenetics also carries out fine karyotype analysis identifying different regions in the chromosomes. The most frequent techniques used are banding and *in-situ* hybridization. Giemsa banding reveals constitutive heterochromatin, and fluorochrome banding, GC- (with chromomycin) and AT-rich (with bisbenzimide) regions. Fluorescent *in situ* hybridization (FISH) permits physical mapping of different regions, including the 18S-5.8S-26S and 5S loci of ribosomal DNA. This has resulted in useful markers for chromosome identification and chromosome evolution and cytotaxonomy. Among the extensive reports of cytogenetic research in *Artemisia* [128, and references therein] several studies dealing with chromosome banding have been published [81,85,86,102,112,114,116,117,128]. Conversely, although FISH has been applied to different plant groups [117 and references therein], to date only a few works concern the tribe Anthemideae; some of them are focused in the *Chrysanthemum* pool of genera [1,2,45,49,50,51,52,53], and some others in *Artemisia* [116,117].

Schweizer and Ehrendorfer [102] coined the expression "banding style" to designate the pattern of bands revealed in the chromosomes of a plant group. *Artemisia* and the scarce other genera of Artemisiinae in which banding techniques have been used show a rather well-defined and constant banding style. Constitutive heterochromatin (Fig. 11.4G) [81, 85, 86, 102] and also AT- and GC-rich DNA (Fig. 11.4A, B) [114, 116, 117, 127] are predominantly terminal, although some intercalary and centromeric bands exist. As reported earlier, a GC-rich centromeric band in A. *vulgaris* is one of the elements that allowed us to postulate a chromosomal fusion as the basis for descending dysploidy in *Artemisia* [127]. rDNA loci are also mainly telomeric or subtelomeric (Fig. 11.4D, E, F), most often colocated with GC-rich regions. 5S loci are scarcer than 18S-5.8S-26S loci and always colocated with one of the latter [116, 117].

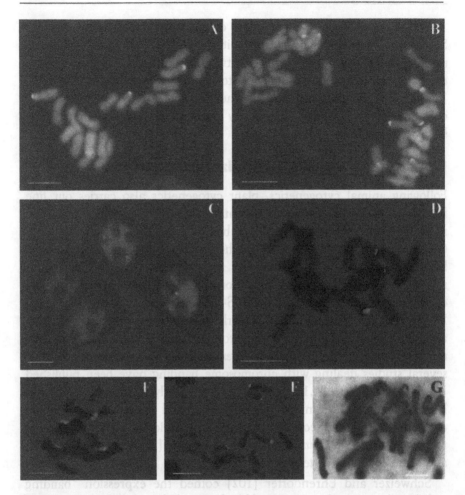

Fig. 11.4: Metaphase plates of various *Artemisia* species after Giemsa or fluorochrome banding or FISH. A-*A. herba-alba* subsp. *valentina*; B-*A. herba-alba* subsp. *herba-alba*; C-*A. barrelieri*; D-*A. caerulescens* subsp. *gallica*; E-*A. annua*; F-*A. tridentata* subsp. *spiciformis*; G-*A. assoana*; A, B, and C: fluorochrome banding with chromomycin; note the large chromocenter in C, corresponding to a B chromosome. D, E, note F: 18S-5.8S-26S loci detected by FISH. G: Giemsa banding. Scale bars = 5 µm [A, B, D, E, F, from 115].

The number of chromomycin-positive bands and 18S-5.8S-26S signals correlate positively with ploidy level (Fig. 11.4A, B).

The comparative fine analysis of molecular cytogenetic data was conclusive regarding the relationships among different taxa. As a case example, the subgenus *Seriphidium*, morphologically very homogeneous and also very constant in rDNA ITS sequences, likewise exhibits a very characteristic and uniform karyotype [117]. The *Seriphidium* banding pattern (four chromomycin-positive bands and four 18S-5.8S-26S rDNA loci per diploid genome; Fig. 11.4D) clearly differs from that of subgenus *Tridentatae* (six chromomycin-positive bands and six 18S-5.8S-26S rDNA loci per diploid genome; Fig. 11.4F), confirming the statements of McArthur et al. [78, 80] that the *Tridentatae* originated independently from *Seriphidium*; thus, it makes no sense to include the North American subgenus in the large subgenus *Seriphidium*, as done by Watson et al. [136]. These results are coincidental with the analysis of ITS sequences [115, 133]. On the other hand, A. *annua*, a taxon belonging to subgenus *Artemisia* but included in molecular phylogeny in the *Seriphidium* clade, presents a karyotype structure (four chromomycin-positive bands and four 18S-5.8S-26S rDNA loci per diploid genome; Fig. 11.4E) which agrees with that of *Seriphidium* but not with that of the subgenus in which it is classically classified.

Nuclear DNA Amount

Genome size, represented by the C-value determined by flow cytometry or by cytodensitometry after Feulgen staining in nuclear DNA pg and expressed in this way or in megabase pairs, is a very well-correlated parameter with many genotypic factors, systematic and evolutionary aspects, ecological questions and characters relevant for plant breeding [9,38,113,143]. Nuclear DNA content has been estimated for seven *Artemisia* species by cytodensitometry after Feulgen staining [10 – from S.R. Band's pers. comm., 23,39,41,83]. Aside from this, we assessed genome size by flow cytometry in about 80 taxa of Artemisiinae, including *Artemisia* and also *Ajania*, *Brachanthemum*, *Dendranthema*, *Filifolium*, *Hippolytia*, *Kaschgaria*, *Lepidolopsis*, *Mausolea*, *Neopallasia*, *Nipponanthemum*, and *Tanacetopsis* [38,113 and unpublished data].

DNA amount is logically very related to ploidy level and other karyological characters, such as chromosome number and karyotype total length. Life cycle and ecology also affect genome size. An increase in DNA amount implies a longer cell cycle, which means that annual plants contain less nuclear DNA than perennial [97]. The annual A. *scoparia* provides a good example of this phenomenon. It has one of the lowest 2C values reported to date in the genus (3.54 pg, the minimum being 3.50 and the maximum 25.65; [38,113]). The studied population inhabits an intermittently dry river bed, and its low genome size promotes a fast life cycle that is rapidly completed before the seasonal summer or fall floods. This case supports the premise that annual species have less nuclear DNA than perennials. In contrast, however, A. *leucodes*, another annual species, has one of the biggest genomes (2C=15.39 pg) of all the diploid species studied; its karyotype comprises large chromosomes [131] and its high C-value, despite its annual life cycle, is supported by the Nagl and Ehrendorfer [83] explanation that large chromosomes could have a higher metabolic rate, which facilitates an increase in RNA synthesis. This condition would provide an increase in synthesis of the proteins necessary for a faster life cycle. Price and Bachmann [94] also found annual plants with a relatively high DNA content. Several authors [8,21,40,96] have suggested an increase in DNA content as an adaptive response to high altitudes. We found no such correlation in Artemisiinae. Conversely, we detected a genome size increase due to other types of ecological selection pressure: the need for adaptation to arid environments, often dominated by *Artemisia* or related genera.

C-value variation also has systematic implications. The five subgenera of *Artemisia* have significantly different mean DNA contents but in some cases a greater variation is observed within than between infrageneric groups, contrary to reports for other genera (*Hypochaeris* [21], *Rosa* [142]). This fact, together with other cytogenetic and molecular data, suggests that the current infrageneric classification of genus *Artemisia* is not organized in completely natural groups. It is specially remarkable that two closely related species belonging to subgenus *Artemisia* (A. *judaica* and A.*lucentica*) have a DNA content closer to that of subgenus *Absinthium* than to subgenus *Artemisia*. This is consistent with the results of molecular phylogeny (Fig. 11.5), in which these two species are placed together with members of subgenus *Absinthium*. Subgenus *Artemisia*, which appears more dispersed in ITS phylogeny, is the one which presents a greater DNA content variation, while the subgenus *Dracunculus* is the most homogeneous in genome size and also the best defined in molecular phylogeny (Fig. 11.5). *Dracunculus*, basal in molecular analysis, is the subgenus with less mean DNA content, supporting that ancestral genomes are relatively small [59]. On the other hand, genome size differs significantly between *Artemisia* and related genera not included in Artemisiinae, while there are no significant differences in DNA amount between *Artemisia* and the remaining Artemisiinae.

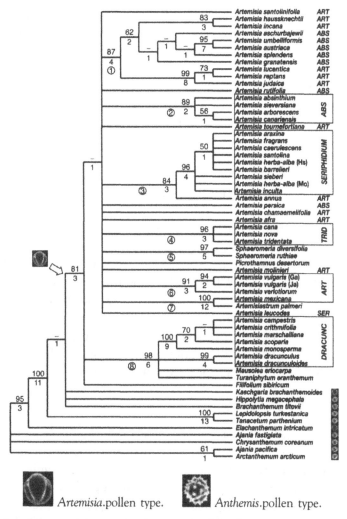

Artemisia.pollen type.   Anthemis.pollen type.

Fig. 11.5: Strict consensus tree showing the hypothetical phylogeny of ITS in *Artemisia* and other genera of Artemisiinae and related subtribes [132]. Numbers above the branches indicate bootstrap percentages; numbers below branches indicate Bremer indices. Circled numbers indicate the eight principal clades. ART = subgenus *Artemisia*; SERIPHIDIUM and SER = subgenus *Seriphidium*; ABS = subgenus *Absinthium*; DRA = subgenus *Dracunculus*; TRID = subgenus *Tridentatae*.

Genetic Variability and Conservation

Conservation biology is a relatively new issue of plant genetic studies, addressed to evaluation of the status of populations of rare and/or endangered plants. One method used frequently in assessing genetic diversity and considered a good indicator of the health situation of a population, is isozymic analysis. We did a study from this viewpoint, complemented with investigation of pollen and achene fertility and cytogenetic research, on one species, namely A. *molinieri*, endemic to a very small area in Southeastern France [114]. It is a narrow endemic species occurring only in two small lakes, with a diameter of 0.5 km and situated 3 km apart. It has a high rate of pollen sterility (pollen fertility only reaches ca. 25%, whereas other species of the genus easily reach ca. 70% [114]), and an extremely low achene germination power (ca. 10% vis-a-vis ca. 90% of other species of the genus [84,114,124 and unpublished data]). In addition, we observed an active vegetative multiplication, via stolons. In spite of all this, the two populations exhibit a high genetic variability: the mean number of alleles per locus was 2.3 in one and 2.2 in the other; the percentage of polymorphic loci with the most common allellic frequency less than 0.95 was 100 and 90.9; the observed heterozygosity 0.401 and 0.433; and the expected heterozygosity 0.426 and 0.371. From these data we deduced that this species is not facing inbreeding depression, so that it is not *a priori* endangered despite its extremely reduced distribution area. The only problems could be anthropic, such as habitat elimination by desiccation of the lakes, which is unlikely as they occur in a protected natural area.

Molecular Phylogeny

A big genus and subtribe such as *Artemisia* and Artemisiinae are logical objects of molecular studies to establish their phylogeny. The first papers on this group focused in the North American endemic subgenus *Tridentatae* [54,55,79,80]. We did a study including a balanced number of representatives of each subgenus of *Artemisia* [115]. Other genera of Artemisiinae were studied from a molecular viewpoint [80,133,136]. Finally, Andrea et al. [6] published a paper on a small number of *Artemisia* species, with more than half the sequences taken from [99] and [115]. Most of these papers are based on the analysis of ITS sequences, whereas those from McArthur [79,80] use RAPDs. Our study [115,133 and unpublished data] includes to date around 50 *Artemisia* species and about 20 genera of Artemisiinae or related subtribes. The ITS provided definitive information that allowed us to draw some conclusions regarding the phylogeny of the group and its infrageneric, generic, and subtribal organization. In many cases these results are consistent with other data, such as cytogenetic, palynological, or chemical. We synthesize them above (Fig. 11.5).

The genus *Artemisia* is clearly monophyletic in the ITS phylogeny, undoubtedly including *Seriphidium* as an infrageneric clade. This result is confirmed by data on karyotypic structure. Also supported by both molecular and cytogenetic data, *Tridentatae* is likewise shown as an infrageneric group within *Artemisia*, at the same level as *Seriphidium*, thus independent from it. Our results agree with those from McArthur et al. [79,80] and Kornkven et al. [54,55]. At this point we must remark that, despite postulating the independence of *Tridentatae* from *Seriphidium* in their first papers, Watson et al. ([135], the principal authors of which are the same as in Kornkven et al. [54,55]), cite species *Artemisia tridentata* as *Seriphidium tridentatum* of the putative genus *Seriphidium*. This fact presents two problems. First, the contradiction of supporting the inclusion of *Seriphidium* in *Artemisia* and, after this, employing it as an independent genus. Second, using as a member of genus *Seriphidium* a species belonging to subgenus *Tridentatae* which, as we have shown, is independent from *Seriphidium*, so that even in the case of segregating *Seriphidium* as a genus, species of *Tridentatae* would not have to be included in it. We state again that the monophyly of *Artemisia* with all its infrageneric groups is clear and that two of these groups, often treated together, *Seriphidium* and *Tridentatae*, appear in separate clades. Several authors [74,75,78,80,105] have already stated that *Seriphidium* and *Tridentatae* are separate groups, which probably evolved from members of subgenera *Artemisia* o r *Dracunculus* and which have undergone convergent evolution. Moreover, Seaman [103], on the basis of chemical characters, also proposed a closer relationship between *Tridentatae* and *Artemisia* than between *Tridentatae* and *Seriphidium*. Phytochemistry is a good complementary tool for the systematics of the genera considered [71, 139].

In summary, genus *Artemisia lato et classico sensu* forms a well-defined and supported clade (Fig. 11.5). The groups within *Artemisia* are as follows:

— A group (clade 1 in Fig. 11.5) constituted by members of subgenera *Artemisia* and *Absinthium*. This clade is supported by other data since some of the species share cytogenetic characters, such as nuclear DNA content.

— Another group formed by species of subgenus *Absinthium* (clade 2), all of them closely related to *A. absinthium*.

— Subgenus *Seriphidium*, already discussed above, united to *A. annua* (clade 3 in Fig. 11.5). Chromosome structure (banding, FISH) supports the closeness of this species belonging to subgenus

Artemisia with subgenus *Seriphidium*. One of the *Seriphidium* species, A. *leucodes*, is not included in this clade. It is an extremely rare annual *Seriphidium* and its annual character may have influenced the apparently anomalous situation (this could also be one of the causes of the A. *annua* situation) [133].

— Subgenus *Tridentatae* (clade 4), already commented on.

— Genus *Sphaeromeria* (clade 5), clearly incardinated in *Artemisia*, has nevertheless been frequently related to *Tanacetum*. Our results support the hypotheses of Holmgren et al. [44] and McArthur et al. [79], who considered it closer to *Artemisia* than to *Tanacetum*.

— Genus *Picrothamnus*, the only species of which was originally described in *Artemisia*, is also included in the mother genus.

— Genus *Filifolium*, also monotypic, whose species was described in *Tanacetum*, appears within *Artemisia* as well, and evidencing no relationship with the *Tanacetum* species analyzed.

— Representatives of the A. *vulgaris* complex are located in two clades (6 and 7), one of which encompasses genus *Artemisiastrum*, originally placed already in *Artemisia*.

— Subgenus *Dracunculus*, including genera *Mausolea* and *Turaniphytum* (clade 8), whose species were originally described in *Artemisia* and linked to subgenus *Dracunculus*. This big group is included in genus *Artemisia*; hence segregation of this subgenus in genus *Oligosporus* as proposed by Cassini [20] makes no sense.

— In addition, five species of subgenus *Artemisia*, two of subgenus *Absinthium* and one of subgenus *Seriphidium* not incorporated in any supported clade.

The remaining genera of subtribe Artemisiinae and other closely related subtribes show in the ITS phylogeny, the following degrees of affinity with *Artemisia*:

— Some genera appear completely included in *Artemisia*, as already stated. All of them belong to subtribe Artemisiinae and have the *Artemisia* pollen type (Fig. 11.5 and Palynology below).

— Three genera are located independently at the same level of the *Artemisia* clade, one (*Kaschgaria*, belonging to subtribe Artemisiinae) with *Artemisia* pollen type, and two (*Hippolytia*, subtribe Tanacetinae, and *Brachanthemum*, subtribe Artemisiinae) with *Anthemis* pollen type.

— One clade including *Lepidolopsis* and *Tanacetum*, both with *Anthemis* pollen type but belonging to two different subtribes, Handeliinae and Tanacetinae respectively.

— Genus *Elachanthemum*, belonging to subtribe Artemisiinae and with *Artemisia* pollen type, but clearly located outside the *Artemisia* clade.

In addition to those of the ITS, we analyzed the sequences of the chloroplast DNA *trn*L-F intron and we verified that they were not informative at the taxonomic level considered. We recently started the study of nuclear ribosomal DNA external transcribed spacer (ETS; unpublished results). The preliminary analyses carried out with 30 species are promising: the ETS are more resolutive than ITS, so that their phylogeny and the combined one of both regions will probably shed light on the subtribal structuration and relationships.

Molecular phylogeny, morphology, and other biological features

We compared the results of molecular analysis with some morphological characters classically used in the systematics of *Artemisia* and its allies (unpublished data). The taxa with ligulate flowers (such as *Arctanthemum* and *Brachanthemum*) are candidates to be excluded from Artemisiinae *stricto sensu* in its revised delineation. Flower head typology (homogamous —subgenera *Seriphidium* and *Tridentatae*—, heterogamous with hermaphrodite peripheral flowers —subgenera *Absinthium* and *Artemisia*—, heterogamous with functionally male peripheral flowers— subgenus *Dracunculus*—), one of the basic characters of classic taxonomy, agrees with some monophyletic groups in the ITS analysis. Features such as number of flowers per head and type of synflorescence define groups in agreement with molecular phylogeny, whereas others are less consistent with the analysis of ITS sequences. Concerning other kinds of biological features, comparison with molecular phylogeny showed that the annual condition residual in the plants studied, appeared at different times in some groups.

Palynology and Its Relationship with Cytogenetic and Molecular Data

Some pollen characters correlate very well with molecular and cytogenetic characters and thus constitute an excellent taxonomic marker. The study of many *Artemisia* species and at least one each of the related genera [72,73,123,125,129 and unpublished data] led to the conclusion that two exine ornamentation patterns occur in *Anthemis* and *Artemisia* types [109,140]. *Artemisia* and its more closely related genera have *Artemisia* pollen type, with slight, echinulate ornamentation. Genera more distant to *Artemisia* have *Anthemis* pollen type, much more ornamented, echinate. These two models are quite consistent with molecular phylogeny (Fig. 11.5). Genome size also accords with separation of the two groups: taxa with *Artemisia* pollen type have a significantly smaller nuclear DNA amount than those with *Anthemis* pollen type [38]. In addition to the fact that this difference could have a biological sense (adaptation to pollination types), it indicates that genome size supports morphology in its relationship with molecular phylogeny.

CONCLUDING REMARKS AND PERSPECTIVES

As reviewed in the preceding pages, knowledge of genome organization and evolution in *Artemisia* and its closely related genera is quite good. Some cytogenetic and molecular characters are consistent with morphological ones and with other biological features. Comparative analysis of all data sets led to the conclusion that both infrageneric classification in the large genus *Artemisia* and subtribal structuring of concerned genera within tribe Anthemideae are still unsolved problems. It is clear that subgenera traditionally considered in *Artemisia* do not represent natural groups. It is also obvious that the currently accepted borders between subtribes Artemisiinae, Chrysantheminae, Handeliinae, Leucantheminae, and Tanacetinae are fuzzy; moreover, members of other subtribes, such as Cancriniinae, will probably have to be considered in further studies. Analysis of ITS sequences having reached the limit of resolution in this group, other regions must be investigated. Cytogenetic studies (in particular molecular cytogenetics and genome size assessment) have to be carried out, especially in genera other than *Artemisia* (many *Artemisia* species belonging to all current subgenera are well known from this viewpoint, whereas data on nuclear DNA content and molecular cytogenetics are available for only a scant number of related genera). These studies will lead to a consistent and stable subtribal, generic, and infrageneric classification in *Artemisia* and its allies. Lastly, research into particular groups or topics is imperative. Foremost is confirmation or not, through molecular cytogenetic researches, of putative *Artemisia* hybrids, in-depth study of the A. *vulgaris* complex and the A. *dracunculus* polyploid series, and a detailed investigation of rare or endangered taxa and of genera such as *Ajania*, with notable taxonomic conflicts.

Acknowledgments

This synthesis, with new data, reflects the research conducted by us over the last 22 years on *Artemisia* and its related genera, which involved collaboration with many people, to whom we are grateful and who, at various levels, must be considered coauthors of this paper: Sònia Garcia, Joan Martín, Marian Oliva, María Sanz, Juan A. Seoane, Maruxa Suárez, Montserrat Torrell (Universitat de Barcelona), Núria Garcia-Jacas, Alfonso Susanna (Institut Botànic de Barcelona), Malika Cerbah, Sonja Siljak-Yakovlev (Université de Paris Sud), E. Durant McArthur (Shrub Research Laboratory, Provo, Utah), Agnieszka Kreitschitz (Universytet Wroctawski) and Aleksandr A. Korobkov (Botanicheskii Institut V.L. Komarova, Saint Petersburg). These colleagues are also thanked for their assistance in plant material collection, as are Carles Benedí, Cèsar Blanché, Maria Bosch, Amelia Gómez, Julián Molero, M. Antonia Ribera, Joan Simon (Universitat de Barcelona), Josep Penuelas (centre de reserca Ecologica i Aplicacions forestals, Bellaterra Anne M. Cauwet-Marc (Universitat de Perpinyà), André Charpin (Conservatoire et Jardin Botaniques, Genève), Joël Mathez (Université de Montpellier), Simonetta Peccenini (Università di Genova), Katsuhiko Kondo (Hiroshima University), Fatime Elalaoui, Aïcha Ouyahya, (Université Mohammed V, Rabat), Georgii Fajvush, Eleonora Gabrielian, Marine Oganesian, Kamilla Tamanian (Botanical Institute, Erevan), Lyuba A. Kapustina, Furkat Khassanov (Botanical Institute, Tashkent), Anna A. Ivaschenko (Botanical Institute, Almaty), Shagdas Tsooj (Botanical Institute, Ulaan Baatar), Shagdas Dariimaa (Education University, Ulaan Baatar), and Valiollah Mozaffarian (Institute of Forests and Rangelands, Tehran). Technical advice and help from Jaume Comas, Màrius Mumbrú (Universitat de Barcelona), Oriane Hidalgo, Miquel Veny, Roser Vilatersana (Institut Botànic de Barcelona), Odile Robin (Université de Paris Sud) and Spencer C. Brown (Institut des Sciences du Végétal, Gif-sur-Yvette) also proved invaluable.

This work was supported by the Spanish government (projects PB-88-0033, PB-93-0032, PB-97-1134, BOS2001-3041-C02-01), the Catalan government (projects AR83-210, 1995SGR00087, 1998BEAI400180, 1999SGR-00332, 2001SGR00125), and the Universitat de Barcelona (projects AR-UB-1985, GRC-UB-3120).

This paper is dedicated with gratitude to Prof. Claude Favarger (Université de Neuchâtel) who devoted most of his 90 active years to the study and teaching of plant cytogenetics.

[1] Abd El-Twab MG, Kondo K. Identification of nucleolar organizing regions and parental chromosomes in F₁ hybrid of Dendranthema japonica and Tanacetum vulgare simultaneously by fluorescence in situ hybridization and fluorescence genomic in situ hybridization. Chrom Sci 1999; 3: 59-62.

[2] Abd El-Twab MG, Kondo K. Molecular cytogenetic identification of the parental genomes in the intergeneric hybrid between Leucanthemella linearis and Nipponanthemum nipponicum during meiosis and mitosis. Caryologia 2001; 54: 109-114.

[3] Adylov TA, Zuckerwanik TI, eds. Opredelitel rasteniy Srednei Azii, tom X. Conspectus florae Asiae Mediae, tomus X. Tashkent: Academy Sciences Republic of Uzbekistan, 1993.

[4] Alexander MP. Differential staining of aborted and nonaborted pollen. Stain Tech 1969; 44: 117-122.

[5] Alexander MP. A versatile stain for pollen, fungi, yeasts and bacteria. Stain Tech 1980; 55: 13-18.

[6] Andrea S d', Caramiello R, Ghignone S, Siniscalco C. Systematic studies on some species of the genus Artemisia: biomolecular analysis. Pl Biosyst 2003; 137: 121-130.

[7] Avetissian EM. Uproschennyi atsotoliznyi metod obrabotki pyltsy. Bot Zh 1950; 35: 385.

[8] Bennett MD. DNA amount, latitude and crop plant distribution. Envir Exper Bot 1976; 16: 93-108.

[9] Bennett MD. Plant genome values: How much do we know? Proc Natl Acad Sci USA 1998; 95: 2011-2016.

[10] Bennett MD, Smith JB. Nuclear DNA amounts in angiosperms. Phil Trans Roy Soc London Biol Sci 1991; 334: 309-345.

[11] Besser WSJG. Synopsis Absinthiorum. Bull Soc Imp Nat Moscou 1829; 1: 219-265.

[12] Besser WSJG. Tentamen de Abrotanis seu de sectione IIda Artemisiarum Linnaei. Bull Soc Imp Nat Moscou 1832; 3: 1-92.

[13] Besser WSJG. De Seriphidiis seu de sectione IIIa Artemisiarum Linnaei. Bull Soc Imp Nat Moscou 1834; 7: 1-46.

[14] Besser WSJG. Dracunculi seu de sectione IVta et ultima Artemisiarum Linnaei. Bull Soc Imp Nat Moscou 1835; 8: 1-95.

[15] Bremer K. Asteraceae. Cladistics and classification. Portland, OR (USA): Timber Press, 1994.

[16] Bremer K, Humphries CJ. Generic monograph of the Asteraceae–Anthemideae. Bull Nat Hist Mus London (Bot) 1993; 23: 71-177.

[17] Bretagnolle F, Felber F, Calame FG, Küpfer Ph. La polyploïdie chez les plantes. Bot Helvet 1998; 108: 5-37.

[18] Camacho JPM, Sharbel TF, Beukeboom LW. B-chromosome evolution. Phil Trans Roy Soc London B 2000; 355: 163-178.

[19] Candolle AP de. Prodromus systematis naturalis regni vegetabilis, pars VI. Paris: Trevttel et Würtz, 1837.

[20] Cassini AHJ. Aperçu des genres formés par M. Cassini dans la famille des Synantherées. Troisième fascicule. Bull Sci Soc Phil Paris 1817; 3: 31-34.

[21] Cerbah M, Coulaud J, Siljak-Yakovlev S. rDNA organization and evolutionary relationships in the genus *Hypochaeris* (Asteraceae). J Hered 1998; 89: 312-318.

[22] Couderc H, Gorenflot R, Moret J, Siami A. Caractéristiques et conséquences de la variation chromosomique chez l'*Ornithogalum divergens* Boureau. Bull Soc Bot France 1985; 132: 63-72.

[23] Dabrowska J. Chromosome number and DNA content in taxa of *Achillea* L. in relation to the distribution of the genus. Prace Bot 1992; 49: 1-84.

[24] Duncan RE. Production of variable aneuploid numbers of chromosomes within the root tips of *Paphiopedilum wardii* Summerhayes. Amer J Bot 1945; 32: 506-509.

[25] Ehrendorfer F. Notizen zur Cytotaxonomie und Evolution der Gattung *Artemisia*. Oesterr Bot Zeitsch 1964; 111: 84-142.

[26] Ehrendorfer F. Polyploidy and distribution. In: Lewis WH, ed. Polyploidy: Biological relevance. New York, NY: Plenum Press, 1980: 45-60.

[27] Erdtman G. Handbook of palynology: morphology, taxonomy, ecology. Copenhagen: Munksgaard, 1969.

[28] Estes JR. Evidence for autoploid evolution in the *Artemisia ludoviciana* complex of the Pacific Northwest. Brittonia 1969; 21: 29-43.

[29] Favarger C. Some recent progress and problems in cytotaxonomy of the higher plants. Nucleus 1981; 24: 151-157.

[30] Favarger C. Cytogeography and biosystematics. In: Grant WF, ed. Plant Biosystematics. London: Academic Press, 1984: 453-475.

[31] Favarger C, Contandriopoulos J. Essai sur l'endémisme. Bull Soc Bot Suisse 1961; 71: 384-408.

[32] Favarger C, Siljak-Yakovlev S. A propos de la classification des taxons endémiques basée sur la cytotaxonomie et la cytogénétique. In: Colloque international de Botanique pyrénéenne, La Cabanasse (Pyrénées-Orientales) 3-5 Juillet 1986. Société Botanique de France/Groupement Scientifique Isard, 1986: 287-303.

[33] Filatova NS. Kariotipy peschanikh vidov polynei podroda *Dracunculus* (Bess.) Rydb. De caryotypis specierum *Artemisii* e subgenere *Dracunculus* (Bess.) Rydb. in arenis vigentium. Bot Mater Gerbar Inst Bot Akad Nauk Kaz SSR, 1971; 7: 46-49.

[34] Filatova NS. K kariosistematicheskomu izucheniyu kazakhstanskikh polynei podroda *Seriphidium* (Bess.) Rouy, Ad stadio de caryotaxonomicum *Artemisii* kazachstaniae e subgenere *Seriphidium* (Bess.) Rouy. Bot Mater Gerbar Inst Bot Akad Nauk Kaz SSR, 1974; 8: 66-75.

[35] Filatova NS. Kariosistematika dvukh vidov polynei podroda *Seriphidium* (Bess.) Rouy. Izvest Akad Nauk Kaz SSR, Ser Biol 1974; 1: 16-20.

[36] Filatova NS. Sistema polynei podroda *Seriphidium* (Bess.) Peterm. (*Artemisia* L., Asteraceae) Evrazii i severnoi Afriki. Systema specierum generis *Artemisia* L. subgeneris *Seriphidium* (Bess.) Perterm. in Eurasia et Africa boreali vigentium. Nov Sist Vysch Rasten 1986; 23: 217-239.

[37] Gabrielian E, Vallès J. New data about the genus *Artemisia* L. (*Asteraceae*) in Armenia. Willdenowia 1996; 26: 245-250.

[38] Garcia S, Sanz M, Garnatje T, Kreitschitz A, McArthur ED, Vallès J. Variation of DNA amount in 47 populations of the subtribe Artemisiinae and related taxa (Asteraceae, Anthemideae): karyological, ecological, and systematic implications. Genome 2004; 47, in press.

[39] Geber G, Hasibeder G. Cytophotometrische Bestimmung von DNA-Mengen: Vergleich einer neuen DAPI-Fluoreszenz Methode mit Feulgen-Absortionsphotometrie. Microsc Acta 1980; suppl. 4: 31-35.

[40] Godelle B, Cartier D, Marie D, Brown SC, Siljak-Yakovlev S. Heterochromatin study demonstrating the non-linearity of fluorometry useful for calculating genomic base composition. Cytometry 1993; 14: 618-626.

[41] Greilhuber J. 'Self-tanning'—a new and important source of stoichiometric error in cytophotometric determination of nuclear DNA content in plants. Plant Syst Evol 1988; 158: 87-96.

[42] Gutermann W. Systematik und Evolution einer alten, dysploid-polyploiden Oreophyten-Gruppe: *Artemisia mutellina* und ihre Verwandten (Asteraceae: Anthemideae). PhD thesis, Universitat Wien, Wien, 1979.

[43] Heywood VH, Humphries CJ. Anthemideae. Systematic review. In: Heywood VH, Harborne JB, Turner BL, eds. The biology and the chemistry of the Compositae, vol. II. London, UK: Academic Press, 1977: 851-898.

[44] Holmgren AH, Shultz LM, Lowrey TK. *Sphaeromeria*, a genus closer to *Artemisia* than to *Tanacetum* (Asteraceae: Anthemideae). Brittonia 1976; 28: 252-262.

[45] Honda Y, Abd El-Twab MH, Ogura H, Kondo K, Tanaka R, Shidahara T. Counting sat-chromosome numbers and species characterization in wild species of *Chrysanthemum sensu lato* by fluorescence *in situ* hybridization using pTa71 probe. Chrom Sci 1997; 1: 77-81.

[46] James CM, Wurzell BS, Stace CA. A new hybrid between a European and a Chinese species of *Artemisia* (Asteraceae). Watsonia 2000; 23: 139-147.

[47] Jones RN, Rees H. B chromosomes. London, UK: Academic Press, 1982.

[48] Kawatani T, Ohno T. Chromosome numbers in *Artemisia*. Bull Natl Inst Hyg Sci 1964; 82: 183-193.

[49] Khaung KK, Kondo K, Tanaka R. Physical mapping of rDNA by fluorescent *in-situ* hybridization using pTa71 probe in three tetraploid species of *Dendranthema*. Chrom Sci 1997; 1: 25-30.

[50] Kondo K, Kokubugata G, Honda Y. Marking and identification of certain chromosomes in wild *Chrysanthemum* and cycads by fluorescence *in situ* hybridization. Genet Polon 1996; 37A: 24-26.

[51] Kondo K, Honda Y, Tanaka R. Chromosome marking in *Dendranthema japonica* var. *wakasaense* and its closely related species by fluorescence *in situ* hybridization using rDNA probe. Kromosomo 1996; II-81: 2785-2791.

[52] Kondo K, Abd El-Twab MH, Tanaka R. Fluorescence *in situ* hybridization identifies reciprocal translocation of somatic chromosomes and origin of extra chromosome by an artificial, intergeneric hybrid between *Dendranthema japonica* x *Tanacetum vulgare*. Chrom Sci 1999; 3: 15-19.

[53] Kondo K, Abd El-Twab MH, Idesawa R, Kimura S, Tanaka R. Genome phylogenetics in *Chrysanthemum sensu lato*. In: Sharma AK, Sharma A, eds. Plant genome.

Biodiversity and evolution. Vol. I, Pt. A: Phanerogams. Enfield (NH, USA)/Plymouth (UK): Science Publishers, 2003: 117-200.

[54] Kornkven AB, Watson LE, Estes JR. Phylogenetic analysis of *Artemisia* sect. *Tridentatae* (Asteraceae) based on sequences from the internal transcribed spacers (ITS) of nuclear ribosomal DNA. Amer J Bot 1998; 85: 1787-1795.

[55] Kornkven AB, Watson LE, Estes JR. Molecular phylogeny of *Artemisia* sect. *Tridentatae* (Asteraceae) based on chloroplast DNA restriction site variation. Syst Bot 1999; 24: 69-84.

[56] Korobkov AA. Polyni Severo–Vostoka SSSR. Leningrad: Nauka, 1981.

[57] Koul MLH. Cytogenetics of polyploids. II. Cytology of polyploid *Artemisia maritima* L. Cytologia 1965; 30: 1-9.

[58] Kreitschitz A, Vallès J. New or rare data about chromosome numbers in several taxa of the genus *Artemisia* L. (*Asteraceae*) in Poland. Fol Geobot 2003; 38: 333-343.

[59] Leitch IJ, Chase MW, Bennett MD. Phylogenetic analysis provides evidence for a small ancestral genome size in flowering plants. Ann Bot 1998; 82 (Suppl A): 85-94.

[60] Lessing CF. Synopsis generum Compositarum. Berlin: Dricker et Humblot, 1832.

[61] Lewis WH. Chromosomal drift, a new phenomenon in plants. Science 1970; 168: 1115-1116.

[62] Ling YR. On the system of the genus *Artemisia* L. and the relationship with its allies. Bull Bot Lab North–East For Inst 1982; 2: 1-60.

[63] Ling YR. The Old World *Seriphidium* (Compositae). Bull Bot Res Harbin 1991; 11(4): 1-40.

[64] Ling YR. The Old World *Artemisia* (Compositae). Bull Bot Res Harbin 1991; 12(1): 1-108.

[65] Ling YR. The genera *Artemisia* L. and *Seriphidium* (Bess.) Poljak. in the world. Comp Newsl 1994; 25: 39-45.

[66] Ling YR. The New World *Artemisia* L. In: Hind DJN, Jeffrey C, Pope GV, eds. Advances in Compositae Systematics. Kew: Roy Bot Gard 1995: 255-281.

[67] Ling YR. The New World *Seriphidium* (Besser) Fourr. In: Hind DJN, Jeffrey C, Pope GV, eds. Advances in Compositae Systematics. Kew: Roy Bot Gard 1995: 283-291.

[68] Linné Kv. Systema naturae. Leiden, Netherlands: T. Haak, 1735, (1st ed.).

[69] Mabberley DJ. The plant–book. Cambridge: Cambridge University Press, 1990.

[70] Malakhova LA. Kariologicheskii analiz prirodnykh populyatsii redkykh i ischezayushchykh rastenii na yuge Tomskoi oblasti. Byull Glavn Bot Sada 1990; 155: 60-66.

[71] Marco JA, Barberá Ó. Natural products from the genus *Artemisia* L. In: Atta-ur-Rahman, ed. Studies in natural products. Chemistry, vol. 7A. Amsterdam: Elsevier, 1990: 201-264.

[72] Martín J, Torrell M, Vallès J. Palynological features as a systematic marker in *Artemisia* L. and related genera (Asteraceae, Anthemideae). Pl Biol 2001; 3: 372-378.

[73] Martín J, Torrell M, Korobkov AA, Vallès J. Palynological features as a systematic marker in *Artemisia* L. and related genera (Asteraceae, Anthemideae), II: Implications for subtribe Artemisiinae delimitation. Pl Biol 2003; 5: 85-93.

[74] McArthur ED. Sagebrush systematics and evolution. In: Sagebrush ecosystem symposium. Logan, UT: Utah State University, 1979: 14-22.

[75] McArthur ED, Plummer AP. Biogeography and management of native western shrubs: A case study, section *Tridentatae* of *Artemisia*. Great Bas Nat Mem 1978; 2: 229-243.

[76] McArthur ED, Pope CL. Karotypes of four *Artemisia* species: *A. carruthii*, *A. filifolia*, *A. frigida*, and *A. spinescens*. Great Bas Nat Mem 1979; 39: 419-426.

[77] McArthur ED, Sanderson SC. Cytogeography and chromosome evolution of subgenus *Tridentatae* of *Artemisia* (Asteraceae). Amer J Bot 1999; 86: 1754-1775.

[78] McArthur ED, Pope CL, Freeman DC. Chromosomal studies of subgenus *Tridentatae* of *Artemisia*: Evidence for autopolyploidy. Amer J Bot 1981; 68: 589-605.

[79] McArthur ED, Buren RV, Sanderson SC, Harper KT. Taxonomy of *Sphaeromeria*, *Artemisia*, and *Tanacetum* (Compositae, Anthemideae) based on randomly amplified polymorphic DNA (RAPD). Great Bas Nat Mem 1998; 58: 1-11.

[80] McArthur ED, Mudge J, Buren RV, et al. Randomly amplified polymorphic DNA analysis (RAPD) of *Artemisia* subgenus *Tridentatae* species and hybrids. Great Bas Nat Mem 1998; 58: 12-27.

[81] Mendelak M, Schweizer D. Giemsa C-banded karyotypes of some diploid *Artemisia* species. Pl Syst Evol 1986; 152: 195-210.

[82] Murín A. Karyotaxonomy of some medicinal and aromatic plants. Thaiszia – J Bot Kosice 1997; 7: 75-88.

[83] Nagl W, Ehrendorfer F. DNA content, heterochromatin, mitotic index, and growth in perennial and annual Anthemideae (Asteraceae). Pl Syst Evol 1974; 123: 737-740.

[84] Oliva M, Vallès J. Contribution to the cytotaxonomical knowledge of the genus *Artemisia* L.: Giemsa C-banded karyotypes of some taxa. Bot Chron 1991; 10: 737-740.

[85] Oliva M, Vallès, J. Karyological studies in some taxa of the genus *Artemisia* (Asteraceae). Can J Bot 1994; 72: 1126-1135.

[86] Oliva M, Torrell M, Vallès J. Data on germination rates and germinative vigour in *Artemisia* (*Asteraceae*). Bocconea 1997; 5: 679-684.

[87] Otto SP, Whitton J. Polyploid incidence and evolution. Ann Rev Genet 2000; 34: 401-437.

[88] Ouyahya A, Viano J. Recherches cytogénétiques sur le genre *Artemisia* L. au Maroc. Bol Soc Brot Sér 2, 1988; 61: 105-124.

[89] Pareto G, ed. Artemisie. Ricerca ed applicazione. Quad Agr suppl 2, 1985: 1-261.

[90] Persson K. Biosystematic studies in the *Artemisia maritima* complex in Europe. Op Bot (Lund) 1974; 35: 1-188.

[91] Podlech D. *Artemisia*. In: Rechinger K, ed. Flora Iranica, vol 6. Graz, Austria: Academische Druck v. Verlagsanstalt, 1986: 159-223.

[92] Poljakov PP. Materialy k sistematike roda polyn – *Artemisia* L. Trudy Inst Bot Akad Nauk Kaz SSR 1961; 11: 134-177.

[93] Poljakov PP. Rod 1550. Polyn – *Artemisia* L. In: Shishkin BK, Bobrov EG, eds. Flora SSSR, vol. 25. Leningrad: Nauka, 1961: 425-631.

[94] Price HJ, Bachmann K. Mitotic cycle time and DNA content in annual and perennial *Microseridinae* (Compositae, Cichoriaceae). Pl Syst Evol 1976; 126: 323-330.

[95] Ramsey J, Schemske DW. Pathways, mechanisms, and rates of polyploid formation in flowering plants. Ann Rev Ecol Syst 1998; 29: 467-501.

[96] Rayburn AL, Auger JA. Genome size variation in *Zea mays* ssp. *mays* adapted to different altitudes. Theor Appl Genet 1990; 79: 470-474.

[97] Rees H, Narayan RKJ. Chromosomal DNA in higher plants. Phil Trans R Soc London B 1981; 292: 569-578.

[98] Reitsma T. Suggestions toward unification of descriptive terminology of angiosperm pollen grains. Rev Paleobot Palynol 1970; 10: 39-60.

[99] Rousi A. Cytogenetic comparison between two kinds of cultivated tarragon (*Artemisia dracunculus*). Hereditas (Lund) 1969; 62: 193-213.

[100] Rouy GCC. Flore de France, tome 8. Asnières/Paris: Rochefort, 1903.

[101] Rydberg PA. *Artemisia* and *Artemisiastrum*. In: Britton NL, Murrill WA, Bamhart JH, eds. North American Flora, vol. 34. New York, NY: New York Botanic Garden, 1916: 244-285.

[102] Schweizer D, Ehrendorfer F. Evolution of C-band patterns in Asteraceae-Anthemideae. Biol Zentralbl 1983; 102: 637-655.

[103] Seaman FC. Sesquiterpene lactones as taxonomic characters in the Asteraceae. Bot Rev 1982; 48: 121-595.

[104] Sharma AK, Sharma A. Chromosome techniques. Theory and practice. London, UK: Butterworths, 1980.

[105] Shultz LM. Taxonomic and geographic limits of *Artemisia* subgenus *Tridentatae* (Asteraceae). In: McArthur ED, Welch BL, eds. Proc. Symp biology of *Artemisia* and *Chrysothamnus*. Ogden, UT: US Dept Agriculture, Forest Service, Intermountain Research Station, 1986: 20-29.

[106] Siljak-Yakovlev, S. La dysploïdie et l'évolution du caryotype. Bocconea 1998; 5: 211-220.

[107] Solbrig OT. Chromosomal cytology and evolution in the family Compositae. In: Heywood VH, Harborne JB, Turner BL, eds. The biology and chemistry of the Compositae, vol. I. London, UK: Academic Press, 1977: 269-281.

[108] Stebbins GL. Chromosomal evolution in higher plants. London, UK: E. Arnold, 1971.

[109] Stix E. Pollenmorphologische untersuchungen an Compositen. Grana Palynol 1960; 2: 41-126.

[110] Swofford DL. PAUP*. Phylogenetic analysis using parsimony (*and other methods). Version 4. Sunderland, MA: Sinauer Assoc. inc., 1999.

[111] Tan RX, Zheng WF, Tang HQ. Biologically active substances from the genus *Artemisia*. Pl Med 1998; 64: 295-302.

[112] Taniguchi K, Tanaka R, Yonezawa Y, Komatsu H. Types of banding patterns of plant chromosomes by modified BSG method. Kromosomo 1975; 100: 3123-3135.

[113] Torrell M, Vallès J. Genome size in 21 *Artemisia* L. species (Asteraceae, Anthemideae): Systematic, evolutionary, and ecological implications. Genome 2001; 44: 231-238.

[114] Torrell M, Bosch M, Martín J, Vallès J. Cytogenetic and isozymic characterization of the narrow endemic species *Artemisia molinieri* (Asteraceae, Anthemideae): Implications for its systematics and conservation. Can J Bot 1999; 77: 51-60.

[115] Torrell M, Garcia-Jacas N, Susanna A, Vallès J. Infrageneric phylogeny of the genus *Artemisia* L. (Asteraceae, Anthemideae) based on nucleotide sequences of nuclear ribosomal DNA internal transcribed spacers (ITS). Taxon 1999; 48: 721-736.

[116] Torrell M, Cerbah M, Siljak–Yakovlev S, Vallès J. Étude cytogénétique de trois taxons du complexe d'*Artemisia campestris* L. (*Asteraceae, Anthemideae*): localisation de l'hétérochromatine et de l'ADN ribosomique. Bocconea 2001; 13: 623-628.

[117] Torrell M, Cerbah M, Siljak–Yakovlev S, Vallès J. Molecular cytogenetics of the genus *Artemisia* (Asteraceae, Anthemideae): Fluorochrome banding and fluorescence *in situ*

hybridization, I. Subgenus *Seriphidium* and related taxa. Pl Syst Evol 2003; 239: 141-153.

[118] Torrell M, Vallès J, Garcia-Jacas N, Mozaffarian V, Gabrielian E. New or rare chromosome counts in the genus *Artemisia* L. (Asteraceae, Anthemideae) from Armenia and Iran. Bot J Linn Soc 2001; 135: 51-60.

[119] Tournerfort, JP de. Institutiones Rei Herbariae. Paris: Typrographia regia, 1700.

[120] Tutin TG, Persson K, Gutermann W. *Artemisia* L. In: Tutin TG, Heywood VH, Burges NA, et al., eds. Flora Europaea, vol. 4. Cambridge, UK: Cambridge University Press, 1976: 178-186.

[121] Urbańska-Worytkiewicz K. Cytological variation within the family of Lemnaceae. Veröff Geobot Inst ETH Stiftung Rübel Zürich 1980; 70: 30-101.

[122] Vallès J. Aportación al conocimiento citotaxonómico de ocho táxones ibéricos del género *Artemisia* L. (*Asteraceae-Anthemideae*). An Jard Bot Madrid 1987; 44: 79-96.

[123] Vallès J. Contribución al estudio de las razas ibéricas de *Artemisia herba-alba* Assoc Bol Soc Brot Sér 2, 1987; 60: 5-27.

[124] Vallès J. Dades sobre la biologia d'espècies ibèrico-baleàriques d'*Artemisia* L. Collect Bot Barcelona 1989; 17: 237-245.

[125] Vallès J, Seoane JA. Étude biosystématique du groupe d'*Artemisia caerulescens* dans la Péninsule Ibérique et les Îles Baléares. Candollea 1987; 42: 365-377.

[126] Vallès J, Oliva M. Contribution à la connaissance du groupe d'*Artemisia umbelliformis* Lam. (*Asteraceae*) dans les Pyrénées. Monog Inst Pir Ecol 1990; 5: 321-330.

[127] Vallès J, Siljak-Yakovlev S. Cytogenetic studies in the genus *Artemisia* L.: fluorochrome banded karyotypes of five taxa, including the Iberian endemic species A. *barrelieri* Besser. Can J Bot 1997; 75: 595-606.

[128] Vallès J, McArthur ED. *Artemisia* systematics and phylogeny: cytogenetic and molecular insights. In: McArthur ED, Fairbanks DJ, eds. Proc Shrubland Ecosystem Genetics and Biodiversity, 2000 June 13-15, Provo, UT. Ogden, UT: US Dept Agriculture Forest Service, Rocky Mountain Research Station, 2001: 67-74.

[129] Vallès J, Suárez M, Seoane JA. Estudio palinológico de las especies ibérico-baleáricas de las secciones *Artemisia* y *Seriphidium* Bess. del género *Artemisia* L. Acta Salmant Cien 1988; 65: 167-174.

[130] Vallès J, Torrell M, Garcia-Jacas N. New or rare chromosome counts in *Artemisia* L. (Asteraceae, Anthemideae) and related genera from Kazakhstan. Bot J Linn Soc 2001; 137: 399-407.

[131] Vallès J, Torrell M, Garcia-Jacas N, Kapustina L. New or rare chromosome counts in the genera *Artemisia* L. and *Mausolea* Bunge (Asteraceae, Anthemideae) from Uzbekistan. Bot J Linn Soc 2001; 135: 391-400.

[132] Vallès J, Garnatje T, Garcia S, Sanz M, Korobkov AA. Chromosome numbers in several taxa of the tribes Anthemideae and Inuleae (Asteraceae). Bot J Linn Soc 2005; in press.

[133] Vallès J, Torrell M, Garnatje T, Garcia-Jacas N, Vilatesana R, Susanna A. The genus *Artemisia* and its allies: phylogeny of the subtribe Artemisiinae (Asteraceae, Anthemideae) based on nucleotide sequences of nuclear ribosomal DNA internal transcribed spacers (ITS). Pl Biol 2003; 5: 274-284.

[134] Watanabe W. Index to chromosome numbers in Asteraceae. http://www-asteraceae.cla.kobe-u.ac.jp/index.html, 2002.

[135] Watson LE, Evans TM, Boluarte T. Molecular phylogeny and biogeography of tribe Anthemideae (Asteraceae) based on chloroplast gene *ndhF*. Mol Phylogenet Evol 2000; 15(1): 59-69.

[136] Watson LE, Bates PL, Evans TM, Unwin MM, Estes JR. Molecular phylogeny of subtribe Artemisiinae (Asteraceae), including Artemisia and its segregate genera. BMC Evol Biol 2002; 2: 17, http://www.biomedcentral.com/1471-2148/2/17.

[137] Weinedel-Liebau F. Zytologische Untersuchungen an *Artemisia*-Arten. Jahrb Wissenschaf Bot 1928; 69: 636-686.

[138] Wiens D, Richter JA. *Artemisia pattersonii*: a 14-chromosome species of alpine sage. Amer J Bot 1966; 53: 981-986.

[139] Williams CA, Greenham J, Harborne JB. The role of lipophilic and polar flavonoids in the classification of temperate members of the Anthemideae. Biochem Syst Ecol 2001; 29: 929-945.

[140] Wodehouse RP. Pollen grain morphology in the classification of the Anthemideae. Bull Torrey Club 1926; 53: 479-485.

[141] Wright CW, ed. *Artemisia*. London, UK: Taylor & Francis, 2002 (series Medicinal and Aromatic Plants–Industrial Profiles).

[142] Yokoka K, Roberts AV, Mottley J, Lewis R, Brandham PE. Nuclear DNA amounts in *Rosa*. Ann Bot 2000; 85: 557-561.

[143] Zoldos V, Papes D, Brown SC, Panaud O, Siljak-Yakovlev S. Genome size and base composition of seven *Quercus* species: inter- and intra-population variation. Genome 1998; 41: 162-168.

Mother of Potato

KAZUYOSHI HOSAKA

Food Resources Education and Research Center, Kobe University
1348 Uzurano, Kasai, Hyogo 675-2103, Japan

ABSTRACT

Potato was domesticated in the Andean highlands of South America. After discovery of the New Continent, potato was brought to Europe and selected for Northern temperate regions. The common potato (*Solanum tuberosum* L. ssp. *tuberosum*), presently grown worldwide, however, is known to have a cytoplasm that differs from Andean potatoes. Chloroplast DNA analysis revealed a unique chloroplast DNA in common potato, which was named T-type chloroplast DNA. Andean highland potatoes exhibit wide variability in chloroplast DNA, which was introduced by successive domestication from ancestral wild species and sexual polyploidization. Meanwhile, the T-type chloroplast DNA was introduced from some populations of a wild diploid species, *S. tarijense* Hawkes. The probable evolutionary pathway of T-type chloroplast DNA could be: first, *S. tarijense* was naturally crossed as female with the most important Andean cultivated potato *S. tuberosum* ssp. *andigena* as male. Second, long-day adapted types were selected from the hybrid progeny and established as *S. tuberosum* ssp. *tuberosum* in southern coastal regions in Chile. Lastly, Chilean ssp. *tuberosum* was used as the genetic backbone in North America and Europe after late-blight epidemics in the mid-nineteenth century, and became the common potato.

Key Words: Potato, T-type chloroplast DNA, maternal ancestor, *Solanum tarijense*

Address for correspondence: Kazuyoshi Hosaka, Food Resources Education and Research Center, Kobe University, 1348 Uzurano, Kasai, Hyogo 675-2103, Japan. Tel: +81-790-49-3121, Fax: +81-790-49-0343, E-mail: hosaka@kobe-u.ac.jp

INTRODUCTION

Potato has a relatively large genetic reservoir compared with other major crops. According to Hawkes [32], seven cultivated species and 226 wild species in genus *Solanum* L. sect. *Petota* Dumortier have been described as the tuber-bearing *Solanum* species (potato and its relatives). Many of these wild as well as cultivated species have proven value in potato breeding as sources of resistance genes and other agronomic traits for cultivar improvement [25, 32, 77].

Wild species occur in ploidy levels from 2x to 6x with a basic chromosome number of 12. These are widely distributed from southwest USA through Central America to southern Chile with two centers of diversity: one in central Mexico and the other in the Andean highlands from Peru through Bolivia to northwest Argentina. These wild and cultivated species were classified into 21 taxonomic series by Hawkes [32]. Cultivated species, all classified in the Series *Tuberosa*, consist of diploid (*S. stenotomum* Juz. et Buk., *S. phureja* Juz. et Buk., and *S. ajanhuiri* Juz. et Buk.), triploid (*S. chaucha* Juz. et Buk. and *S. juzepczukii* Buk.), tetraploid (*S. tuberosum* L. ssp. *andigena* Hawkes and ssp. *tuberosum*), and pentaploid (*S. curtilobum* Juz. et Buk.) species [32]. All these species are grown in the Andes of South America except *S. tuberosum* ssp. *tuberosum*, which is grown in southern Chile (referred to as Chilean ssp. *tuberosum*) and worldwide (referred to as the common potato). Species relationships among cultivated species have been morphologically investigated by various authors [10, 15, 32, 68], leading to a serious controversy on the taxonomic treatment of cultivated species (reviewed in [52]). In the latest taxonomic treatment by Huamán and Spooner [52], all cultivated potatoes were classified into a single species *S. tuberosum*, and divided into nine cultivar groups. In this paper, however, Hawkes' [32] classification system is tentatively adopted.

There have long been arguments on the origin of the first domesticated diploid species *S. stenotomum* and the most important Andean cultivated potato *S. tuberosum* ssp. *andigena*. Based on morphological and phytogeographic evidence, Hawkes [28] suggested *S. leptophyes* Bitt. and *S. canasense* Hawkes as the ancestral species of *S. stenotomum* and later favored *S. leptophyes* because it is distributed at the same altitudes as *S. stenotomum* and in the same phytogeographic region [31, 32, 34]. *S. stenotomum* is highly polymorphic [10, 27, 32, 68] and Ugent [95] proposed its ancestor to be a single superspecies, the '*S. brevicaule* complex', which included *S. brevicaule* Bitt., *S. bukasovii* Juz., *S. canasense*, *S. coelestipetalum* Vargas, *S. gourlayi* Hawkes, *S. leptophyes*, *S. multidissectum* Hawkes, *S. multiinterruptum* Bitt., and *S. spegazzinii* Bitt. Most of these wild species are closely related to each other and there are many controversies concerning their taxonomy [10, 12, 32, 68]. For the origin of *S. tuberosum* ssp. *andigena*, there are several different hypotheses: it originated via polyploidization from an intervarietal or interspecies cross within cultivated diploid potatoes [42, 66, 93], from an interspecies cross between *S. stenotomum* and a wild diploid species *S. sparsipilum* (Bitt.) Juz. et Buk. (Fig. 12.1) [13, 27, 32], or directly from a wild species *S. vernei* Bitt. et Wittm. [4]. The origin of the other cultivated species is generally undisputed. *S. phureja* derived as a nontuber-dormancy variant from *S. stenotomum* [31, 32]. *S. ajanhuiri* originated from natural hybrids between *S. stenotomum* and a wild frost-resistant diploid species *S. megistacrolobum* Bitt. [54]. *S. chaucha* is a triploid hybrid between tetraploid *S. tuberosum* ssp. *andigena* and diploid *S. stenotomum* [28, 34]. The most frost-resistant

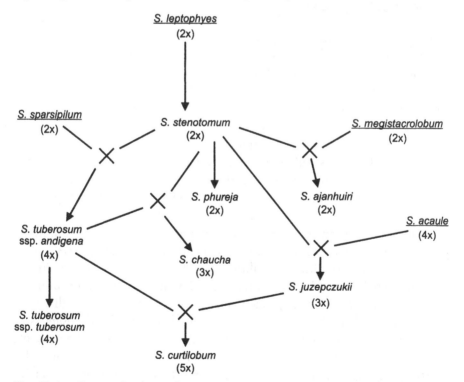

Fig. 12.1: Origin of cultivated potato species suggested by Hawkes [32]. Wild species underlined.

species, *S. juzepczukii*, is a triploid hybrid between a wild frost-resistant tetraploid *S. acaule* Bitt. and *S. stenotomum* [28, 29, 86, 97]. A pentaploid species, *S. curtilobum*, arose by fertilization between a normal gamete from *S. tuberosum* ssp. *andigena* and a 2n gamete from *S. juzepczukii* [8, 28, 29, 86]. In the origin of all the Andean cultivated species, *S. stenotomum* was involved as the backbone germplasm (Fig. 12.1).

The most primitive cultivated species, *S. stenotomum*, was possibly domesticated around Lake Titicaca 10,000 to 7,000 years ago [32]. This species and its derived cultivated taxa became staple crops to support the Incas. Apart from these Andean potatoes, there are two cultigens, Chilean ssp. *tuberosum* and the common potato, which have the same botanical name *S. tuberosum* ssp. *tuberosum*. In this chapter, we trace the maternal history of the common potato and the Andean potatoes by means of chloroplast DNA analysis.

RECOGNITION OF UNIQUE CYTOPLASM IN THE COMMON POTATO

Sterility problems have long been recognized for potato breeders. J. R. Livermore, an erstwhile potato breeder at Cornell University, described: "Nine times out of ten one of the two clones that you wish to cross will not flower, but if they both do flower, nine times out of ten they will both be male sterile. If one is male sterile, nine times out of ten the other one will be female sterile. If a berry is set, nine times there are very few seed will be set, and they will not germinate. If they do germinate, nine times out of ten the plants will be extremely weak, but if they are good strong plants that form good tubers, nine times out of ten they will be sterile." [23]. Among the many factors probably involved in sterility, cytoplasmic aspects are some of the known genetic abnormalities. The common potato and Chilean ssp. *tuberosum* shared at least seven different cytoplasmic sterility factors ([ASH^s], [Fm^s], [In^s], [Sm^s], [Sp^s], [TA^s], and [VSA^s]) that condition sterilities in the presence of dominant chromosomal genes (*ASF, Fm, In, Sm, Sp, TA,* and *VSA*) that occur in *S. tuberosum* ssp. *andigena* [24]. Such genetic-cytoplasmic male sterility was often observed in the progeny from a cross between *S. tuberosum* ssp. *tuberosum* as female and Andean cultivated species as male. The cross of *S. phureja* × a haploid of *S. tuberosum* ssp. *tuberosum* resulted in male fertile progeny, while the reciprocal cross resulted in a high level of male sterility [78]. Forty-three percent of the progeny from 4x × 2x crosses (the common potato × diploid hybrids between *S. phureja* or *S. stenotomum* and haploid *S. tuberosum* ssp. *tuberosum*) were male sterile, while only 4% of the progeny from the reciprocal crosses were male sterile [26].

Cytoplasmic differences for several agronomic traits were also recognized by reciprocal crossing. The cytoplasm of *S. tuberosum* ssp. *tuberosum* was associated with high percent tuberization, high tuber yield, many tubers, poor flowering, early vine maturity, low pollen stainability, poor pollen shedding in the reciprocal hybrids between haploid *S. tuberosum* ssp. *tuberosum* and a mixed population of *S. phureja* and *S. stenotomum* [84]. Hilali et al. [39] studied yield components in 28 exact reciprocal hybrid families between 12 haploid clones of *S. tuberosum* ssp. *tuberosum* and 12 clones of *S. phureja*. Hybrids with ssp. *tuberosum* cytoplasm were superior to those with *S. phureja* cytoplasm by 18% for tuber yield, 21% for tuber number, and 9% for vine maturity (or earlier maturity). Also, at the tetraploid level, large yield differences were found between reciprocal families of *S. tuberosum* ssp. *tuberosum* and ssp. *andigena* with an average of 45% from reciprocal seedlings and 46% from tuber-grown plants [85]. Maris [65] investigated seven clones of *S. tuberosum* ssp. *andigena* (A) and three varieties of *S. tuberosum* ssp. *tuberosum* (T) in an incomplete diallel cross. Among the four groups of crosses (T×T, T×A, A×T, and A×A), T×A populations gave the highest tuber yield and A×T came second. The T×A populations outyielded their exact reciprocals by an average of 10.7%. Contrarily, the *S. tuberosum* ssp. *andigena* cytoplasm was preferable for high male fertility [65].

The uniqueness of *S. tuberosum* ssp. *tuberosum* cytoplasm was disclosed by isoelectric focusing patterns of Fraction 1 protein (= Rubisco), of which large and small subunits are encoded in the chloroplast and nuclear genomes respectively. Gatenby and Cocking [17] reported that *S. sparsipilum*, *S. stenotomum* and *S. tuberosum* ssp. *andigena* had an identical large subunit polypeptide composition but *S. tuberosum* ssp. *tuberosum* shared only one of its three large subunit polypeptides with the other species.

UNIQUE CHLOROPLAST AND MITOCHONDRIAL DNA IN THE COMMON POTATO

Chloroplast DNA is inherited through the maternal parent in most angiosperms and evolves relatively slowly; thus, it is a reliable indicator to trace maternal ancestry of crops [69]. The potato chloroplast DNA is approximately 155-156 kilo base pair (kbp) in size, which is slightly smaller than that of other Solanaceae species (156-160 kbp), and consists of two large inverted repeat regions of 23-27 kbp separated by two single-copy regions of approximately 19 kbp and 109-113 kbp respectively (Fig. 12.2) [37]. Although complete sequences of potato chloroplast DNA are not yet available, the structural features and sequences of potato chloroplast DNA seem to be essentially very similar to those of tobacco [5, 37, 60].

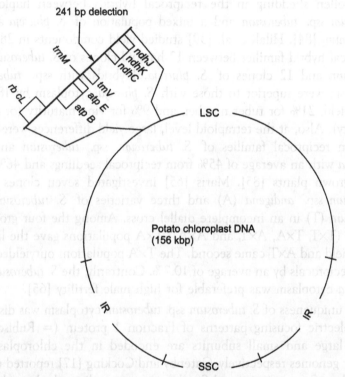

Fig. 12.2: Restriction map of Potato chloroplast DNA (Solarun tuberosum ssp. tuberosum cv. katah din), cited from [37] with kind permission of the authors and the publisher. The location of a 241 bp deletion characterizing T-type chloroplast DNA is shown on this map.

In May 1983, I started a restriction fragment length polymorphism (RFLP) analysis of potato chloroplast DNA. It was found that cv. May Queen, the most leading cultivar in western Japan, and Chilean ssp. *tuberosum* shared an identical chloroplast DNA, but differed from all Andean potatoes [50]. The chloroplast DNA extracted from the common potato was digested with a restriction endonuclease *Bam*HI, *Bgl*II, *Eco*RI, *Hind*III, *Kpn*I, *Pst*I, *Pvu*II, *Xba*I, or *Xho*I and separated on agarose gels by electrophoresis. The fragment patterns generated by digestion with *EcoRI*, *HindIII*, *KpnI*, *PvuII*, and *XhoI* showed specific single fragments smaller than those generated from other species. The *BamHI* pattern showed a fragment change caused likely by base substitution in the recognition site of this enzyme [50]. Later, it became clear that all these fragment differences were caused by one physical deletion occurring in the chloroplast DNA of common potato [48]. The size of the deleted region was determined by sequencing to be 241 bp [60] and approximate position is shown in Figure 12.2. This chloroplast DNA was named T-type chloroplast DNA after *S. tuberosum* ssp. *tuberosum*. Among Andean cultivated and closely related wild species, four other chloroplast DNA types were identified; A type (after *S. tuberosum* ssp. *andigena*), C type (after *S. canasense*), S type (after *S. stenotomum*), and W type (after wild species) [40]. These chloroplast DNA types were distinguished by single differences detected on restriction fragment patterns of chloroplast DNA digested by *BamHI* or *HindIII* (Fig. 12.3). W1, W2, and W3 types, found only in wild species, were distinguished by additional restriction patterns with *PvuII* [47]. The evolutionary directions between these chloroplast DNA types, i.e., primitive vs. advanced (or derived), were determined by shared mutations, that is, in the sense that the state in the characteristic *BamHI* or *PvuII* recognition sites of W-type chloroplast DNA is similar to that of distantly related wild species and/or tomato [50], and the absence of the 241 bp deletion is a common state in the potato relatives, the W-type is considered primitive compared to C, T, W1, W2, and W3.

For mitochondrial DNA analysis, Lössl et al. [64] used 11 mitochondrial DNA probes to reveal variability among 180 haploid

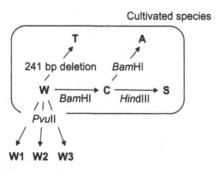

Fig. 12.3: Relationships of chloroplast DNA types. Each type is distinguished by a specific fragment change of the restriction endonuclease *BamHI*, *HindIII*, or *PvuII*, or by the 241 bp deletion.

clones of *S. tuberosum* ssp. *tuberosum* and 144 tetraploid German potato cultivars. Five mitochondrial DNA types—Alpha, Beta, Gamma, Delta and Epsilon, based on their RFLP patterns, were distinguished. The mitochondrial DNA type Beta was always associated with T-type chloroplast DNA, reflecting a coevolution of both organelles and a common uniparental inheritance [64].

DISTRIBUTION OF T-TYPE CHLOROPLAST DNA IN THE COMMON POTATO

The T-type chloroplast DNA was predominantly found among Japanese potato cultivars and some representative European and American cultivars (Table 12.1) [41, 46]. Ten historically important cultivars (Garnet Chili, Beauty of Hebron, Irish Cobbler, Early Rose, Burbank, Early Ohio, Bliss Triumph, Green Mountain, Rural New Yorker, and White Rose), had T-type chloroplast DNA [16]. The T-type chloroplast DNA was also found in 44 of 56 old and modern European cultivars [71], 151 of 178 European cultivars [73] or 47% of German cultivars [63].

Table 12.1: Chloroplast DNA types in commercial varieties of common potato

Chloroplast DNA type	Place	Variety
T	Europe	May Queen, Up-to-date
	America	Early Rose, Garnet Chili, Irish Cobbler, Katahdin, Kennebec, Norland, Pike, Russet Burbank
	Japan	Beni-akari, Benimaru, Bifukabeni, Bifukashiro, Bihoro, Chitose, Dejima, Eniwa, Hatsufubuki, Hokkai-aka, Hokkaikogane, Hokkaishiro, Kita-akari, Konafubuki, Myojo, Nemurobeni, Niseko, Nishiyutaka, Norin 1, Norin 2, Oojiro, Shimabara, Tachibana, Tarumae, Tokachikogane, Touya, Toyoshiro, Unzen, Waseshiro, Yukijiro, Yukirasha
W	Europe	Greta, Tunika
	Japan	Chijiwa, Ezo-akari, Meiho, Rishiri, Setoyutaka, Shiretoko, Toyo-akari, Yoraku

Compiled from [41, 46].

However, some other chloroplast DNA types were also found in the modern potato cultivars (Table 12.1). One apparent source donating W-type chloroplast DNA is *S. demissum* Lindl. In Table 12.1, cv. Tunika and its derived cultivars Toyo-akari and Ezo-akari had W type chloroplast

DNA in common, because the cytoplasm of cv. Tunika came from a so-called W-race, which was a German breeding line derived from *S. demissum* [77]. The cytoplasm of cvs. Chijiwa, Meiho, Rishiri, Setoyutaka, and Yoraku were all derived from the same parent 41089-8, an important breeding line conferring late-blight resistance, which was bred by the Japanese using *S. demissum*. The origin of W-type chloroplast DNA of cv. Shiretoko is not known since pedigree information on the grandmother of cv. Shiretoko (HLT-4 from USDA) is not available [41]. The W-type chloroplast DNA of *S. demissum* (see [50]) was apparently inherited through backcrossing of *S. demissum* (6x) by *S. tuberosum* (4x) for introduction of late-blight resistance from the former species. For this backcrossing, successful crosses were made only from *S. demissum* as female, and the resultant pentaploid hybrid was female. Thus, the *S. demissum* cytoplasm was preferentially transmitted to the bred varieties. The W-type chloroplast DNA is more frequent in German cultivars (51%) [63], of which those having Alpha-type mitochondrial DNA (40% of German cultivars) seem to be introduced from *S. demissum* since Alpha-type mitochondrial DNA was classified closely with that of *S. demissum* [64]. Extensive use of *S. demissum* was reported in that 70% [76] or more than 83% [77] of German cultivars carry genes from *S. demissum*.

A second possible source donating a different chloroplast DNA type could be old European cultivars grown before late-blight epidemics in the 1840s, which were descended from early introductions of *S. tuberosum* ssp. *andigena* brought from South America to Europe.

Thirdly, various chloroplast DNA types might have been incorporated from various cultivated and wild species to modern cultivars by crossbreeding in order to introduce specific traits. The utility of exotic germplasms in modern breeding has been repeatedly emphasized and past records have indicated successful use of related cultivated and wild species, in particular *S. tuberosum* ssp. *andigena*, *S. acaule*, *S. chacoense* Bitt., and *S. phureja* [70, 77]. However, there seems to be no particular reason for preferential transmission of cytoplasm from these cultivated and wild species.

Differences between *S. tuberosum* ssp. *andigena* and Chilean ssp. *tuberosum*

S. tuberosum ssp. *andigena* is the most important and widely grown potato in the Andes of Venezuela, Colombia, Ecuador, Peru, Bolivia, and northwest Argentina, and also sparingly in Guatemala and Mexico [32]. Tubers are formed in the high Andes (2,000 – 4,000 m) only under short-day conditions. Ssp. *andigena* is distinguished from ssp. *tuberosum* by somewhat narrower, more numerous leaflets, and leaves set at a more acute angle on the stem, more dissected and only slightly arched, with pedicels not thickened at apex [32]. Very wide morphological variability has been recognized. Ochoa [68] described 13 varieties (not cultivated varieties, but botanical ones) and 35 forms only for Bolivian *S. tuberosum* ssp. *andigena*. In fact, 2,667 accessions have been maintained clonally in the gene bank at the International Potato Center (CIP), Lima, Peru [51]. Within *S. tuberosum* ssp. *andigena*, some degree of geographical trends has been recognized. Salaman [80] examined leaf characters of 139 varieties of *S. tuberosum* ssp. *andigena* and noted that incidence of *S. tuberosum* ssp. *tuberosum*-like forms steadily increased from 5% in Bolivia and Argentina, to 27% in Peru, to 65% in Ecuador, and to 70% in Colombia. Glendinning [18] analyzed seedling progenies of 189 accessions of *S. tuberosum* ssp. *andigena* and found that the proportion of the terminal leaflet in the whole leaf increased northward of the Andes (more ssp. *tuberosum*-like), while actual leaf length and relative breadth of the terminal leaflet increased both northward and southward from south-central Peru.

Chilean ssp. *tuberosum* is grown in the coastal regions of Chiloé island and adjacent mainland in south-central Chile. It forms tubers under long days, or short days in the tropics, at lower altitudes only.

Some degree of genetic differentiation between the two subspecies has been recognized by Raker and Spooner [75], who analyzed 35 accessions of Chilean ssp. *tuberosum* and 35 accessions of *S. tuberosum* ssp. *andigena* using nuclear DNA microsatellite (or simple sequence repeat) markers, and successfully discriminated them from each other. However, from the morphological point of view the two subspecies are distinguished with difficulty by overlapping character states such as number of stems, angle of insertion of leaves, lateral leaflet length/width ratio, etc. [11, 32, 52].

After Hosaka et al. [50] reported unique chloroplast DNA for the common potato, Buckner and Hyde [6] found that *S. tuberosum* ssp. *tuberosum* (cv. Kennebec) had the same chloroplast DNA as *S. tuberosum* ssp. *andigena* (WRF2288). To solve this inconsistency, chloroplast DNA types were determined for 286 accessions of *S. tuberosum* ssp. *andigena* from the whole range of the distribution area and 28 accessions of Chilean ssp. *tuberosum* (Table 12.2). Type A was the major chloroplast

Table 12.2: Number of accessions with different chloroplast DNA types in *S. tuberosum* ssp. *andigena* (and Chilean ssp. *tuberosum* in parentheses) in various countries

Country	T	A	S	C	W
Mexico	0	8	0	0	0
Guatemala	0	1	0	0	1
Costa Rica	0	1	0	0	0
Venezuela	0	3	0	0	0
Colombia	0	19	0	1	0
Ecuador	1	13	0	0	0
Peru	0	50	9	21	2
Bolivia	0	34	10	20	10
Argentina	7	45	9	11	9
Chile	1 (24)	0 (3)	0	0	0 (1)

DNA in *S. tuberosum* ssp. *andigena* (60.8%) and type T in Chilean ssp. *tuberosum* (85.7%). The frequencies of chloroplast DNA types showed geographical cline from north to south in the Andes [46]. Most accessions from Mexico, Central America, and northern parts of the Andes have A-type chloroplast DNA. The incidence of A-type chloroplast DNA steadily decreased southward, while C-type, and then S- and W-type chloroplast DNA increased. Maximum heterogeneity was found in Argentine accessions of *S. tuberosum* ssp. *andigena*, which contained A-, S-, C-, W-, and T-type chloroplast DNA. In Chile, T-type chloroplast DNA predominated in both subspecies.

Evolutionary Relationship between *S. tuberosum* ssp. *andigena* and Chilean ssp. *tuberosum*

Andean cultivated potatoes were most likely domesticated in Peru, because A-, S-, C-, and W-type chloroplast DNA found in Andean cultivated tetraploid (*S. tuberosum* ssp. *andigena*) and diploid (*S. stenotomum*) potatoes were shared with a group of mostly Peruvian wild diploid species [42]. On the other hand, Chilean ssp. *tuberosum* is well isolated geographically from Andean potatoes by severe desert regions between the western slopes of the Andes and the southern coastal regions in Chile. Thus, an independent origin for Chilean ssp. *tuberosum* from a wild species in Chile may be suggested. Bukasov [7, 9] suggested that a tetraploid wild species *S. leptostigma* Juzepczuk or *S. molinae* Juzepczuk, both collected on Chiloé Island, was an ancestral species of Chilean ssp.

tuberosum. Sykin [94] insisted that S. *leptostigma* or S. *molinae* on the islands of Chiloé and Chonos are obviously wild, since these were collected from rather inaccessible areas, far from human colonies, and had long stolons (up to 90 cm) and quite small, tasteless tubers. However, Hawkes [27] regarded both S. *leptostigma* and S. *molinae* as escaped forms of Chilean ssp. *tuberosum* because these species possessed a gene for red tuber color not found in any other wild species, but which did occur in both diploid and tetraploid cultivated potatoes. Brücher [3] explored thoroughly Chiloé and concluded that no true wild species existed in Chile except S. *maglia* Schlechtd.

S. *maglia* is the only wild tuber-bearing species grown in Chile, which is unique in possessing a loose, barrel-shaped anther column with anthers and filaments not well differentiated from each other; hence it is classified alone in Series *Maglia* [32]. S. *maglia* is distributed mostly from central to north Chile, generally near the seacoast, and all collections but two are sterile triploids. Of the two diploids one was from the lower slopes of the Andes in Argentina, Province of Mendoza, the other from near Valparaiso in Chile [32]. Ugent et al. [96] examined 13,000-year-old tuber skins recovered from the archaeological site of Monte Verde in south-central Chile, and identified them as those of S. *maglia* based on the comparative morphology of starch grains. Further observation that Chilean ssp. *tuberosum* and S. *maglia* are remarkably alike in having reddish-purple-skinned tubers (tubers with similarly colored skins had not been discovered in any wild species, as Hawkes insisted above) led them to propose that Chilean ssp. *tuberosum* arose as a result of spontaneous chromosome doubling of diploid S. *maglia*, followed by many years of human selection [96]. Spooner et al. [90] argued that their identification of extant populations of S. *maglia* from Chiloé Island was based on a probable misidentification in the field of S. *tuberosum* as S. *maglia*, and that the starch data were inadequate because statistical presentation of comparative data was lacking. Irrespective of the validity in identifying the ancient potato remains, S. *maglia* could not be ancestral to Chilean ssp. *tuberosum* because neither Chilean triploid (PI 245087) nor Argentine diploid (PI 407408) clones of S. *maglia* has T-type but rather A-type chloroplast DNA [40]. Grun [22] also described one Chilean (PI 210813) and one Argentine (PI 407408) accession of S. *maglia* as having resistant cytoplasmic factors, i.e., $[In^r]$, $[Sm^r]$ and $[Sp^r]$, $[TA^r]$, similar to those of S. *tuberosum* ssp. *andigena*. He, however, also mentioned that a third clone of S. *maglia*, from Concon, Province of

Valparaiso, had cytoplasmic factors similar to those of Chilean ssp. *tuberosum*. This statement appears dubious since no accession number was provided by Grun for the third clone [22].

Therefore, no wild progenitor can be set up with certitude for Chilean ssp. *tuberosum*. Alternatively, it has been suggested that the Chilean ssp. *tuberosum* originated by selection from *S. tuberosum* ssp. *andigena* [27]. This selection hypothesis has been reinforced by the facts that: (1) early European potatoes brought from South America in the late sixteenth century were actually *S. tuberosum* ssp. *andigena*, from which long-day adaptive *S. tuberosum* ssp. *tuberosum* was artificially selected in Europe [27, 35, 41, 46, 82, 83], and (2) *S. tuberosum* ssp. *tuberosum*, or "Neo-Tuberosum", was experimentally recreated by recurrent mass selection from *S. tuberosum* ssp. *andigena* populations [19, 89]. The continuous geographical cline in the frequency of different chloroplast DNA types from north to south of the Andes toward Chile (Table 12.2) supported the selection hypothesis from *S. tuberosum* ssp. *andigena* to Chilean ssp. *tuberosum* [46].

HISTORY OF THE COMMON POTATO

Potato was first brought from the New to the Old World in the late sixteenth century. Similar to many New World plants, potatoes were brought to the Canary Isles before they arrived in Spain. According to Hawkes and Francisco-Ortega [36], there are records in the archives of public notaries that barrels of potatoes were exported from Gran Canaria to Antwerp, Belgium in November 1567 and from Tenerife via Gran Canaria to Rouen, France in April 1574. Thus, the potato was obviously grown as a crop in Gran Canaria and Tenerife in 1567 and 1574 respectively. As some five years would have been needed to bulk it sufficiently as an export crop, Hawkes and Francisco-Ortega [36] suggested that the original introduction from South America to the Canary Isles could well have been about 1562.

So, there were two introductions into continental Europe, the first into Spain ca. 1570 and the second into England in 1590 [32, 81]. Account books of the Hospital de la Sangre in Seville during the period 1546 to 1601 were studied for purchases of potatoes by Hawkes and Francisco-Ortega [35]. Potatoes were bought regularly in the Seville market from 1580 onwards, with the first record appearing in 1573, agreeing thereby with Salaman's [81] conclusion that potatoes became established in Spain by about 1570. Since purchases were confined almost entirely to December and January each year, these potatoes were short-day adapted *S. tuberosum* ssp. *andigena*, actually grown in Spain and forming tubers in the short days at the end of the year.

Juzepczuk and Bukasov [59] believed that the early European potato came from Chile because Chilean potatoes were adapted to form tubers under long-day conditions of southern latitudes and would have immediately adapted to the similar daylength of Europe. Salaman [80] disagreed because the length of time and the number of transshipments needed to get a potato from Chile to Spain would have resulted in tuber death and suggested that the potato left some northern port in South America, possibly Cartagena, Colombia. Herbarium specimens preserved in Europe during ca. 1600 and 1750 were examined by morphological comparison of leaf shape [80, 83]. The earliest types of potato grown in western Europe and England were typical of *S. tuberosum* ssp. *andigena*. New varieties were continually raised from true seeds and by selection *S. tuberosum* ssp. *tuberosum* was evolved rather more rapidly and at an earlier date in England (in the early decades of the eighteenth century) than in western Europe [83]. Simmonds [87] analyzed leaf morphology of a whole series of potato clones covering the earliest European potatoes that Salaman and Hawkes [83] examined to modern cultivars. He provided strong evidence that the European potatoes changed little from *S. tuberosum* ssp. *andigena* until the nineteenth century, and changed very rapidly, transforming into modern potatoes in about 100 years.

Further spread in Europe is well described by Hawkes [32]. Potatoes were first received in North America from Bermuda in 1621, where they had been grown after an initial introduction from England in 1613. Potatoes were brought into India either by the Portuguese or later by the British in the early seventeenth century [92]. The first introduction of potato into Japan is generally recognized as during the Keicho period (1596-1614) [61] or between 1609 and 1615 [62]. The Dutch may have brought potatoes via Java (Indonesia) to Nagasaki, Japan.

A great turnover occurred in the 1840s in Europe. Almost all previously existing varieties were lost by late-blight attack (the Irish Famine). Thus analysis of chloroplast DNA of early European potatoes seemed impossible. Fortunately, P. Grun, Pennsylvania State University, gifted me some hybrid seeds whose female parent had descended from cv. Myatt's Ashleaf. Salaman [79] had written: "The oldest variety now in general use is cv. Myatt's Ashleaf which may possibly be over a hundred-

and-fifty years old." It had A-type chloroplast DNA, strongly suggesting that potatoes grown in Europe before the late-blight epidemic were *S. tuberosum* ssp. *andigena* [46]. Distinctiveness of cv. Myatt's Ashleaf from modern European cultivars in chloroplast DNA was highlighted again in RFLP analysis of chloroplast DNA by Powell et al. [71] and by microsatellite marker analysis of chloroplast DNA by Provan et al. [73].

Another evidence for early European potatoes being of *S. tuberosum* ssp. *andigena* origin came from old Japanese potatoes. Since the first introduction during the Keicho period, the potato gradually spread via a succession of famines as a hardy crop in mountain regions or cool areas in central and northern Japan. Potatoes were introduced into Hokkaido, a northern island in 1706 [1]. Under strictly restricted conditions of transportation for farmers by governmental policies, it is amazing that potatoes reached from the south to the north in about 100 years. Figure 12.4 shows a drawing by Kazan Watanabe that appeared in the book "Kyuko-Nibutsu-Ko" [Two Hardy Crops] written by Choei Takano in 1836. Long thin stems and leaves set at acute angles on the stem give the impression of *S. tuberosum* ssp. *andigena* grown in temperate latitudes. (A potato flower was incorrectly drawn imaginatively or sketched from some other source, probably because the plant was not flowering.)

During an approximate 200-year period between 1639 and 1854, Japan was closed to the outside world except for the Chinese and Dutch who traded at Nagasaki, the only permitted trading port. Once Japan opened its doors to the outer world in the mid-nineteenth century, western varieties were successively introduced and modern breeding programs initiated in the early twentieth century. As a result (Table 12.1), Japanese cultivars bred through modern breeding have either T- or W-type chloroplast DNA. However, among local or old cultivars of unknown origin obtained from local farmers or experiment stations, four cultivars were found to have A-type chloroplast DNA; Nemuro-murasaki, Rankoku 3, Rankoku 5 (from northern regions) and Murasaki-imo (from a central region in Japan) [41]. Nuclear DNA variation was also analyzed by random amplified polymorphic DNA (RAPD) analysis for these A-type chloroplast DNA cultivars, most Japanese registered cultivars and breeding lines, some representative European and North American cultivars, and *S. tuberosum* ssp. *andigena* [49]. On the dendrogram by cluster analysis, the four old cultivars were first clustered together and then clustered with *S. tuberosum* ssp. *andigena* accessions, strongly supporting that these old cultivars are relict potatoes of early European

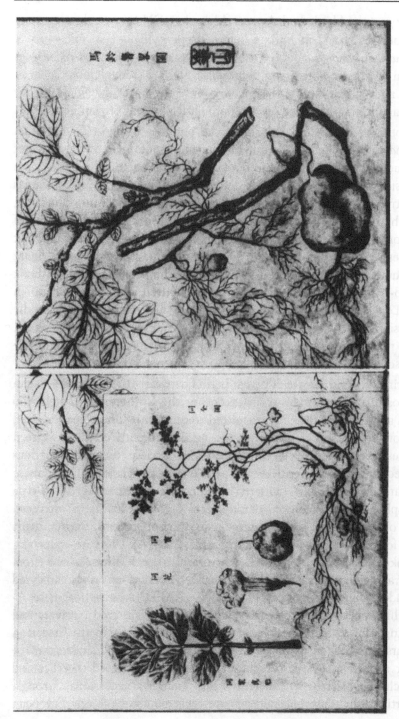

Fig. 12.4: Old potato in Japan. A drawing by Kazan Watanabe appeared in "Kyuko-Nibutsu-Ko" written by Choei Takano in 1836. Photographed by the author in April 1985, through the courtesy of the Takano Choei Memorial Hall.

introduction and have characteristics of *S. tuberosum* ssp. *andigena* in common [49].

After the late-blight epidemics in the 1840s, the most influential clone has undoubtedly been Rough Purple Chili [30, 70]. C.E. Goodrich of Utica, New York obtained this clone in the Panama market in 1851. Since it purportedly came from Chile, he named the clone Rough Purple Chili. He raised seedlings from it, selecting one which he named Garnet Chili. Among seedlings raised from open-pollinated berries of Garnet Chili possibly by selfing, Early Rose was selected by Albert Bresee in 1861, which in turn became a founder of modern potato cultivars. The pedigree of almost all modern cultivars can be traced back to Early Rose [70]. Rough Purple Chili has become extinct while Garnet Chili and Early Rose still exist. Both Garnet Chili and Early Rose have T-type chloroplast DNA, which clearly explains why most modern cultivars have T-type chloroplast DNA [16, 46].

EVOLUTION OF ANDEAN CULTIVATED POTATOES BASED ON CHLOROPLAST DNA TYPES

None of the chloroplast DNA types were species specific but the frequencies differed among accessions of different species. As described earlier, *S. tuberosum* ssp. *andigena* had A-type chloroplast DNA in many accessions but also four other chloroplast DNA types with different frequencies varied from north to south of the Andes (Table 12.2) [46]. The most primitive cultivated species *S. stenotomum* had four of five chloroplast DNA types with S-type the most frequent (Table 12.3) [42]. A considerable overlap was found in the chloroplast DNA type frequencies between *S. stenotomum* and wild diploid species *S. bukasovii* (A, S, C and W), *S. canasense* (S and C), *S. candolleanum* Berth. (S and C), *S. leptophyes* (C and W) and *S. multidissectum* (S and C) (Table 12.3) [42]. Based on chloroplast DNA types, the most probable domestication process of Andean potatoes was inferred by Hosaka [42]: cultivated potato tubers were first taken from "ancestral species" and presumably should have had the W-type chloroplast DNA common in other South American wild species. During the evolutive course of the "ancestral species", C-type chloroplast DNA derived from the W type and was domesticated and incorporated into the stock of cultivated forms. When the inferiority of old stocks did not prevail over the newly domesticated C-type chloroplast DNA genotypes, both could be vegetatively maintained by farmers. At this stage, some genotypes of the "ancestral species" that adapted well

Table 12.3: Number of accessions with different chloroplast DNA types in cultivated and some wild species

Species	T	A	S	C	W	W1	W2	W3
Cultivated species								
S. tuberosum ssp. andigena	9	174	28	53	22	0	0	0
S. tuberosum ssp. tuberosum	24	3	0	0	1	0	0	0
S. stenotomum	0	27	92	2	2	0	0	0
S. phureja	0	7	41	0	0	0	0	0
Wild species								
S. bukasovii	0	1	1	23	1	0	0	0
S. canasense	0	0	7	18	0	0	0	0
S. multidissectum	0	0	5	5	0	0	0	0
S. candolleanum	0	0	1	3	0	0	0	0
S. leptophyes	0	0	0	4	12	1	0	0
S. brevicaule	0	0	0	0	15	0	0	0
S. sparsipilum	0	0	0	0	33	0	2	0
S. chacoense	0	0	0	0	13	28	0	1

Compiled from [42] with additional new data.

to a particular environment became S. leptophyes. After S-type chloroplast DNA occurred in the "ancestral species", it was introduced to the cultivated gene pool. On the other hand, some genotypes differentiated into S. canasense and S. multidissectum in Peru and S. candolleanum in Bolivia. The most recent event occurring in the "ancestral species" was the derivation of A-type chloroplast DNA in central Peru giving rise to cultivated forms with A-type chloroplast DNA. S. bukasovii might have differentiated at this stage from the "ancestral species". This suggested a multiple origin of S. stenotomum by successive domestication from time to time and place to place from the presumed "ancestral species" [42], known as the successive domestication hypothesis.

This successive domestication over a long period of time might have accumulated different genotypes maintained vegetatively as cultivated stocks, resulting in a highly polymorphic diploid cultigen. Subsequent polyploidization occurred many times in the field of diploid cultivars [93] and formed a series of chloroplast DNA variation in S. tuberosum ssp. andigena similar to that in diploid cultivars [47]. It is highly probable that sexual polyploidization by the union of 2n gametes contributed to the origin and evolution of the cultivated tetraploid potatoes [55, 98, 99]. Various combinations of diploid genotypes through sexual polyploidization could result in tremendous genetic diversity in *S. tuberosum* ssp. *andigena* [14].

Chloroplast DNA Differentiation in Andean Potatoes

Recent advancement in new types of molecular markers has made it possible to evaluate chloroplast DNA variability in even greater detail. Chloroplast DNA microsatellite markers detect polymorphisms in repeated numbers of mononucleotides in chloroplast DNA [72]. Such mononucleotide repeated regions were searched using complete sequence of tobacco chloroplast DNA, and primers flanking the repeated regions (NTCP markers) were designed by Provan et al. [73]. These NTCP markers revealed much higher levels of diversity than chloroplast DNA RFLPs in potato [5, 44, 73].

The H3 marker was originally found as a restriction site difference in the *Dra*I restriction fragments of potato chloroplast DNA, and converted to a polymerase chain reaction (PCR)-based marker, which amplifies from the base number 63082 to 63490 in tobacco chloroplast DNA (including coding regions of *ycf4* and *ycf10*). Polymorphisms were detected after restriction digestion of PCR products by *Dra*I [44].

To compare differences of chloroplast DNA types as defined by restriction site analysis with those of chloroplast DNA high-resolution markers (seven NTCP and H3 markers) and with those of nuclear DNA RFLPs, Sukhotu et al. [91] investigated polymorphisms in 76 accessions of seven cultivated species, 17 accessions of putative ancestral wild species *S. acaule* (acl), *S. brevicaule* (brv), *S. bukasovii* (buk), *S. canasense* (can), *S. leptophyes* (lph), *S. megistacrolobum* (mga), *S. sparsipilum* (spl), and *S. vernei* (vrn), and two accessions of a distantly related wild species *S. chacoense* (chc). Overall, 33 fragments were detected using chloroplast DNA microsatellites, while the H3 marker provided two banding patterns. Consequently, chloroplast DNA high-resolution markers identified 25 different chloroplast DNA haplotypes, as shown in Figure 12.5. The A- and S-type chloroplast DNA were discriminated as unique haplotypes from 12 haplotypes having C-type chloroplast DNA, and T-type chloroplast DNA from 10 haplotypes having W-type chloroplast DNA. Differences among chloroplast DNA types correlated strongly with those of chloroplast DNA high-resolution markers ($r = 0.822$). Thus simple determination of chloroplast DNA types is apt to provide useful information pertaining to phylogenetic relationships, or more specifically

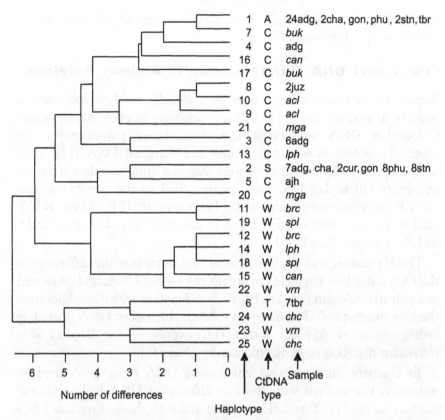

Fig. 12.5: Relationships of chloroplast DNA haplotypes, modified from [91]. Corresponding chloroplast DNA types and sample names are also denoted with haplotype numbers. Sample names are represented by species abbreviations (see text). Wild species are shown in Italic.

maternal lineages among cultivated species and their closely related species.

Using 35 single-copy RFLP probes, a total of 111 polymorphic bands were scored. A UPGMA dendrogram was constructed, which identified three large clusters (Clusters 1, 2 and 3) (Fig. 12.6). The most distant cluster (Cluster 3) consisted of S. *juzepczukii*, S. *curtilobum*, and S. *acaule*, strongly supporting common ancestry of these species [29, 86, 97].

Fig. 12.6: A UPGMA dendrogram constructed using nuclear DNA RFLPs among cultivated potatoes and their relatives, cited from [91]. Sample names are represented by species abbreviations (see text) with accession identity numbers. The chloroplast DNA type and haplotype number (ND = not determined) are also denoted for each accession.

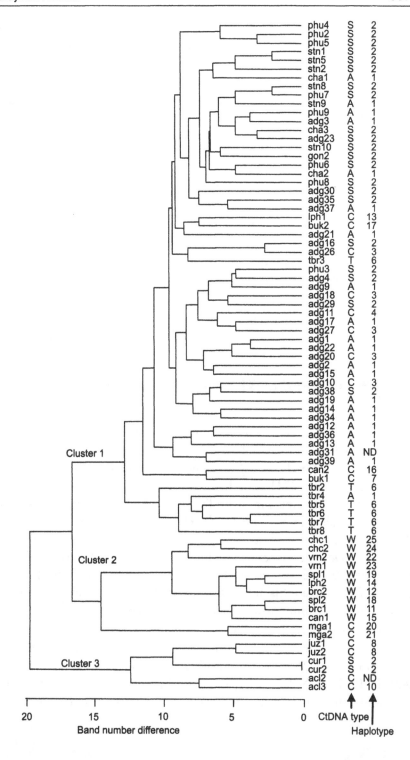

Cluster 2 was formed by exclusively wild species accessions having W-type chloroplast DNA and S. *megistacrolobum* with C-type chloroplast DNA. The remaining cultivated species, one accession each of S. *leptophyes* (lph1) and S. *canasense* (can2) and two accessions of S. *bukasovii* (buk1 and buk2) were classified in Cluster 1 with chloroplast DNA type A, S, C or T. Differentiation between W-type chloroplast DNA and A-, S-, and C-type chloroplast DNA was supported by nuclear DNA RFLPs in most species except for those of recent or immediate hybrid origin. However, differentiation among A-, S-, and C-type chloroplast DNA was not clearly supported by nuclear DNA RFLPs, suggesting that frequent genetic exchange occurred among them and/or they shared the same gene pool due to common ancestry.

As suggested by the dendrograms in Figures 12.3 and 12.5, the Andean potatoes and Chilean ssp. *tuberosum*, or the A-, S-, and C-type chloroplast DNA group and T-type chloroplast DNA might have passed independent maternal pathways. Andean potatoes were possibly differentiated from the hypothesized "ancestral species" [42], while T-type chloroplast DNA of Chilean ssp. *tuberosum* should have had a different maternal ancestor.

T-TYPE CHLOROPLAST DNA IN WILD SPECIES

Since S. *chacoense* f. *gibberulosum* (Juz. et Buk.) Corr., a wild species of Series *Yungasensa* from the Province of Córdoba in Argentina (PI 133073) shares at least three cytoplasmic sterility factors with Chilean ssp. *tuberosum* ([In^s], [Sm^s] and [Sp^s], unknown for [ASH^s], [Fm^s], [TA^s], and [VSA^s]) [20], Grun [21] once hypothesized that Chilean ssp. *tuberosum* evolved from interspecific hybridization between S. *tuberosum* ssp. *andigena* as male and S. *chacoense* f. *gibberulosum* as female. Accession of PI 133073 was revealed to have W1-type chloroplast DNA [47] and other S. *chacoense* accessions chloroplast DNA of either W (11 accessions), W1 (27 accessions) or W3 (one accession) type [47]. So Grun [22] changed his mind and posited the maternal ancestor of Chilean ssp. *tuberosum* as an unidentified species.

T-type chloroplast DNA had never been found in any wild species until 1997 and it was suggested that the deletion event characterizing T-type chloroplast DNA had occurred after potatoes were domesticated [42]. In 1997, K. Nakagawa, working for her master thesis on the origin of a wild tetraploid species S. *acaule*, observed a restriction fragment pattern specific to T-type chloroplast DNA in an *S. neorossii* Hawkes et Hjerting accession [67].

At that moment T-type chloroplast DNA was found in South America in most Chilean ssp. *tuberosum*, some northwest Argentine collections of *S. tuberosum* ssp. *andigena*, and one accession of *S. neorossii* from the Department of Salta, Argentina. This implied that T-type chloroplast DNA originated somewhere in northern Argentina or adjacent areas in Bolivia, where approximately one-fourth of the 226 wild tuber-bearing *Solanum* species are distributed.

A pair of primers, 5'-GGAGGGGTTTTTCTTGGTTG-3' and 5'-AAGTTTACTCACGGCAATCG-3', flanking the 241 bp deleted region of potato chloroplast DNA were designed as H1 marker, based on the sequence information published by Kawagoe and Kikuta [60]. The H1 marker amplifies by PCR a fragment approximately 200 bp from the accession with T-type chloroplast DNA and a fragment approximately 440 bp from that of the other type chloroplast DNA (Fig. 12.7). Based on the distribution areas (Bolivia, Argentina, or Chile), morphological

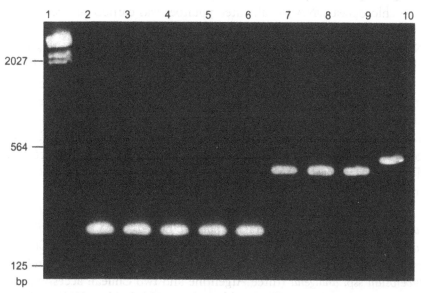

Fig. 12.7: Polymorphic fragments, using H1 marker, separated in a 3% NuSieve 3:1 agarose gel, cited from [43]. Lane 1, *Hind*III-digested lambda DNA; lanes 2 and 3, *S. tuberosum* ssp. *tuberosum* cv. Konafubuki and CIP 703254 respectively; 4, *S. neorossii* (PI 473428); 5, *S. tarijense* (PI 414152); 6, *S. berthaultii* (PI 568918); 7, *S. phureja* (1.22); 8, *S. boliviense* (PI 265861); 9, *S. maglia* (PI 407408); 10, *S. vernei* (PI 230468).

similarities to cultivated potatoes (the taxonomic series *Yungasensa*, *Megistacroloba*, *Maglia* or *Tuberosa*) and the availability in gene banks, 566 accessions of 35 wild species were surveyed for the presence or absence of the 241 bp deletion by the simple PCR [43]. In all species except *S. berthaultii* Hawkes, *S. neorossii*, *S. tarijense*, and *S. vernei*, only a 440 bp fragment was observed, indicating no deletion in the amplified region of the chloroplast DNA. Sixteen of 80 accessions of *S. tarijense*, *S. berthaultii*, and *S. neorossii* showed a 200 bp fragment. Sequencing of these fragments revealed that the same 241 bp was deleted at the same position in these accessions [43]. Three of 29 accessions of *S. vernei* showed a higher molecular weight fragment with approximately 500 bp. The sequencing result indicated that 65 bp were tandemly duplicated.

MATERNAL ANCESTOR OF POTATO

The 241 bp deletion characterizing T-type chloroplast DNA in the common potato and Chilean ssp. *tuberosum* was found in *S. berthaultii*, *S. neorossii*, and *S. tarijense*. Further search for which species conferred T-type chloroplast DNA to cultivated potatoes was carried out using high-resolution chloroplast DNA markers (H1, H2, H3, NTCP6, NTCP7, NTCP14, and NTCP18) [44]. The H1, H3, and NTCP markers have already been described above. The H2 marker was first identified as a restriction fragment length polymorphism detected in *Hae*III restriction fragments among T-type chloroplast DNA holders. The H2 marker amplifies from the base number 57851 to 58184 in tobacco chloroplast DNA within a coding region for large subunit polypeptides of Rubisco gene (*rbcL*). After restriction digestion by *Hae*III, the PCR product from T-type chloroplast DNA of cultivated potatoes was cleaved into two fragments.

Among *S. berthaultii* (30 accessions), *S. neorossii* (five accessions), *S. tarijense* (62 accessions), putative natural hybrids of *S. berthaultii* × *S. tarijense* (25 accessions), *S. tuberosum* ssp. *tuberosum* (six Chilean primitive accessions and one Japanese advanced cultivar), and *S. tuberosum* ssp. *andigena* (three Argentine and two Chilean accessions, all previously identified as T-type chloroplast DNA holders [46]) and one accession each of *S. phureja* (a cultivated diploid clone with S-type chloroplast DNA) and *S. chacoense* (a wild diploid clone with W-type chloroplast DNA), seven chloroplast DNA markers revealed 26 banding pattern types or bands (H1, two types; H2, two types; H3, four types; NTCP6, six bands; NTCP7, four bands; NTCP14, four bands; NTCP18, four bands) (Fig. 12.8). All the cultivated materials with T-type chloroplast DNA showed the same polymorphism types in all the markers. Thus, this chloroplast DNA type was named haplotype 1. Likewise, 25 haplotypes were distinguished by combination of marker phenotypes. The number of accessions in each species group is tabulated for each haplotype in Table 12.4. The chloroplast DNA possessing the 241 bp deletion appeared in five haplotypes (haplotypes 1 to 5). Haplotype 1 was also found in 14 accessions of S. *tarijense* (12 accessions from northern Argentina and two accessions from southern Bolivia) and

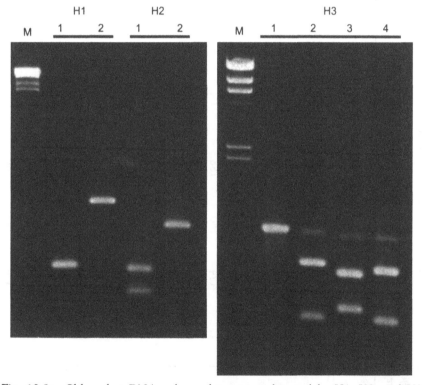

Fig. 12.8: Chloroplast DNA polymorphism types detected by H1, H2, and H3 markers, cited from [43]. H1 marker bands are detected using undigested PCR products (a primer pair of 5'-GGAGGGGTTTTTCTTGGTTG-3' and 5'-AAGTTTACTCACGGCAATCG-3'). H2 (5'-GCATCGAGCGTGTTGTTGGA-3' and 5'-AGTCCACCGCGAAGACATTC-3') and H3 (5'-CAGGGGTCCATT CCCTTGAC-3' and 5'-AGAAAGAAATCCACCAGGGC-3') marker bands are detected after digestion of PCR products with restriction endonucleases *Hae*III and *Dra*I respectively. All cultivated materials with T-type chloroplast DNA showed Type 1 fragment patterns in all markers. M, *Hind*III-digested lambda DNA.

Table 12.4: Chloroplast DNA haplotypes and their distribution among *S. berthaultii* (ber), putative hybrid of *S. berthaultii* × *S. tarijense* (hybrid), *S. tarijense* (tar), *S. neorossii* (nrs), *S. tuberosum* ssp. *tuberosum* (tbr), ssp. *andigena* (adg), *S. stenotomum* (stn), *S. chacoense* (chc), *S. phureja* (phu)

Haplotype	Marker phenotype (type or fragment length in bp)							No. of accessions				
	H1	H2	H3	NTCP6	NTCP7	NTCP14	NTCP18	ber	hybrid	tar	nrs	Others
1	1	1	1	173	173	149	188		4	14		6tbr, 5adg, 1stn
2	2	2	1	173	172	150	188			1		
3	2	1	1	174	173	149	188			1		
4	2	1	1	174	173	150	188			1		
5	2	4	1	173	173	149	188				2	
6	2	2	1	172	173	152	186			2		
7	2	1	1	172	174	149	188	4	7	11		
8	2	1	1	172	174	149	189	2		1		
9	2	1	1	172	174	150	188	1		1		
10	2	1	1	172	175	149	188	1				
11	2	1	1	173	174	149	188	1				
12	2	1	1	173	174	150	188			9		
13	2	1	1	173	175	150	187		1			
14	2	1	1	173	174	151	187	1	1	8		
15	2	1	1	174	174	149	188		1	1		
16	2	1	1	174	174	150	189		1			
17	2	1	1	174	174	150	187		1			
18	2	1	1	174	175	150	188		1	1		
19	2	1	1	174	175	151	187	1	1			
20	2	1	1	174	174	149	188		2			
21	2	1	1	175	174	151	187				3	
22	2	1	1	175	174	152	187					
23	2	2	1	175	174	149	188	17	3			1chc
24	2	2	2	172	174	149	188			4		
25	2	3	3	173	173	150	186			2		1tbr, 1phu

four accessions of *S. berthaultii-S. tarijense* hybrids (all from central Bolivia). Haplotypes 2, 3, and 4 were found in one accession each of *S. tarijense*. Although two accessions of *S. neorossii* had the 241 bp deletion, these had chloroplast DNA of haplotype 5, differing thereby from the T-type chloroplast DNA of cultivated potatoes. *S. berthaultii* accessions, once determined as having T-type chloroplast DNA [43], were all classified into the group of natural hybrids because previously misclassified [44]. Therefore, it was suggested that some populations of *S. tarijense* initially conferred haplotype-1 chloroplast DNA on cultivated potatoes [44].

Introduction of T-type chloroplast DNA from *S. tarijense* into Chilean ssp. *tuberosum*

Since chloroplast haplotype 1 in *S. tuberosum* ssp. *andigena* and *S. tarijense* was found mainly in the Bolivia-Argentine boundary area, and partly from central Bolivia, the haplotype-1 chloroplast DNA was likely introduced from some populations of *S. tarijense* to cultivated potatoes somewhere in this region.

To elucidate a possible evolutionary link of haplotype-1 chloroplast DNA between *S. tarijense* (diploid) and Chilean ssp. *tuberosum* (tetraploid), the first domesticated and most important diploid species *S. stenotomum* was extensively investigated by Hosaka [45]. One accession of *S. stenotomum* (CIP 704089), collected from the southernmost distribution area of this species in the Department of Potosi, Bolivia (southern boundary with Argentina), was reported to have T-type chloroplast DNA [42]. But this accession was probably misidentified since several plants of this accession grown for re-evaluation, morphologically resembled *S. tuberosum* ssp. *tuberosum* and their ploidy level was tetraploid [45].

Out of 529 accessions (11 *S. stenotomum* ssp. *goniocalyx*, 204 *S. stenotomum* ssp. *stenotomum*, 286 *S. tuberosum* ssp. *andigena* and 28 Chilean ssp. *tuberosum*) surveyed, nine accessions of *S. tuberosum* ssp. *andigena* (PI 133667, PI 208563, PI 209421, PI 234592, PI 245816, PI 246979, PI 280936, PI 473257, and PI 558141) and 24 Chilean ssp. *tuberosum* accessions had the T-type chloroplast DNA (Table 12.5). The T-type accessions of *S. tuberosum* ssp. *andigena* were mostly from northern Argentina. One Ecuadorian *S. tuberosum* ssp. *andigena* (PI 246979) showed T-type chloroplast DNA, which was possibly a later introduction or a hybrid with *S. tuberosum* ssp. *tuberosum* because no T-type chloroplast

Table 12.5: Number of accessions surveyed and those with T-type chloroplast DNA (in parentheses) in Andean and Chilean cultivated potatoes

Country of origin	S. stenotomum ssp.		S. tuberosum ssp.	
	goniocalyx[1]	stenotomum	andigena	tuberosum
Mexico			8	
Guatemala			2	
Costa Rica			1	
Venezuela			3	
Colombia		1	20	
Ecuador			14 (1)	
Peru	10	127	82	
Bolivia	1	76	74	
Argentina			81 (7)	
Chile			1 (1)	28 (24)
Total	11	204	286 (9)	28 (24)

[1]Peruvian variant
Cited from [45].

DNA has been found in surrounding areas, and because the introduction of high-yielding S. *tuberosum* ssp. *tuberosum* and hybridization with ssp. *andigena* are frequent in modern breeding efforts in South America.

None of the S. *stenotomum* accessions had T-type chloroplast DNA [45]. The S. *stenotomum* accessions used covered most of the existing collections for this species (240 accessions maintained in CIP [51]). Since the T-type chloroplast DNA has never been found in the other diploid cultivated species S. *phureja* [47] and S. *ajanhuiri* [91], it was concluded that T-type chloroplast DNA was not present in any cultivated diploid species. Therefore, it is likely that T-type chloroplast DNA was first introduced into the cultivated potato gene pool not at the diploid, but at the tetraploid level.

Thus the most probable evolutionary pathway could be that some populations of S. *tarijense* with T-type chloroplast DNA were naturally crossed as female with S. *tuberosum* ssp. *andigena* from which Chilean ssp. *tuberosum* was predominantly selected [45]. This 2x × 4x cross could be verified because S. *tarijense* produced a 2n egg with relatively high frequency, resulting in tetraploid progeny having the S. *tarijense* cytoplasm [99]. Although no evidence of natural hybridization occurring between S. *tarijense* and S. *tuberosum* ssp. *andigena* could be found in the monographs [33, 34, 68], there may have been natural hybridization between them since the distribution area of *S. tarijense* (central Bolivia to northwest Argentina at altitudes of 2,400-2,600 m) overlaps with that of *S. tuberosum* ssp. *andigena* (mostly from Venezuela to northwest Argentina at altitudes of 2,500-4,000 m). Some of the *S. tarijense* accessions contain very low levels of foliar glycoalkaloid [2] and have been shown to be superior parents giving tuberization ability under long daylength to hybrid progenies with *S. tuberosum* haploids [38]. *S. tarijense* is a plant of comparatively dry regions [34] and is probably drought tolerant, which might have facilitated range extension over severe desert regions between the western slopes of the Andes and the southern coastal regions in Chile. These features would support the likelihood of *S. tarijense* germplasm having been incorporated in Chilean ssp. *tuberosum*. Although initial hybridization could have occurred between *S. tarijense* female and *S. tuberosum* ssp. *andigena* male, it remains unknown whether Chilean ssp. *tuberosum* was selected from immediate hybrids or those with further hybridization with *S. tuberosum* ssp. *andigena*.

CONCLUSION

Since T-type chloroplast DNA had not been found in any Andean highland potatoes, the author entitled his 1986 paper "Who is the mother of the potato?—restriction endonuclease analysis of chloroplast DNA of cultivated potatoes" [40]. Twenty years have elapsed since recognition of the unique T-type chloroplast DNA in common potato. The T-type chloroplast DNA was ultimately discovered in some populations of *S. tarijense*. It was hypothesized that hybridization between *S. tarijense* female and *S. tuberosum* ssp. *andigena* male and selection from the hybrids gave rise to the Chilean ssp. *tuberosum* [45]. Based on this collective knowledge, the probable evolutionary pathway of chloroplast DNA in potato is illustrated in Figure 12.9, which could indicate maternal or cytoplasmic ancestry of potato since evolution of chloroplast DNA and mitochondrial DNA appears to be highly correlated [64]. The evolutionary pathway shown in Figure 12.9 indicates the main stream of chloroplast DNA introductions from wild species or presumed ancestral species, to cultivated diploid potato, to Andean cultivated tetraploid potato and Chilean cultivated tetraploid potato. Other minor streams might be possible by introduction of chloroplast DNA through introgressive hybridization from wild species to cultivated potatoes after cultivated forms arose [10, 22, 47, 95]. It was demonstrated that gene flow occurred from both wild species *S. sparsipilum* [74] and *S.*

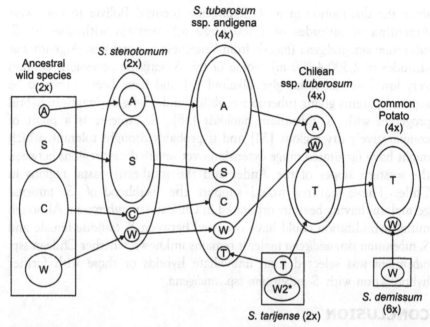

Fig. 12.9: Evolutionary pathways of chloroplast DNA. *S. tarijense* has several different W-type-derived chloroplast DNA [44], of which W2 is a major type.

megistacrolobum [58] to cultivated diploid potatoes. Ongoing introgressive hybridization has been reported between *S. stenotomum* and *S. megistacrolobum* [53, 57]. Grun [22] described *S. tuberosum* ssp. *andigena* as a genetic sponge, absorbing genes via introgression from any wild and cultivated species with which it hybridized via 2x × 4x or 4x × 2x crosses. At least some W-type chloroplast DNA in *S. tuberosum* ssp. *andigena* was possibly introduced by introgression because corresponding haplotypes were not found in *S. stenotomum* and showed closer relationships to some other species than to the wild species derived from the presumed ancestral species (Sukhotu and Hosaka, unpubl. data).

As already mentioned, the *S. tuberosum* ssp. *tuberosum* cytoplasm shows profound effects on various agronomic traits such as yield, maturity, and sterility. Functional differences of chloroplast DNA as well as mitochondrial DNA are to be exploited for full understanding of the selection process of Chilean ssp. *tuberosum* and for agronomic utility of cytoplasmic genome.

[1] Asama K. Jagaimo 43 wa [43 stories on potato]. Sapporo: Hokkaido-Shinbunsha, 1978.

[2] Bamberg JB, Martin MW, Schartner JJ. Elite selections of tuber-bearing *Solanum* species germplasm. Wisconsin: Inter-Regional Potato Introduction Station, 1994.

[3] Brücher H. Untersuchungen über die *Solanum (Tuberarium)*—Cultivare der Insel Chiloé. Z Pflanzenzücht 1963; 49:7-54.

[4] Brücher H. El origen de la papa (*Solanum tuberosum*). Physis 1964; 24: 439-52.

[5] Bryan GJ, McNicoll J, Ramsay G, Meyer RC, De Jong WS. Polymorphic simple sequence repeat markers in chloroplast genomes of Solanaceous plants. Theor Appl Genet 1999; 99: 859-67.

[6] Buckner B, Hyde BB. Chloroplast DNA variation between the common cultivated potato (*Solanum tuberosum* ssp. *tuberosum*) and several South American relatives. Theor Appl Genet 1985; 71: 527-31.

[7] Bukasov SM. The potatoes of South America and their breeding possibilities. Bull Appl Bot Leningrad (Suppl) 1933; 58: 1-192.

[8] Bukasov SM. The origin of potato species. Physis 1939; 18: 41-6.

[9] Bukasov SM. Die Kulturarten der Kartoffel und ihre wildwachsenden Vorfahren. Z Pflanzenzücht 1966; 55: 139-64.

[10] Bukasov SM. Systematics of the potato. Bull Appl Bot Genet Breed 1978; 62(1): 1-42.

[11] Contreras A, Ciampi L, Padulosi S, Spooner DM. Potato germplasm collecting expedition to the Guaitecas and chonos Archipelagos, Chile, 1990. Potato Res 1993; 36: 309-16.

[12] Correll DS. The potato and its wild relatives. Renner, TX: Texas Research Foundation, 1962.

[13] Cribb PJ, Hawkes JG. Experimental evidence for the origin of *Solanum tuberosum* subspecies *andigena*. In: D'Arcy WG, ed. Solanaceae: biology and systematics. New York: Columbia University Press. 1986: 383-404.

[14] den Nijs TPM, Peloquin SJ. 2n gametes in potato species and their function in sexual polyploidization. Euphytica 1977; 26: 585-600.

[15] Dodds KS. Classification of cultivated potatoes. In: Correll DS, ed. The potato and its wild relatives. Renner, TX: Texas Research Foundation, 1962: 517-539.

[16] Douches DS, Ludlam K, Freyre R. Isozyme and plastid DNA assessment of pedigrees of nineteenth century potato cultivars. Theor Appl Genet 1991; 82: 195-200.

[17] Gatenby AA, Cocking EC. Fraction 1 protein and the origin of the European potato. Plant Sci Lett 1978; 12: 177-81.

[18] Glendinning DR. Regional variation in leaf form and other characters of *Solanum tuberosum* Group Andigena. Europ Potao J 1968; 11: 277-80.

[19] Glendinning DR. Neo-Tuberosum: new potato breeding material. 1. The origin, composition, and development of the Tuberosum and Neo-Tuberosum gene pools. Potato Res 1975; 18: 256-61.

[20] Grun P. Cytoplasmic sterilities that separate the cultivated potato from its putative diploid ancestors. Evolution 1970; 24: 750-8.

[21] Grun P. Evolution of the cultivated potato: a cytoplasmic analysis. In: Hawkes JG, Lester RN, Skelding AD, eds. The biology and taxonomy of the Solanaceae. London: Academic Press; 1979: 655-665.

[22] Grun P. The evolution of cultivated potatoes. Econ Bot 1990; 44: 39-55.

[23] Grun P, Staub J. Evolution of tetraploid cultigens from the view of cytoplasmic inheritance. In: Report of the planning conference on the Exploration, Taxonomy and Maintenance of Potato Cream Plasm III. Lima, Peru: International Potato center; 1979: 141-152.

[24] Grun P, Ochoa C, Capage D. Evolution of cytoplasmic factors in tetraploid cultivated potatoes (Solanaceae). Amer J Bot 1977; 64: 412-20.

[25] Hanneman RE Jr. The potato germplasm resources. Amer Potato J 1989; 66: 655-67.

[26] Hanneman RE Jr, Peloquin SJ. Genetic-cytoplasmic male sterility in progeny of 4x-2x crosses in cultivated potatoes. Theor Appl Genet 1981; 59: 53-5.

[27] Hawkes JG. Taxonomic studies on the tuber-bearing Solanums. 1. Solanum tuberosum and the tetraploid complex. Proc Linn Soc London 1956; 166: 97-144.

[28] Hawkes JG. Kartoffel. 1. Taxonomy, cytology and crossability. In: Kappert H, Rudorf W, eds. Handbuch der Pflanzenzüchtung, vol 3. Berlin: Paul Parey, 1958: 1-43.

[29] Hawkes JG. The origin of Solanum juzepczukii Buk. and S. curtilobum Juz. et Buk. Z Pflanzenzüchtg 1962; 47: 1-14.

[30] Hawkes JG. The history of the potato. J Roy Hort Soc 1967; 92: 207-302.

[31] Hawkes JG. The evolution of cultivated potatoes and their tuber-bearing wild relatives. Kulturpflanze 1988; 36: 189-208.

[32] Hawkes JG. The potato—evolution, biodiversity and genetic resources. London: Belhaven Press, 1990.

[33] Hawkes JG, Hjerting JP. The potatoes of Argentina, Brazil, Paraguay, and Uruguay. Oxford, UK: Oxford University Press, 1969.

[34] Hawkes JG, Hjerting JP. The potatoes of Bolivia; their breeding value and evolutionary relationships. New York, NY: Oxford University Press, 1989.

[35] Hawkes JG, Francisco-Ortega J. The potato in Spain during the late 16th century. Econ Bot 1992; 46: 86-97.

[36] Hawkes JG, Francisco-Ortega J. The early history of the potato in Europe. Euphytica 1993; 70: 1-7.

[37] Heinhorst S, Gannon GC, Galun E, Kenschaft L, Weissbach A. Clone bank and physical map of potato chloroplast DNA. Theor Appl Genet 1988; 75: 244-51.

[38] Hermundstad SA, Peloquin SJ. Germplasm enhancement with potato haploids. J Heredity 1985; 76: 463-7.

[39] Hilali A, Lauer FI, Veilleux RE. Reciprocal differences between hybrids of Solanum tuberosum Groups Tuberosum (haploid) and Phureja. Euphytica 1987; 36: 631-9.

[40] Hosaka K. Who is the mother of the potato?—restriction endonuclease analysis of chloroplast DNA of cultivated potatoes. Theor Appl Genet 1986; 72: 606-18.

[41] Hosaka K. Similar introduction and incorporation of potato chloroplast DNA in Japan and Europe. Jpn J Genet 1993; 68: 55-61.

[42] Hosaka K. Successive domestication and evolution of the Andean potatoes as revealed by chloroplast DNA restriction endonuclease analysis. Theor Appl Genet 1995; 90: 356-63.

[43] Hosaka K. Distribution of the 241 bp deletion of chloroplast DNA in wild potato species. Amer J Potato Res 2002; 79: 119-23.

[44] Hosaka K. T-type chloroplast DNA in *Solanum tuberosum* L. ssp. *tuberosum* was conferred from some populations of *S. tarijense* Hawkes. Amer J Potato Res 2003; 80: 21-32.

[45] Hosaka K. Evolutionary pathway of T-type chloroplast DNA in potato. Amer J Potato Res 2004; 81: 153-158.

[46] Hosaka K, Hanneman RE Jr. The origin of the cultivated tetraploid potato based on chloroplast DNA. Theor Appl Genet 1988; 76: 172-6.

[47] Hosaka K, Hanneman RE Jr. Origin of chloroplast DNA diversity in the Andean potatoes. Theor Appl Genet 1988; 76: 333-40.

[48] Hosaka K, de Zoeten GA, Hanneman RE Jr. Cultivated potato chloroplast DNA differs from the wild type by one deletion—evidence and implications. Theor Appl Genet 1988; 75: 741-5.

[49] Hosaka K, Mori M, Ogawa K. Genetic relationships of Japanese potato cultivars assessed by RAPD analysis. Amer Potato J 1994; 71: 535-46.

[50] Hosaka K, Ogihara Y, Matsubayashi M, Tsunewaki K. Phylogenetic relationship between the tuberous *Solanum* species as revealed by restriction endonuclease analysis of chloroplast DNA. Jpn J Genet 1984; 59: 349-69.

[51] Huamán Z. *Ex-situ* conservation of potato genetic resources at CIP. CIP Circular 1994; 20(3): 2-7.

[52] Huamán Z, Spooner DM. Reclassification of landrace populations of cultivated potatoes (*Solanum* sect. *Petota*). Amer J Bot 2002; 89: 947-65.

[53] Huamán Z, Hawkes JG, Rowe PR. *Solanum ajanhuiri*: an important diploid potato cultivated in the Andean Altiplano. Econ Bot 1980; 34: 335-43.

[54] Huamán Z, Hawkes JG, Rowe PR. A biosystematic study of the origin of the cultivated diploid potato, *Solanum* × *ajanhuiri* Juz. et Buk. Euphytica 1982; 31: 665-76.

[55] Iwanaga M, Peloquin SJ. Origin and evolution of cultivated tetraploid potatoes via 2n gametes. Theor Appl Genet 1982; 61: 161-9.

[56] Jackson MT, Hawkes JG, Rowe PR. The nature of *Solanum* × *chaucha* Juz. et Buk., a triploid cultivated potato of the South American Andes. Euphytica 1977; 26: 775-83.

[57] Johns T, Keen SL. Ongoing evolution of the potato on the Altiplano of western Bolivia. Econ Bot 1986; 40: 409-24.

[58] Johns T, Huamán Z, Ochoa C, Schmiediche E. Relationships among wild, weed, and cultivated potatoes in the *Solanum* × *ajanhuiri* complex. Syst Bot 1987; 12: 541-52.

[59] Juzepczuk SW, Bukasov SM. A contribution to the question of the origin of the potato. Proc USSR Cong Genet Pl & Animal Breed 1929; 3: 593-611.

[60] Kawagoe Y, Kikuta Y. Chloroplast DNA evolution in potato (*Solanum tuberosum* L.). Theor Appl Genet 1991; 81: 13-20.

[61] Kawakami K. Bareisho-Tsuron [Textbook on potato]. Tokyo: Yokendo, 1948.

[62] Laufer B. The American plant migration. Part 1: The potato. The potato in Japan and Korea. Anthropological Series, Field Museum of Natural History 1938; 28: 80-3.

[63] Lössl A, Götz M, Braun A, Wenzel G. Molecular markers for cytoplasm in potato: male sterility and contribution of different plastid-mitochondrial configurations to starch production. Euphytica 2000; 116: 221-30.

[64] Lössl A, Adler N, Horn R, Frei U, Wenzel G. Chondriome-type characterization of potato: mt α, β, γ, δ, ε and novel plastid-mitochondrial configurations in somatic hybrids. Theor Appl Genet 1999; 99: 1-10.

[65] Maris B. Analysis of an incomplete diallel cross among three ssp. *tuberosum* varieties and seven long-day adapted ssp. *andigena* clones of the potato (*Solanum tuberosum* L.). Euphytica 1989; 41: 163-82.

[66] Matsubayashi M. Phylogenetic relationships in the potato and its related species. In: Tsuchiya T, Gupta PK, eds. Chromosome engineering in plants: genetics, breeding, evolution. Part B. Amsterdam, Netherlands: Elsevier, 1991, 93-118.

[67] Nakagawa K, Hosaka K. Species relationships between a wild tetraploid potato species, *Solanum acaule* Bitter, and its related species as revealed by RFLPs of chloroplast and nuclear DNA. Amer J Potato Res 2002; 79: 85-98.

[68] Ochoa CM. The potatoes of South America: Bolivia. Cambridge, UK: Cambridge University Press, 1990.

[69] Palmer JD, Jansen RK, Michaels HJ, Chase MW, Manhart JR. Chloroplast DNA variation and plant phylogeny. Ann Mo Bot Gard 1988; 75: 1180-206.

[70] Plaisted RL, Hoopes RW. The past record and future prospects for the use of exotic potato germplasm. Amer Potato J 1989; 66: 603-27.

[71] Powell W, Baird E, Duncan N, Waugh R. Chloroplast DNA variability in old and recently introduced potato cultivars. Ann Appl Biol 1993; 123: 403-10.

[72] Provan J, Powell W, Hollingsworth PM. Chloroplast microsatellites: new tools for studies in plant ecology and evolution. Trends Ecol Evol 2001; 16: 142-7.

[73] Provan J, Powell W, Dewar H, et al. An extreme cytoplasmic bottleneck in the modern European cultivated potato (*Solanum tuberosum*) is not reflected in decreased levels of nuclear diversity. Proc Roy Soc Lond B 1999; 266: 633-9.

[74] Rabinowitz D, Linder CR, Ortega R, et al. High levels of interspecific hybridization between *Solanum sparsipilum* and *S. stenotomum* in experimental plots in the Andes. Amer Potato J 1990; 67: 73-81.

[75] Raker C, Spooner DM. The Chilean tetraploid cultivated potato, *Solanum tuberosum*, is distinct from the Andean populations; microsatellite data. Crop Sci 2002; 42: 1451-8.

[76] Ross H. The use of wild *Solanum* species in German potato breeding of the past and today. Amer Potato J 1966; 43: 63-80.

[77] Ross H. Potato breeding- problems and perspectives. Berlin: Verlag Paul Parey, 1986.

[78] Ross RW, Peloquin SJ, Hougas RW. Fertility of hybrids from *Solanum phureja* and haploid *S. tuberosum* matings. Eur Potato J 1964; 7: 81-9.

[79] Salaman RN. Potato Varieties. Cambridge, UK: Cambridge University Press, 1926.

[80] Salaman RN. The early European potato: its character and place of origin. J Linn Soc Bot 1946; 53: 1-27.

[81] Salaman RN. The history and social influence of the potato. Cambridge, UK: Cambridge University Press, 1949.

[82] Salaman RN. The origin of the early European potato. J Linn Soc Bot 1954; 53: 185-90.

[83] Salaman RN, Hawkes JG. The character of the early European potato. Proc Linn Soc Lond 1949; 161: 71-84.

[84] Sanford JC, Hanneman RE Jr. Reciprocal differences in the photoperiod reaction of hybrid populations in *Solanum tuberosum*. Amer Potato J 1979; 56: 531-40.

[85] Sanford JC, Hanneman RE Jr. Large yield differences between reciprocal families of *Solanum tuberosum*. Euphytica 1982; 31: 1-12.

[86] Schmiediche PE, Hawkes JG, Ochoa CM. Breeding of the cultivated potato species *Solanum × juzepczukii* Buk. and *Solanum × curtilobum* Juz. et Buk. 1. A study of the natural variation of *S. × juzepczukii*, *S. × curtilobum* and their wild progenitor, *S. acaule* Bitt. Euphytica 1980; 29: 685-704.

[87] Simmonds NW. Studies of the tetraploid potatoes. 2. Factors in the evolution of the Tuberosum Group. J Linn Soc Bot 1964; 59: 43-56.

[88] Simmonds NW. Studies of the tetraploid potatoes. 3. Progress in the experimental re-creation of the Tuberosum Group. J Linn Soc Bot 1966; 59: 279-88.

[89] Simmonds NW. Change of leaf size in the evolution of the Tuberosum potatoes. Euphytica 1968; 17: 504-6.

[90] Spooner DM, Contreras MA, Bamberg JB. Potato germplasm collecting expedition to Chile, 1989, and utility of the Chilean species. Amer Potato J 1991; 68: 681-90.

[91] Sukhotu T, Kamijima O, Hosaka K. Nuclear and chloroplast DNA differentiation in Andean potatoes. Genome 2004; 47: 46-56.

[92] Swaminathan MS. The origin of the early European potato—evidence from Indian varieties. Indian J Genet Pl Breed 1958; 18: 8-15.

[93] Swaminathan MS, Magoon ML. Origin and cytogenetics of the commercial potato. In: Advances in genetics, vol 10. London, UK: Academic Press, 1961: 217-256.

[94] Sykin AG. Zur Frage der Abstammung und der wildwachsenden Vorfahren chilenischer Kulturkartoffeln. Z Pflanzenzüchtg 1971; 65: 1-14.

[95] Ugent D. The potato. Science 1970; 170: 1161-6.

[96] Ugent D, Dillehay T, Ramirez C. Potato remains from a late Pleistocene settlement in south-central Chile. Econ Bot 1987; 41: 17-27.

[97] van den Berg RG, Zevenbergen MJ, Kardolus JP, Groendijk-Wilders N. The origin of *Solanum juzepczukii*. In: Andrews S, Leslie AC, Alexander C, eds. Taxonomy of cultivated plants: 3rd int'l symp. Kew: Royal Botanic Gardens, 1999: 369-370.

[98] Watanabe K, Peloquin SJ. Occurrence of 2n pollen and *ps* gene frequencies in cultivated groups and their related wild species in tuber-bearing *Solanums*. Theor Appl Genet 1989; 78: 329-36.

[99] Werner JE, Peloquin SJ. Occurrence and mechanisms of 2n egg formation in 2x potato. Genome 1991; 34: 975-82.

Authors Index

Detailed Contents